HOW TO SOLVE PHYSICS PROBLEMS

HOW TO SOLVE PHYSICS PROBLEMS

Robert M. Oman

Daniel M. Oman

McGraw-Hill
New York San Francisco Washington, D.C. Auckland Bogatá
Caracas Lisbon London Madrid Mexico City Milan
Montreal New Delhi San Juan Singapore
Sydney Tokyo Toronto

Library of Congress Cataloging-in-Publication Data

Oman, Robert M.
How to solve physics problems / Robert M. Oman, Daniel M. Oman.
p. cm.
Includes index.
ISBN 0-07-048166-0
1. Physics—Problems, exercises, etc. I. Oman, Daniel M.
II. Title.
QC32.053 1996
530'.076—dc20 96-32639
CIP

McGraw-Hill

A Division of The ***McGraw-Hill*** *Companies*

1 2 3 4 5 6 7 8 9 0 MAL/MAL 9 0 1 0 9 8 7 6

ISBN 0-07-048166-0

The sponsoring editor for this book was Arthur Biderman, the editing supervisor was Paul R. Sobel, and the production supervisor was Pamela A. Pelton. It was set in Times New Roman by the authors.

This book is printed on recycled, acid-free paper containing a minimum of 50% recycled de-inked fiber.

CONTENTS

	How to Use This Book	vii
	How to Excel in Your Physics Course	ix
	Preface	xvii
Introduction	Mathematical Background	1
1.	Vectors	19
2.	Motion in One Dimension	28
3.	Falling Body Problems	37
4.	Projectile Motion	47
5.	Forces (including friction)	56
6.	Apparent Weight	71
7.	Work and the Definite Integral	76
8.	Work-Energy Problems	82
9.	Momentum Analysis	93
10.	Collision and Impulse	102
11.	Rotational Motion	110
12.	Rotational Dynamics	120
13.	Equilibrium	130
14.	Gravity	139
15.	Simple Harmonic Motion	145
16.	Fluids	159
17.	Temperature and Calorimetry	171
18.	Kinetics and the Gas Laws	181
19.	First Law of Thermodynamics	191
20.	Second Law of Thermodynamics	202
21.	Mechanical Waves	209
22.	Standing Waves (Strings and Pipes)	216
23.	Sound	222
24.	Charge and Coulomb's Law	230
25.	The Electric Field	234
26.	Gauss' Law	243
27.	Electric Potential	250
28.	Capacitance	260
29.	Conductivity	271
30.	Resistors in D.C. Circuits	279
31.	Kirchhoff's Laws	284
32.	R-C Circuits	294
33.	Magnetic Fields	301
34.	Magnetic Forces	312
35.	Ampere's Law	318
36.	Biot-Savart Law	326
37.	Faraday's Law	331
38.	Inductance	341
39.	R-L Circuits	346

40. Oscillating L-C Circuits 352
41. Series R-L-C Circuits and Phasors 356
42. Maxwell's Equations 368
43. Electromagnetic Waves 372
44. Reflection, Refraction, and Polarization 380
45. Mirrors and Lenses 386
46. Diffraction and Interference 392
47. Special Relativity 396
Physical Constants 403
Index 405

HOW TO USE THIS BOOK

This book will teach you how to do physics problems. The explanation of not only how to do a problem but why we do it a certain way teaches you not just a collection of solved problems, but a collection of methods that can be used, modified, and built upon to do other physics problems. As researchers and teachers, we know that the key to solving new and challenging problems is contained within the collection of techniques already learned for solving simpler problems. Seeing a problem solved and knowing why it was done in a certain manner is the best way to learn how to solve related, more difficult problems.

This book is not a presentation of every problem you are going to encounter on a test. It is a presentation of the methods that we have found to work for large groups of problems. If you develop the techniques we describe for solving problems then you will know how to successfully attack the problems you will encounter on the tests. This is the book you should have as a reference when you are doing your homework problems. It will show you how to work the problems and explain why they are being done the way they are.

The topics in this book are in the order of most physics texts. Each chapter begins with a theoretical discussion. Problems are mixed in with the discussion as soon as possible. These problems follow the development of the theory. In this way you do not have to assimilate a large amount of conceptual material before begining to work problems.

A "standard" route is followed for problems wherever possible. In this way you will learn that broad categories of problems worked in a standard "logical" way always produce correct solutions. Our emphasis is on logic and order in solving problems. We avoid methods that may be quick and have limited application to problem solving in favor of possibly longer solutions that have broad applications and always work. We believe that a lot of good physics can be taught in problems so we use problems to illustrate and expand a topic and sometimes introduce new concepts. For this reason problems and text are integrated with a minimum of artificial barriers between them.

The book is intended as a complement to either the calculus-based or the non-calculus-based elementary physics course. It has been our experience that calculus concepts can be introduced into the traditional non-calculus course and used in the development of concepts. Conceptually, calculus is not difficult and when it is introduced in the context of a physics problem it is even easier. We use calculus concepts to explain theory, but calculus is rarely used in problems. Even those students who are taking calculus concurrent with their physics course usually learn calculus concepts in physics before they see them in their calculus course.

(C) In those instances where calculus is needed, the problems and paragraphs are marked with a calculus icon. Even the student without formal calculus training should read these sections. They are often explained in a simple manner so that the calculus does not present a problem.

The chapters on electricity and magnetism are also excellent background chapters for someone taking an undergraduate course in Electricity and Magnetism.

We have used two significant figures for the physical constants and most of the numbers in the problems. Results are given to two, and occasionally three, significant figures. Using two significant figures cuts down on the clutter in the problems, allowing the technique to receive greater exposure. Do not be concerned in working through the problems if your answers do not agree exactly with ours. This is no doubt due to when, or if, intermediate calculations were rounded off. SI units are used nearly universally throughout the book.

HOW TO EXCEL IN YOUR PHYSICS COURSE

Most students realize that putting off studying until the day before the exam and then cramming at the last minute is not efficient. Some students do this anyway, because so far they have gotten away with it. Perhaps most of the other students you previously competed with had poor study skills. This may have allowed you to adopt poor or non-existent study habits and still keep up, or even get good grades if you are naturally a better student. Now that you are in college, the courses will be more difficult and it is to your advantage to develop a more organized approach to handling your course work.

Successful people generally have three things in common. They make effective use of their time, they set goals for themselves, and they have a positive attitude. Physics is a challenging course for most students. It will take a well-organized consistent effort to do well in this course, but success in a challenging area is a worthwhile goal.

General Approach for Studying Physics

Many people believe the following: more work and more study results in higher grades. This is not necessarily so. You certainly must be willing to make a certain commitment of time and energy to this course, but the key to academic success is concentrating your efforts on the right things at the right times. You may have noticed that those students who receive the highest grades are not necessarily the ones who work the greatest number of hours. Some students may boast that they have studied all night for an exam, but don't be impressed by this habit. "Allnighters" and the like are almost always the result of procrastination and bad study habits. Getting no sleep before an exam is foolish and it usually takes several days to recover from this kind of activity. By taking advantage of the study techniques that follow you can achieve higher grades with less effort.

The most efficient way of learning Physics by attending lectures, problem solving sessions, and performing supplementary readings is to:

1. Do a quick reading on the topics to be covered in the lecture before attending class. Ten or fifteen minutes may be sufficient for a one hour lecture. The purpose here is to generally familiarize yourself with the topics to be discussed. Perhaps you can identify one or two questions or key points to listen for during the lecture.

2. Attend class and take notes. Attend all of the classes. Someone is paying for these classes so BE THERE! Be on the alert for any indication by the instructor of possible test questions. If the professor says something like "This is very important, you may be seeing this again," make a special note of this in your notebook.

3. Review your lecture notes. Don't save this step until a few days before the exam. It is far more efficient to review your notes a little bit at a time during the semester than to try and do it

all at once. At this point you should also do a more detailed reading of the text to fill in any gaps in your class notes.

4. This may be the most important step. Do the homework problems regularly. In other courses it may be sufficient to read the text and review your notes, but in Physics you must be able to work the problems. You don't learn problem solving skills by just reading examples of solved problems, you must do the problems yourself. By doing the homework problems on a regular basis you will be able to identify areas that you need more work on well in advance of the test. Physics problems can be difficult. Therefore, when you set out to work problems do not set yourself the task of working a certain number of problems, but rather set out a certain amount of time to work on problems.

5. Compile a formal set of notes and prepare a detailed outline. The general strategy here is that a number of short exposures to manageable pieces of the course is more efficient than one long exposure to a large amount of material. As you progress through the course, you first get your information in an initial reading of the material, then again in the lecture, then again in a second reading, and yet again in an organizing session where you prepare a detailed outline. The detailed outline is essential to success on the exams. It contains the examination questions. Your main preparation for the exam will be to extract the questions and prepare to answer them. Notice we did not say "study for the exam;" the studying for the exam has been going on all along. That is what you have been doing as you make up your formal notes, outline, etc. What you have done with this systematic approach is to reproduce the notes and outline that the instructor is using. If you are reasonably good at it, you will have as good a source of exam questions as the instructor.

How to Prepare for a Physics Test

Examine the shelves of any bookstore catering to career oriented students and you will find books with titles such as: How to Pass the Real Estate Licensing Exam, or How to Succeed on the S.A.T. Examining these books will help you to develop your personal exam-taking program. One common thread in all books on how to pass particular exams is to know the questions in advance. Most writers of these types of books are in the business of training people in their particular areas, so they are close to the people who are making up the exams. This gives them a ready source of test questions, and knowing the questions (or at least the type of questions) is half way to knowing the answers. Therefore we make the following suggestions :

1. Almost all instructors in physics will place some problems on the test that are very similar to examples that they have done in class. Many times you may encounter the same problem with different numbers. This makes it very important to attend every class so as not to miss the opportunity to see possible test questions. If you do miss class, always get the notes from a friend.

2. Another frequent occurrence is for slight modifications of homework problems to appear on the test. Join a study group that does homework problems together. This can be more efficient than grinding away on your own. Don't waste too much time with a study group unless it is productive. Your final preparations for a test should be done privately so that you can concentrate on developing a plan for taking the test.

3. Find sample physics tests given by your instructor for the past few years. It is a good bet that most of the questions for the exams in the near future will be very much like those of the immediate past.

4. Some physics problems involve mathematics that can be deceptively easy. For example, if you expect problems involving the manipulation of logarithms or exponents be sure you practice the mathematical operations and entering the numbers into your calculator so you don't have to stop and figure out how to take exponents during the test. Practice any unfamiliar mathematical operations before the test.

Timing and the Use of the Subconscious

Have you ever experienced the frustration of having a conversation with someone and forgetting momentarily a name or fact that is very familiar to you? Usually, shortly after such an experience, the name or fact will come to you when you are not consciously trying to recall it. Another variation of this same phenomenon is when a person doesn't feel right about making a decision immediately upon receiving or defining a problem. They like to "sleep on it." Both of these situations have a common characteristic - the use of the subconscious. The fact that solutions are often presented to us in the absence of active work on the problem at the moment we receive the solution indicates that another part of the brain was analyzing the pertinent information and providing a solution. We call this part of the brain the subconscious, and this part of the brain is very effective at solving problems.

Here are some tips for effectively using the subconscious:

1. Your subconscious will not work without information. You must consciously sort out all of the facts or information for a particular problem. If you are having difficulty with a problem, try to get straight in your mind what you *do know* about the problem. Then also define in your mind what specifically you don't know or don't understand about the problem.

2. Put conscious effort into the problem up to the point of confusion. Many people grind and grind on a problem after this point and accomplish very little. It is more efficient for you to plan your study time so that you do not put yourself in a situation where your only choice is to grind on a problem.

3. After you have done all you can consciously on the problem, "Put it in the back of your mind." Don't keep worrying about it. It is important that you clear your mind so that you can accept the solution when it comes.

4. Be sure you have a deadline for the solution.

5. When a solution comes, be sure to act on it quickly, so you can go on to something else. Sometimes instead of a solution to the problem you will receive a request for more information. The problem may still be unanswered, but will be clearer to you. What could be happening here is that your subconscious has analyzed the problem and found an essential piece of information missing and is asking you for it.

The study program that we have outlined, consisting of regular review of lecture notes, frequent working of homework problems, and periodic updates of your formal notes and outline, makes maximum use of your subconscious. The periodic intake of new material and the required conscious review serves to keep you subconsciously analyzing and fitting new information into the body of knowledge you are accumulating.

Here would be a good approach to practicing for a Physics test:

ED - 4: (Exam day minus four) Prepare a sample exam from your outline. This may consist of questions from previous exams given by the instructor and variations of homework problems or examples done in class. Keep in mind that this is probably the same way that the professor is making up your exam.

ED - 3: Study for your first sample exam. Go over your notes, text, and homework problems.

ED - 2: Take your first sample exam. As soon as possible after the exam, do a detailed review concentrating on the weaker areas. Make up your final sample exam.

ED - 1: Take your final sample exam. Again review the difficult points of this sample exam. Get a good night's sleep tonight.

ED: Do as little as possible on the day of the exam. You may want to quickly review your outline or a couple of difficult points.

You will notice that the bulk of the work in preparing for a test this way consists of writing and taking sample tests. It is planned that way. One of the common fallacies in preparing for exams is to prepare for the wrong thing. Many students will prepare for a Physics exam by reading the text or by reading solutions to problems. A Physics exam, however, is not a reading exam but a writing and problem-solving exam. If you have not practiced writing solutions to typical problems, you have not prepared as well as you might for the exam.

The second advantage to taking sample tests is that it increases your speed in writing solutions to types of problems that are likely to be on the test. This will allow you more time during the test to spend on unexpected or more troublesome problems.

Strategies to Use During a Physics Test

You are now entering the test room. You are well prepared to take the test. You have taken practice tests and know what to expect on the exam. You have gotten a good night's sleep the night before and eaten a healthy breakfast that will provide you with the energy needed for good concentration. You have a positive attitude. At this point worrying about how you will do on the exam is useless. Study time is over. You now need to concentrate on the strategies that will get you the highest possible score on the test. Here are some suggestions:

1. It is usually a good idea to take a minute or two at the beginning of the exam to look over all the questions. Look for the type of questions that you expected and have practiced and do these first. Save the hardest questions for last. It can be very frustrating to run out of time working on question # 4 only to realize that you didn't even get a chance to start question #5 that was much easier.

2. Have a rough idea of how much time you should be spending on each question. Sometimes certain questions will count for more points than others and the instructor should provide that information on the test.

3. If you are required to memorize a lot of formulas you may want to take the time at the beginning of the test to write down a few of the more complicated ones next to problems that involve those formulas as you are glancing over the test. Later during the test, your mind may be cluttered with formulas and it may be harder to correctly recall one of the more complicated ones.

4. Always include the units of your answer (miles per hour if the answer is a velocity for example). Don't make the mistake of not including units. This is very important to almost all physics teachers.

5. Write your work clearly when you are solving a problem. It is easier for the professor to give you partial credit if he can clearly see that you did the problem correctly and just made a minor computational error.

6. Think about your answer to a problem. Does the answer make sense? For example, if you are solving for the length of one side of a right triangle and you are given the hypotenuse, your answer better not be a length greater than the hypotenuse. It is very important to be able to think like this on a test. This will help you to catch a lot of mistakes like missing a minus sign.

7. Unfortunately some instructors give tests that are much too long for a given period of time. It seems as if they are more interested in measuring how *fast* you can do physics than how well you can do physics. Try to find out in advance of the test if your professor's tests are like this. If the cutoff for an A is usually 75% instead of 90% then you need to be aware of this. This will save you from panicking as you run out of time on the test. Remember that you may be able to work for partial credit on that last answer. On these kinds of tests it is very important to keep your cool and try to get as many points as you possibly can. Stay positive all the way through and give it your best shot!

8. Make sure you know the difference between radian mode and degree mode on your calculator when taking a test that includes trigonometry (See the Mathematical Background Section).

9. Avoid prolonged contact with other students immediately before the exam. Many times the nervous tension, frustration, defeatism, and perhaps wrong information expressed by fellow students can be harmful to your performance.

10. Multiple Choice Tests: Find out if there is any penalty for a wrong answer. If not, don't leave any question unanswered. Find out if there is any partial credit for showing your work

on a separate sheet of paper. One thing to think about for multiple choice tests is how the professor is generating the choices other than the correct answer. Here are some typical wrong choices on a multiple choice Physics test:

a) A formula requires the input of length in meters. In the problem the length is specified in centimeters. The wrong answer is off by a factor of 100.

b) A formula requires the input of a radius. Diameter is given in the problem. The wrong answer is off by a factor of two.

c) A question asks for a velocity. Choice A is 10 lbs. This is the correct number, but the wrong units. Choice D is 10 miles per hour, the correct answer. The lesson here is to look carefully at all the choices.

Your Self Image as a Student

To a large extent, many people perform at the level of their own self image. One thing to get straight in your mind at the beginning of the course is that you are capable of mastering the material in your Physics course. Some students get stuck in the mode of saying something like, "I have always been a C student." There is a simple logical argument that will show you that the C student in physics or mathematics or any subject where skill is built from course to course, is not getting C's because of their understanding of the material, but because that is how they view themselves, consciously or unconsciously. In a series of three to five sequential mathematics courses, for example, it is virtually impossible to go from one course to the next, let alone a sequence of several, without eventually mastering the material in each previous course. Think back to your first math course where you were taught how to add, subtract, multiply, and divide. At some point in that course you may have thought that you *couldn't* understand certain concepts. By now you have mastered those skills. College Physics is the same way. You are mentally capable of understanding and even mastering basic physics. Now it is true that different people learn at different speeds. You may need to spend a little extra time on physics or, more likely, make more effective use of your time.

At this point you need to set a goal for yourself in your Physics course. The first question is how important is Physics in your academic program. If you are a Biology major and you are taking Physics only because it is a general requirement, then your primary goals should be to get the best grades in your Biology courses, since that is your major. If one of your goals is to have a high G.P.A., then you should strive for an A or at least a B. If your major is Physics or Engineering then you should definitely go for an A in this course. Write down your goals and check them off as they are accomplished. Your goals for the first part of a Physics course may look something like this:

Main Goal: An A in Physics I

week 1: establish a schedule for reading text, reviewing notes and doing homework problems week 2: investigate the possibility of joining a study group

week 3: find out if past exams from this professor are available: find out how many points it will take to make an A on the first test

week 4: prepare and take sample exams for first test

The purpose of writing down your goals is not to create more work, but to keep you focused on the most important things that you need to accomplish as the semester progresses. Please remember that all of the study techniques outlined in this chapter are designed to make achieving higher grades *easier* for you. The sooner you become more organized and focused on your goals, the sooner you will begin to realize that you are capable of impressive accomplishments with a reasonable amount of effort.

Perhaps Physics is a favorite area of study that you may wish to pursue in the future or perhaps you are primarily interested in the most efficient way to make it through this course. Whatever you choose for your major area of study, find something you enjoy and pursue excellence. Give it your best today, and better tomorrow. We wish you success.

PREFACE

The purpose of this book is to show you how to do physics problems. It is only through application of concepts to solving problems that we can know for certain that we understand something. Nowhere is this more true than in a physics course where performance is measured almost exclusively by your ability to do problems.

This book is not a collection of problems. Neither is it a text. It is an attempt to strike a balance between theory and problem solving with heavy emphasis on the problem solving. As such it is intended to complement your course text. Generations of physics students, the authors included, have often lamented, concerning their physics courses, "I understood everything in lecture and the text but I can't do the problems." This book will help you do the problems.

Learning physics is different from most other disciplines. Most disciplines can be learned by reading and listening, with mastery demonstrated by writing. Physics is not like that. Reading and listening are the first step, but mastery is demonstrated by doing problems. Writing comes easy to most people. Working problems in mathematical symbolism is not so easy to most, and it is not something we do regularly. Learning physics requires learning to do the problems of physics not by writing about them but by manipulating mathematical symbols in the correct manner.

The book was started around 1980 (by RMO) and was provided in rough form to his students in the elementary physics sequence to help them understand concepts and give them practice and confidence in working problems. The favorable response from those students provided motivation to continue to expand the number and extent of the topics. In 1984 the problems were used (by DMO) as an aid in the elementary physics courses he was taking. Since then the collection of problems and text has been expanded by both authors and refined through further use by their students.

It is the sincere desire of the authors that this book help you to better understand physical concepts and work the associated problems. We would like to thank the many students who have contributed to this work by using the material and offering their suggestions. Also the fine staff at McGraw-Hill, especially our editor, Arthur Biderman, have contributed greatly to the clarity of presentation.

Robert M. Oman
St. Petersburg, Florida

Daniel M. Oman
Orlando, Florida

HOW TO SOLVE PHYSICS PROBLEMS

INTRODUCTION

MATHEMATICAL BACKGROUND

The purpose of this chapter is to provide you with a review of and reference for the mathematical techniques you will need in working the physics problems in this book. Some topics may be familiar to you while others may not. Depending on the mathematical level of your physics course, some topics may not be of interest to you. Each topic is covered in sufficient depth to allow you to perform the mathematical manipulations necessary for a particular problem without getting bogged down in lengthy derivations. It is not our intention to teach mathematics, but to show you how to apply specific mathematical procedures to physics problems.

The most efficient use of this chapter is for you to do a brief review of the chapter, spending time on those sections that are unfamiliar to you and that you know you will need in your course, then refer to specific topics as they are encountered in the solution to problems. With this reference you should be able to perform all the mathematical operations necessary to complete the problems in your physics course. If you need or desire more depth in a particular topic go to an algebra or calculus text.

Solving Equations

The simplest equations to solve are the linear equations of the form $ax+b=0$ which have as solution $x=-b/a$. You should be very familiar with these.

The next most complicated equations are the quadratics. The simplest quadratic is the type that can be solved by taking square roots directly, without any other manipulations.

An example is $4x^2=36$, which is first divided by 4 to read $x^2=9$ and square roots taken to produce $x=\pm 3$. Both plus and minus values are legitimate solutions. The reality of the physical problem producing the equation may dictate that one of the solutions be discarded.

The next complication in quadratic equations is the factorable equations such as $x^2-x-6=0$, which can be factored to $(x-3)(x+2)=0$. The solutions, the values of x that make each parentheses equal to zero and satisfy the factored equation, are $x=3$ and $x=-2$.

If the quadratic cannot be solved by factoring, the most convenient solution is by quadratic formula, a general formula for solution of any quadratic equation in the form $ax^2+bx+c=0$.

The solution according to the quadratic formula is

$$x = \frac{-b \pm \sqrt{b^2 - 4ac}}{2a}$$

See any algebra book for a derivation of this formula.

The physics problems you are doing should not produce square roots of negative numbers. If your solution to a quadratic produces any square roots of negative numbers, you are probably doing something wrong in the problem.

Certain cubic equations such as $x^3 = 8$ can be solved directly producing the single answer $x = 2$. Cubic equations with quadratic (x^2) and linear (x) terms can be solved by factoring (if possible) or approximated using graphical techniques. You most likely will not encounter cubic equations in your early physics courses.

Another category of equations you will encounter is simultaneous equations: two independent equations in two unknowns and later three equations in three unknowns. We'll start with two equations in two unknowns. Take two equations

$$1)\quad 2x + 3y = 7$$

$$2)\quad x - 4y = -3$$

The most direct way of solving these equations is by substitution, solving one equation for one unknown and substituting in the other equation. Looking at these two equations the easiest variable to solve for is x in the second equation

$$1)\quad 2x + 3y = 7$$

$$2)\quad x = 4y - 3$$

Now substitute equation 2) in equation 1) and solve

$$2(4y - 3) + 3y = 7$$

$$8y - 6 + 3y = 7$$

$$11y = 13 \quad \text{or} \quad y = 13/11$$

Now put this value of y in either original equation and solve for x

$$x = 4\frac{13}{11} - 3 = \frac{52}{11} - \frac{33}{11} = \frac{19}{11}$$

These answers can be checked by substituting into both the original equations.

Another method, often involving less manipulation, is addition and subtraction where the equations are multiplied in such a way that upon addition or subtraction one of the variables adds away leaving one equation in one unknown. Start with the equations used previously and write equation 1) and -2 times equation 2), and add

$$\begin{aligned} 1) \quad & 2x+3y=7 \\ -2\times 2) \quad & -2x+8y=6 \\ 11y=13 \quad & \text{or} \quad y=13/11 \end{aligned}$$

This is the same value obtained above and by substitution in either original equation will produce the value for x. The equations could be handled differently by making the y terms add away. Multiply equation 1) by 4 and equation 2) by 3, and add

$$\begin{aligned} 4\times 1) \quad & 8x+12y=28 \\ 3\times 2) \quad & 3x-12y=-9 \\ 11x=19 \quad & \text{or} \quad x=19/11 \end{aligned}$$

The use of determinants in solving simultaneous equations is discussed in the next section.

Determinants

A **determinant** is a square array of numbers. Determinants are very convenient for solving two equations in two unknowns and three equations in three unknowns. The determinant technique for solving equations, called Cramers Rule, can be derived from the addition and subtraction method of solving simultaneous equations. Use as an example the equations of the previous section.

$$\begin{aligned} 2x+3y&=7 \\ x-4y&=-3 \end{aligned}$$

For the master, or main, determinant the array is the coefficients of the variables.

$$D=\begin{vmatrix} 2 & 3 \\ 1 & -4 \end{vmatrix}=-8-3=-11$$

The numeric equivalent of the determinant is found by multiplying 2 times -4 and subtracting the multiplication of 3 times 1. The numeric equivalence of a 2 by 2 determinant is this first diagonal multiplication minus the second diagonal multiplication. With a little practice this goes very quickly.

Now form the x associated determinant by replacing the x coefficients with the constants.

$$D_x = \begin{vmatrix} 7 & 3 \\ -3 & -4 \end{vmatrix} = -28 + 9 = -19$$

Perform the same diagonal multiplication minus diagonal multiplication operation: multiply 7 times –4 and subtract the multiplication of 3 times –3.

The y associated determinant is formed by replacing the y associated coefficients with the constants and multiplying and subtracting.

$$D_y = \begin{vmatrix} 2 & 7 \\ 1 & -3 \end{vmatrix} = -6 - 7 = -13$$

The solutions are

$$x = D_x / D = 19/11 \qquad \text{and} \qquad y = D_y / D = 13/11$$

If you need practice with determinants write down some sets of equations and solve them by substitution and determinants. After a few manipulations with determinants you will be able to solve simultaneous equations very quickly. Some calculators that solve systems of equations with Cramers Rule ask you to enter the numbers in a determinant format.

Three by three determinants require a little more manipulation. Consider three equations with the master determinant

$$\begin{aligned} 2x + y - z &= 2 \\ x - 3y + 2z &= 5 \\ x - 2y + z &= 3 \end{aligned} \qquad D = \begin{vmatrix} 2 & 1 & -1 \\ 1 & -3 & 2 \\ 1 & -2 & 1 \end{vmatrix}$$

There are several ways to find the values of this determinant. We'll look at one simple method called expanding the determinant, using the first row and the associated determinants obtained by crossing off the rows and columns associated with each number in the top row. This is easier to see than explain.

$$D = 2\begin{vmatrix} -3 & 2 \\ -2 & 1 \end{vmatrix} - 1\begin{vmatrix} 1 & 2 \\ 1 & 1 \end{vmatrix} - 1\begin{vmatrix} 1 & -3 \\ 1 & -2 \end{vmatrix}$$

Look at the top row of (the 3 by 3) D and write each term times the determinant obtained by blocking off the row and column associated with that term. Also, alternate the signs of the three 2 by 2 determinants so the second number, 1, is changed to –1. The two by two determinants are evaluated as before.

$$D = 2[-3+4] - 1[1-2] - 1[-2+3] = 2[1] - 1[-1] - 1[1] = 2$$

The x associated determinant is (again replacing the x coefficients with the constants)

$$D_x = \begin{vmatrix} 2 & 1 & -1 \\ 5 & -3 & 2 \\ 3 & -2 & 1 \end{vmatrix} = 2\begin{vmatrix} -3 & 2 \\ -2 & 1 \end{vmatrix} - 1\begin{vmatrix} 5 & 2 \\ 3 & 1 \end{vmatrix} - 1\begin{vmatrix} 5 & -3 \\ 3 & -2 \end{vmatrix}$$

$$D_x = 2[-3+4] - 1[5-6] - 1[-10+9] = 2+1+1 = 4$$

so

$$x = D_x / D = 4/2 = 2$$

The y associated determinant is

$$D_y = \begin{vmatrix} 2 & 2 & -1 \\ 1 & 5 & 2 \\ 1 & 3 & 1 \end{vmatrix} = 2\begin{vmatrix} 5 & 2 \\ 3 & 1 \end{vmatrix} - 2\begin{vmatrix} 1 & 2 \\ 1 & 1 \end{vmatrix} - 1\begin{vmatrix} 1 & 5 \\ 1 & 3 \end{vmatrix}$$

$$D_y = 2[5-6] - 2[1-2] - 1[3-5] = -2+2+2 = 2$$

so

$$y = D_y / D = 2/2 = 1$$

As an exercise find the z associated determinant and calculate z. The value of $z = 3$ can be verified from any of the original equations.

With a little practice determinants can be a very quick way of solving multiple equations in multiple unknowns.

Determinants as applied to vector products are discussed in the chapter on vectors.

Binomial Expansions

Squaring $(a+b)$ is done so often that most would immediately write $a^2 + 2ab + b^2$.

Cubing $(a+b)$ is not so familiar but easily accomplished by multiplying $(a^2 + 2ab + b^2)$ by $(a+b)$ to obtain $a^3 + 3a^2b + 3ab^2 + b^3$.

There is a simple procedure for finding the nth power of $(a+b)$. Envision a string of $(a+b)$'s multiplied together, $(a+b)^n$. Notice that the first term has coefficient 1 with a raised to the nth power, and the last term has coefficient 1 with b raised to the nth power. The terms in between contain a to progressively decreasing powers, $n, n-1, n-2, \ldots$, and b to progressively increasing powers. The coefficients can be obtained from an array of numbers or more conveniently from the binomial expansion or binomial theorem

$$(a+b)^n = \frac{a^n}{0!} + \frac{na^{n-1}b}{1!} + \frac{n(n-1)a^{n-2}b^2}{2!} + \ldots$$

The factorial notation may be new to you. The definitions are

$$0! = 1,\ 1! = 1,\ 2! = 2 \cdot 1,\ 3! = 3 \cdot 2 \cdot 1, \text{ etc.}$$

As an exercise use the binomial expansion formula to verify $(a+b)^3$.

The real utility of the binomial expansion in physics problems is in finding approximations to expressions where a is equal to 1 and b is less than 1. For this case the expansion looks like

$$(1+b)^4 = \frac{1}{0!} + \frac{4b}{1!} + \frac{4 \cdot 3b^2}{2!} + \frac{4 \cdot 3 \cdot 2b^3}{3!} + \ldots$$

The terms of the series decrease depending on the value of b. Two or three terms is usually a good approximation. Also the "next" term in the expansion is a good measure of the error in using a fixed number of terms of the binomial expansion.

The classic use of this expansion is in special relativity where the expression $\left(1 - v^2/c^2\right)^{1/2}$ regularly occurs.

$$\left(1 - \frac{v^2}{c^2}\right)^{1/2} = \frac{1}{0!} + \frac{(1/2)}{1!}\frac{v^2}{c^2} + \frac{(1/2)(-1/2)}{2!}\left(\frac{v^2}{c^2}\right)^2 + \ldots$$

In special relativity v/c is always less than one so this approximation, whether used algebraically or with numbers, is very convenient.

Coordinate Systems

The standard two dimensional coordinate system works well for most physics problems. In working problems in two dimensions do not hesitate to arrange the coordinate system for your convenience in doing a problem. If a motion is constrained to move up an incline, it may be more convenient to place one axis in the direction of the motion rather than in the traditional horizontal direction. If a projectile is dropped from an airplane, it may be more convenient to place the origin of the coordinate system at the place where the projectile was dropped and have the positive directions down, since the projectile and possibly the distances in the problem are given in reference to the airplane.

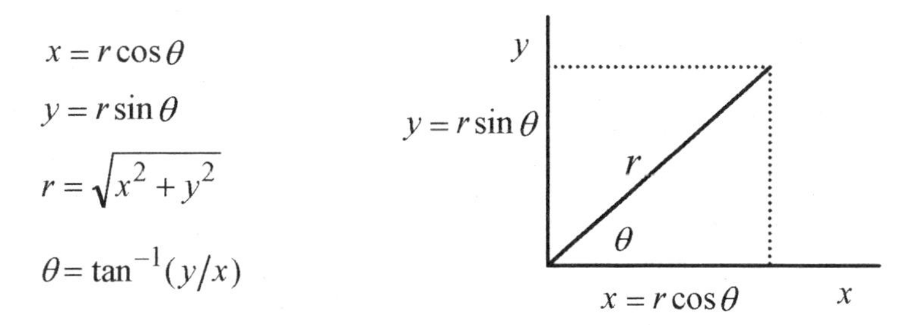

Fig. I-1

Positions in the standard right angle coordinate system are given with two numbers. In a polar coordinate system positions are given by a number and an angle. In the accompanying diagram it is clear that any point (x,y) can also be specified by (r,θ). Rather than moving distances in mutually perpendicular directions, the r and θ locate points by moving a distance r from the origin along what would be the $+x$ direction, then rotating through an angle θ. The relationship between rectangular and polar coordinates is also shown in Fig. I-1.

Three dimensional coordinate systems are usually right-handed. In Fig. I-2 imagine your right hand positioned with fingers extended in the $+x$ direction closing naturally so that your fingers rotate into the direction of the $+y$ axis while your thumb points in the direction of the $+z$ axis. It is this rotation of x into y to produce z with the right hand that specifies a right-handed coordinate system. Points in the three dimensional system are specified with three numbers (x,y,z).

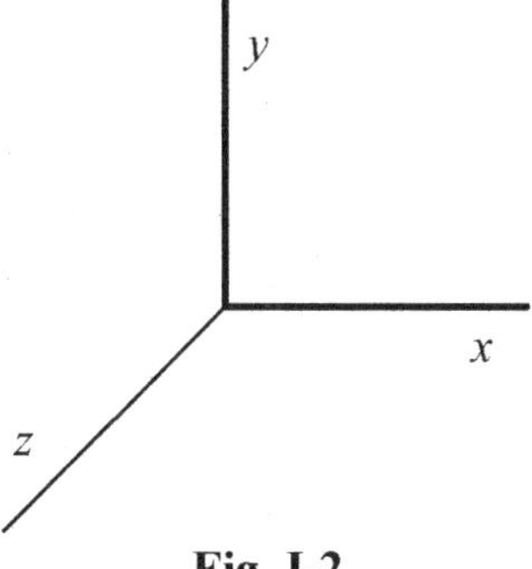

Fig. I-2

For certain types of problems, locating a point in space is more convenient with a cylindrical coordinate system. Construct a cylinder with the central axis on the z-axis of a right-handed coordinate system.

A point is located by specifying a radius measured out from the origin in the $+x$ direction, an angle in the x-y plane measured from the x-axis, and a height above the x-y plane. Thus the coordinates in the cylindrical system are (r,θ,z). The relation of these coordinates to x,y,z is given in Fig. I-3.

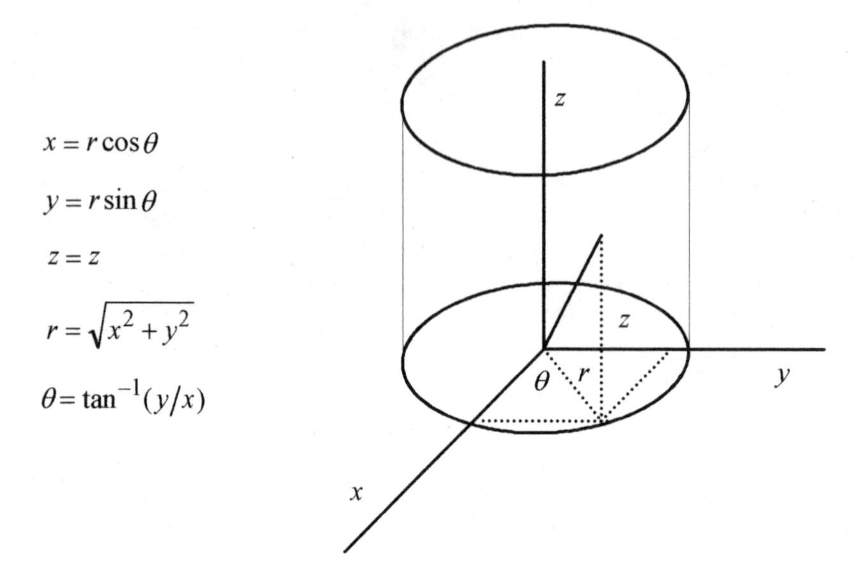

Fig. I-3

Spherical coordinates are also convenient in some problems. As the name suggests, points are located on a sphere centered on the origin of an (x,y,z) system. The radius of the sphere is the distance from the origin (to the sphere). The angle between this radius and the z-axis is one angle, and the angle between the x-axis and the projection of r on the x-y plane is the other angle. Thus, the coordinates in the spherical system are (r,θ,ϕ). The relation of these coordinates to x,y,z is given in Fig. I-4.

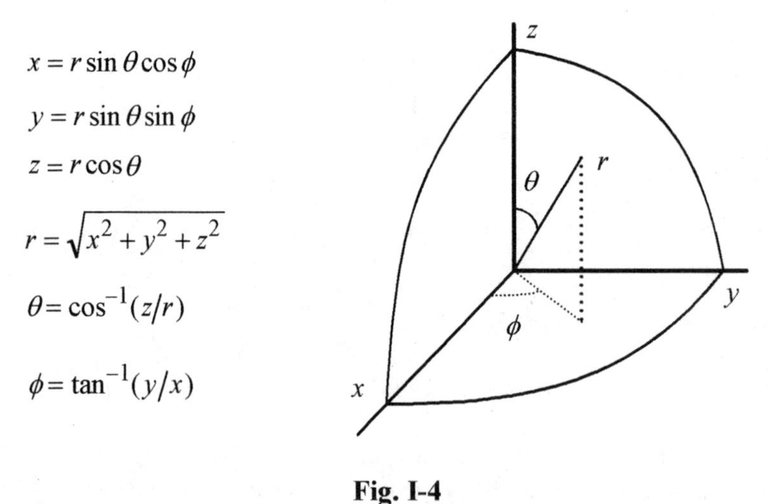

Fig. I-4

Trigonometry

The trigonometric relations can be defined in terms of right angle trigonometry or through their functions. The basic trigonometric relations, as they relate to right triangles, are shown in the box.

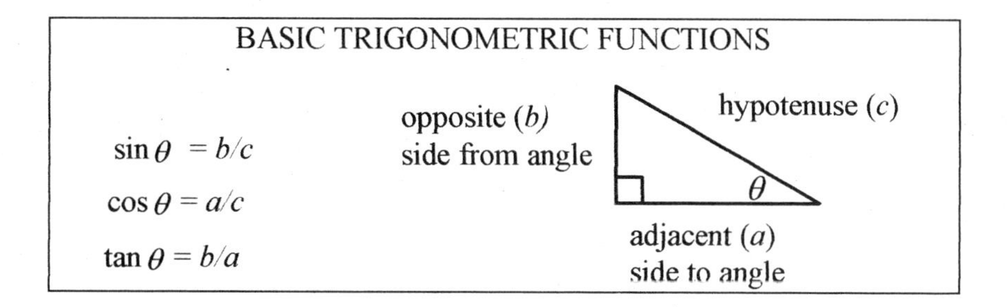

Graphs of the trigonometric relations are shown in the Fig. I-5.

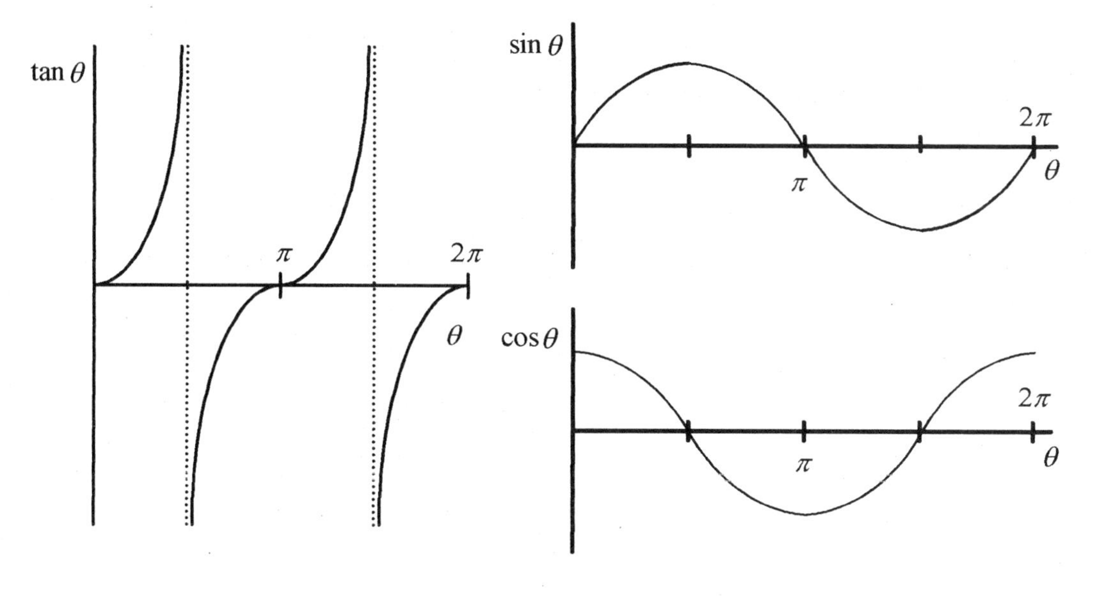

Fig. I-5

Angles are measured in radians. Radian measure is a pure number, the ratio of arc length to radius to produce the desired angle. Figure I-6 shows the relationship of arc length to radius to define the angle.

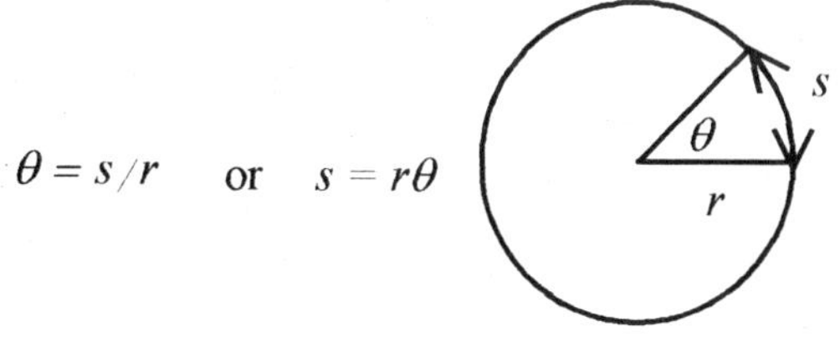

Fig. I-6

The $s = r\theta$ is the basic relation in rotational motion.

The relation between radians and degrees is $2\pi\,\mathrm{rad} = 360^o$

The sine of small angles can be approximated with the radian measurement of angles. Figure I-7 shows the sine of a very small angle and the radian measure of the angle. Take the two sides of the triangle as equal to r. For small angles this is nearly an isosceles triangle. The sine of the angle is

$$\sin\theta = \frac{h}{r} \text{ and the angle in radians is } \theta = \frac{s}{r}$$

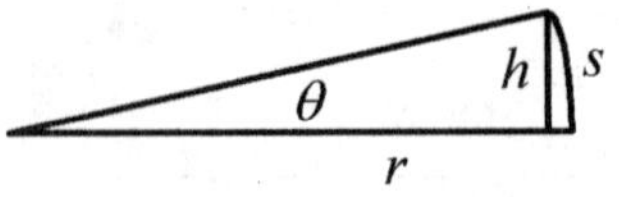

Fig. I-7

For small angles s is approximately h and the sine of the angle is nearly equal to the angle (measured in radians). Take a small angle, $\pi/30$, which is equal to 6 degrees, and as an exercise find the $\sin 6^o$ and $\pi/30$ and verify that the error in using the radian rather than the sine is 2 parts in 1000 or 0.2%.

There are a large number of trigonometric identities that can be derived using geometry and algebra. Several of the more common are in the box below.

TRIGONOMETRIC IDENTITIES

$$a^2 + b^2 = c^2 \qquad \sin^2\theta + \cos^2\theta = 1$$

$$\sin\theta = \cos(90^o - \theta) \qquad \cos\theta = \sin(90^o - \theta)$$

$$\sin(\alpha \pm \beta) = \sin\alpha\cos\beta \pm \cos\alpha\sin\beta \qquad \tan\theta = 1/\tan(90^o - \theta)$$

$$\cos(\alpha \pm \beta) = \cos\alpha\cos\beta \mp \sin\alpha\sin\beta$$

$$\tan(\alpha \pm \beta) = \tan\alpha \pm \tan\beta / 1 \mp \tan\alpha\tan\beta$$

Functions

It is helpful in visualizing problems to know what certain functions look like. The linear algebraic function (see Fig. I-8) is $y = mx + b$, where m is the slope of the straight line and b is the intercept, the point where the line crosses the x-axis.

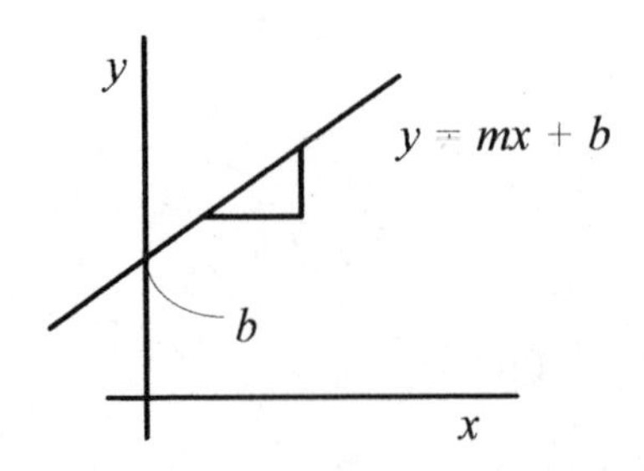

Fig. I-8

The next most complicated function is the quadratic (see Fig. I-9), and the simplest quadratic is $y = x^2$, a curve of increasing slope symmetric about the y-axis. Quadratics are also called parabolas. Adding a constant a in front of the x^2 either sharpens ($a > 1$) or flattens ($a < 1$) the graph. Adding a constant to obtain $y = ax^2 + c$ serves to move the curve up or down the y-axis. Adding a linear term, producing the most complicated quadratic, moves the curve up and down and sideways. If a quadratic is factorable then the places where it crosses the x-axis are obtained directly in factorable form. This discussion of parabolas is continued in the chapter on projectile motion.

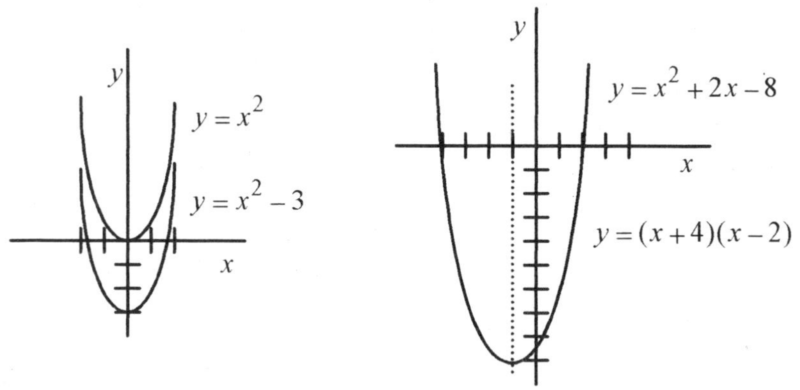

Fig. I-9

With a little experience you should be able to look at a function $y = x^2 + 2x - 8$ (see Fig. I-9) and say that the x^2 means it is a parabola, the coefficient of 1 means it has standard shape, and the other two terms mean that it is moved up and down and sideways. Factor to $y = (x + 4)(x - 2)$, and the curve crosses the x-axis at $x = 2$ and $x = -4$. Because it is a parabola the curve is symmetric about $x = -1$.

Cubic curves have the general shape shown in Fig. I-10. Adding a constant term moves the curve up or down the y-axis. A negative in front of the x^3 term produces a mirror image about the x-axis. Quadratic and linear terms in a cubic produce peaks and troughs in the curve.

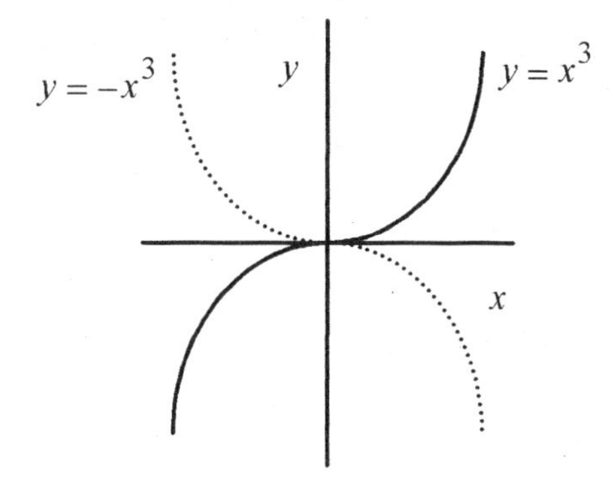

Fig. I-10

Logarithms and Exponents

Logarithms and exponents are used to describe several physical phenomena. The exponential function $y = a^x$ is a unique one with the general shape shown in Fig. I-11.

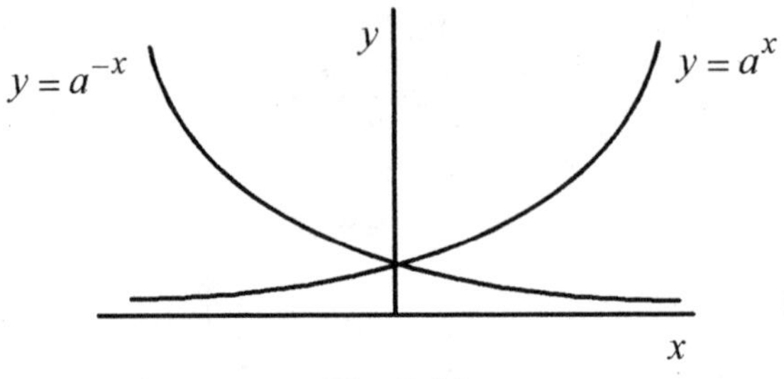

Fig. I-11

This exponential equation $y = a^x$ cannot be solved for x using normal algebraic techniques. The solution to $y = a^x$ is one of the definitions of the logarithmic function: $x = \log_a y$.

The language of exponents and logarithms is much the same. In exponential functions we say "a is the base raised to the power x." In logarithm functions we say "x is the logarithm to the base a of y." The laws for the manipulation of exponents and logarithms are similar. The manipulative rules for exponents and logarithms are summarized in the box.

The term "log" is usually used to mean logarithms to the base 10, while "ln" is used to mean logarithms to the base e. The terms "natural" (for base e) and "common" (for base 10) are frequently used.

LAWS OF EXPONENTS AND LOGARITHMS	
$(a^x)^y = a^{xy}$	$y \log_a x = \log_a x^y$
$a^x a^y = a^{x+y}$	$\log_a x + \log_a y = \log_a xy$
$\dfrac{a^x}{a^y} = a^{x-y}$	$\log_a x - \log_a y = \log_a \dfrac{x}{y}$

Derivatives

There are numerous definitions of the derivative, but the one that fits most physics problems best is that the derivative of a function is another function that gives the slope of the original function at any point. Consider a function $f(x)$, often written as $y = f(x)$, over an interval Δx. The notation $y = f(x)$ is mathematical symbolism that says "a variable y is going to be described by certain operations on another variable x."

Using the Δ notation the general expression for slope is

$$\text{slope} = \frac{\Delta f}{\Delta x} = \frac{f(x+\Delta x) - f(x)}{\Delta x}$$

This equation says that the slope of a function is the value of the function at a point $x + \Delta x$ minus the value of the function at x all divided by the Δx. This assumes the function is linear between x and $x + \Delta x$; an approximation that gets better as Δx gets smaller. The slope defined this way is an average slope between x and $x + \Delta x$. The derivative is the general expression for the slope at any point, thus, it is a function that gives the slope of another function at every point. The derivative, df/dx or f' is the limiting case of the slope where $\Delta x \to 0$

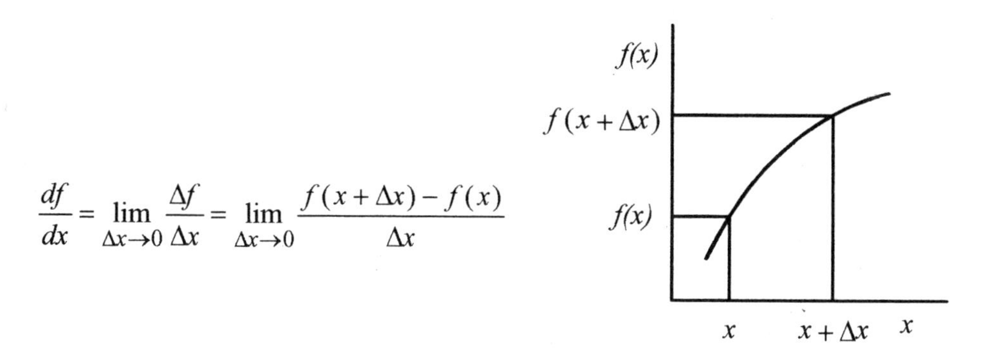

$$\frac{df}{dx} = \lim_{\Delta x \to 0} \frac{\Delta f}{\Delta x} = \lim_{\Delta x \to 0} \frac{f(x+\Delta x) - f(x)}{\Delta x}$$

Fig. I-12

Now apply this procedure to several functions.

$$f(x) = y = 3$$

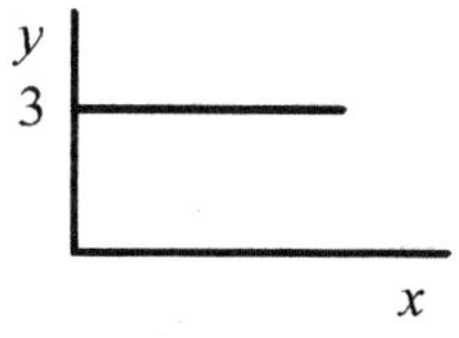

Fig. I-13

The function is a constant so $f(x+\Delta x) = f(x)$ and the slope is zero as is evident from the graph.

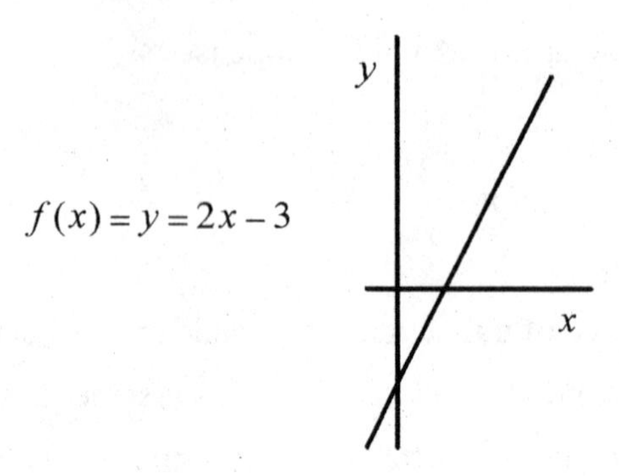

Fig. I-14

Form $f(x+\Delta x) - f(x) = [2(x+\Delta x) - 3] - [2x - 3]$ so that $\dfrac{df}{dx} = \lim\limits_{\Delta x \to 0} \dfrac{2\Delta x}{\Delta x} = 2$.

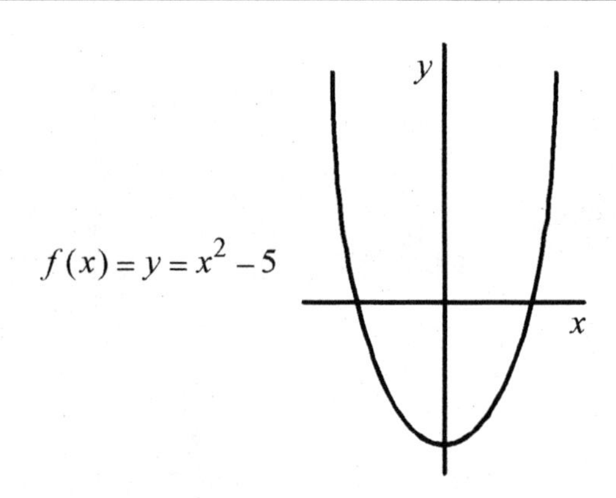

Fig. I-15

Form $f(x+\Delta x) - f(x) = \left[(x+\Delta x)^2 - 5\right] - \left[x^2 - 5\right] = 2x\Delta x + (\Delta x)^2$ so that

$$\frac{df}{dx} = \lim_{\Delta x \to 0} \frac{2x\Delta x + (\Delta x)^2}{\Delta x} = 2x$$

The slope of the curve $y = x^2 - 5$ is $2x$. Just pick a value of x, and the slope is two times this value.

As an exercise verify that the derivative of $y = x^3$ is $3x^2$.

The derivative of power law functions is very easy with the procedure described above. After performing a few of these, we can come to the conclusion that for any power law $y = cv^n$, the general expression for the slope (derivative) is $y' = cnv^{n-1}$. Listed below are the derivatives for power laws as well as some trigonometric, exponential, and logarithmic functions. All of these can be derived using the procedures employed above.

DERIVATIVES

Function	Derivative
x^n	nx^{n-1}
$\sin \alpha x$	$\alpha \cos \alpha x$
$\cos \alpha x$	$-\alpha \sin \alpha x$
$\tan \alpha x$	$\alpha \sec^2 \alpha x = \alpha / \cos^2 \alpha x$
$e^{\alpha t}$	$\alpha e^{\alpha t}$
$\ln \alpha x$	$1/x$

One other useful rule of differentiation is the **chain rule**. If y is written in terms of x and x is written terms of t, it is possible to write dy/dt through the simple expediency of a chain derivative.

$$\frac{dy}{dt} = \frac{dy}{dx}\frac{dx}{dt}$$

If $y = x^4 + 3x^2$ and $x = 2t^2 - t$, then $\dfrac{dy}{dx} = 4x^3 + 6x$ and $\dfrac{dx}{dt} = 4t - 1$ so

$$\frac{dy}{dt} = (4x^3 + 6x)(4t - 1)$$

and since x is written in terms of t, the derivative dy/dt can be written in terms of x or t.

Integrals

Integrals can be viewed two ways, as the area under a curve or as the inverse operation to the derivative. Look upon the derivative as an operation performed on a function. If $y = 3x^2 + 2x - 1$ is the function, then the derivative is d by dx of y or

$$\frac{d}{dx}(y) = 6x + 2$$

The inverse of this operation is called integration. The actual operation of integration is seen by writing the d/dx operation as a total derivative $dy = (6x + 2)dx$, and the integral is

$$\int dy = \int (6x + 2)dx$$

so $\quad y = \dfrac{6x^2}{2} + 2x + \text{const} = 3x^2 + 2 + \text{const}$ $\quad$ the original function plus a constant.

Just as the derivative of a power function $y = cv^n$ is $dy = cnv^{n-1}dv$ the integral of

$$\int dy = \int cnv^{n-1}dv \text{ is } y = cv^n + \text{const}.$$

The constant is necessary because constants are lost in differentiation! Evaluating the constant requires some knowledge of the physical problem.

The other definition of the integral is as the area under a curve. This definition is most convenient in many physical problems, especially those involving work. The integrals of several curves are done below. The integrals are represented by the area under the curves between two specific values.

$y = 3$

$x = 0$ to $x = 2$

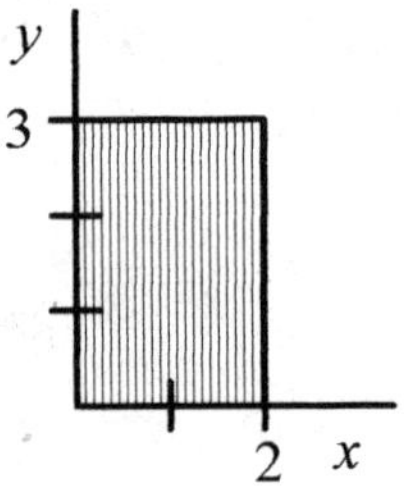

y 3 2 x

Fig. I-16

The area under the curve is $= \int_0^2 3dx = 3x\big|_0^2 = 6 - 0 = 6$. A quick glance at the graph of $y = 3$ confirms this calculation.

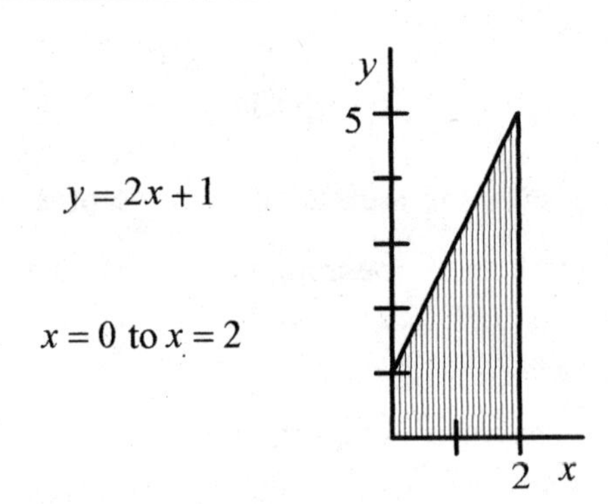

Fig. I-17

The area under the curve is $= \int_0^2 (2x+1)dx = \dfrac{2x^2}{2} + x\Big|_0^2 = \left[2^2 + 2\right] - \left[0^2 + 0\right] = 6$

The shaded area consisting of a rectangle and triangle is equal to 6.

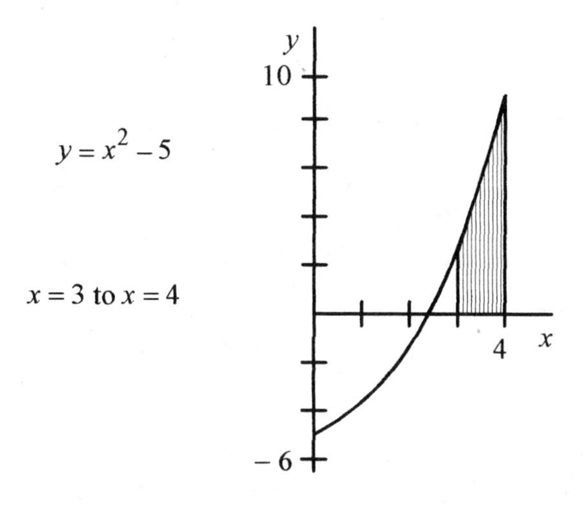

Fig. I-16

$$\text{Area} = \int_3^4 (x^2 - 5)dx = \frac{x^3}{3} - 5x\Big|_3^4 = \left[\frac{64}{3} - 20\right] - [9 - 15] = \frac{64}{3} - 20 + 6 = \frac{64}{3} - \frac{42}{3} = \frac{22}{3}$$

As an exercise approximate the area under the curve. The area between $x = 3$ and $x = 4$ can be approximated with a rectangle and triangle. Find the value of y at $x = 3$ and the area of the rectangle. Find the value of y at $x = 4$ and find the approximate area of the triangle. These two areas are very close to the area found from the integral.

The integrals of these three curves are what is known as definite integrals, ones that have specific limits. As such they do not need constants. The indefinite integrals, those without limits, need the constant. Listed below are the integrals of some common functions.

INTEGRALS

Function	Integral
$x^n\ (n \neq -1)$	$(n+1)^{-1}x^{n+1}$
$1/x$	$\ln x$
$\sin \alpha x$	$-\alpha^{-1}\cos \alpha x$
$\cos \alpha x$	$\alpha^{-1}\sin \alpha x$
$\tan \alpha x$	$\alpha^{-1}\ln\lvert\cos \alpha x\rvert$
$e^{\alpha x}$	$\alpha^{-1}e^{\alpha x}$
$\ln \alpha x$	$x\ln \alpha x - x$

Average Value of a Function

In Fig.I-18 the shaded area is the value of the integral. This area could be represented by a rectangle with one side of the rectangle equal to 1, the length of the integral, from 3 to 4, and the other side, the average height of the function between 3 and 4. This average height is the average value of the function over the interval from 3 to 4. From the geometry then we can say that the average value of the function times the length of the integral equals the area or value of the integral. Rearranging then, the average value of a function over a particular range is the value of the integral over the range divided by the range. Applying this to the function $y = x^2 - 5$ the average value of the function between 3 and 4 is

$$(x^2 - 5)_{\text{avg}3-4} = \frac{1}{4-3}\int_3^4 (x^2 - 5)dx = \frac{22}{3}$$

At $x = 3$ the function has value 4, and at $x = 4$ it has value 11, so 22/3 is a reasonable value for the average.

Likewise, the average value of the function $y = x^3 - 3$ between 1 and 3 is

$$(x^3 - 3)_{\text{avg}1-3} = \frac{1}{3-1}\int_1^3 (x^3 - 3)dx = \left.\frac{x^4}{4} - 3x\right|_1^3$$

$$(x^3 - 3)_{\text{avg}1-3} = \frac{1}{2}\left\{\left[\frac{81}{4} - 9\right] - \left[\frac{1}{4} - 3\right]\right\} = \frac{1}{2}\left[\frac{45}{4} + \frac{11}{4}\right] = 7$$

CHAPTER 1

VECTORS

This chapter serves a dual purpose. First, it will help you to do the addition and subtraction of vectors in the chapter in your text on vectors and the vector problems in the chapters on motion and forces. Second, it will serve as a reference for those topics involving vector products, especially the definitions of work and torque encountered in mechanics.

Certain physical quantities such as mass or temperature are described with a number called a **scalar**. Other quantities, such as displacement, velocity, or force, have a direction associated with them and are called **vectors**. Operationally, a vector is an arrow oriented in space with the length (of the arrow) representing the number and the orientation, the direction. Vectors can be placed anywhere on a coordinate system so long as they maintain their required length and orientation.

Number Plus Angle and Components

A vector can be described with a number and an angle as $A = 23\angle 37^o$. In performing the basic mathematical operations of addition, subtraction, multiplication, and division, it is more convenient to write vectors in component form. If A were placed with the tail (of the vector) at the origin of a coordinate system, then the x and y components could be written as shown in Fig. 1-1.

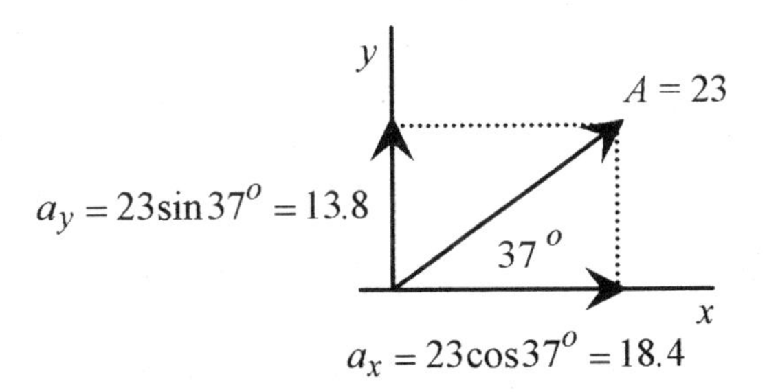

Fig. 1-1

It is very important in the use of vectors to be able to go from the number plus angle format to component format quickly and accurately. Before going any further in this chapter review the basic trigonometric relations and the formulas for going from number plus angle to components and vice versa (Fig. 1-2). And if you are at all unsure of yourself make up a 3 x 5 card with figure and formulas and review it several times a day until you can perform the operations without hesitation.

A vector $\boldsymbol{A} = A\angle\theta$ has components as shown in Fig. 1-2. (The bold type represents the vector, and the normal type represents the number associated with that vector.) If the components, a_x and a_y, are given, then the number plus angle form can be obtained with the trigonometric relations on the left side of Fig. 1-2.

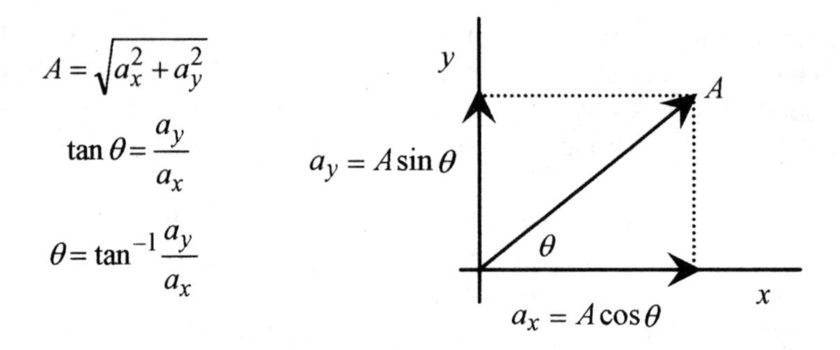

Fig. 1-2

The vector $\boldsymbol{A}$ (see Fig. 1-1) can be reconstructed from the components $a_x = 18.4$ and $a_y = 13.8$.

$$A = \sqrt{18.4^2 + 13.8^2} = 23 \quad \text{and} \quad \theta = \tan^{-1}\frac{13.8}{18.4} = 37^o$$

1-1 Diagram the vector $\boldsymbol{C} = 47\angle 193^o$ and write the components.

Solution: The components are $c_x = 47\cos 193^o = -45.8$ and $c_y = 47\sin 193^o = -10.6$.

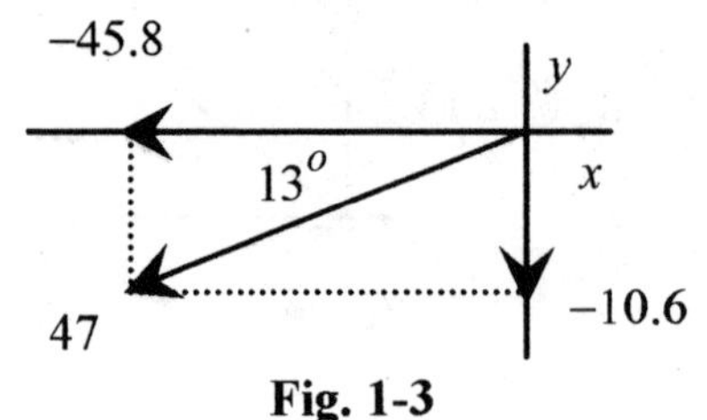

Fig. 1-3

Taking the sine or cosine of the 193° angle will produce the appropriate negative numbers, but drawing a figure and using the principle angle is a better procedure. Less mistakes are made from figures than from the readout of calculators. When the vector and its components are drawn on the coordinate system there can be no mistake that 45.8 and 10.6 are both negative.

Unit Vectors in Two Dimensions

The use of unit vectors simplifies the mathematical operations on vectors. In two dimensions **unit vectors** are vectors of unit value directed in the $+x$ and $+y$ directions.

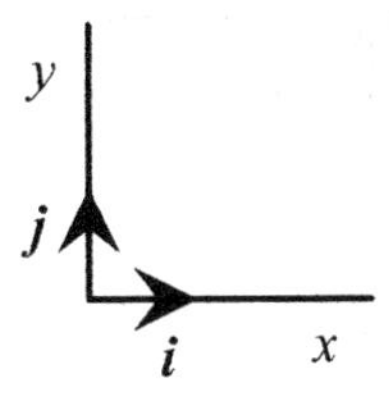

Fig. 1-4

The vector A (Fig. 1-1) would be written as $A = 18.4\boldsymbol{i} + 13.8\boldsymbol{j}$
and C (problem 1-1) would be written as $C = -45.8\boldsymbol{i} - 10.6\boldsymbol{j}$

1-2 Add the vectors A and C.

Solution: The addition of A and C is now accomplished by adding the components

$$S = A + C = -27.8\boldsymbol{i} + 3.2\boldsymbol{j}$$

1-3 Subtract **C** from **A**.

Solution: $T = A - C = 64.2\boldsymbol{i} + 24.4\boldsymbol{j}$

1-4 Diagram $T = A - C$ and write in number plus angle form.

Solution: The diagram is started by drawing the components on the coordinate system. With the components the magnitude and angle can be calculated $T = 68.7\angle 20.8^o$.

$$T = \sqrt{64.2^2 + 24.4^2} = 68.7$$

$$\theta = \tan^{-1}(24.4/64.2) = 20.8^o$$

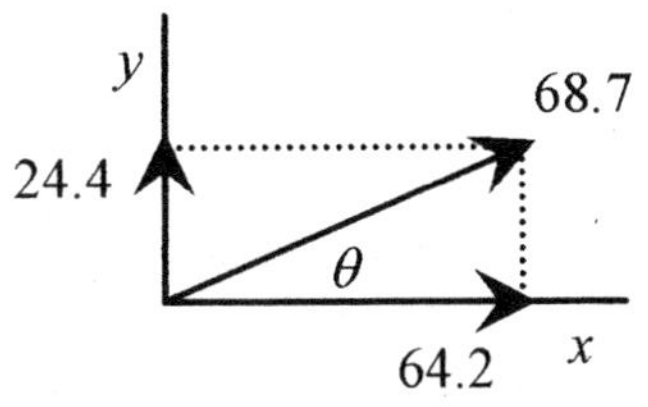

Fig. 1-5

1-5 Add the vectors $A = 13\angle 50^o$, $B = 15\angle -60^o$, and $C = 17\angle 20^o$.

Solution: Diagram each vector along with the components.

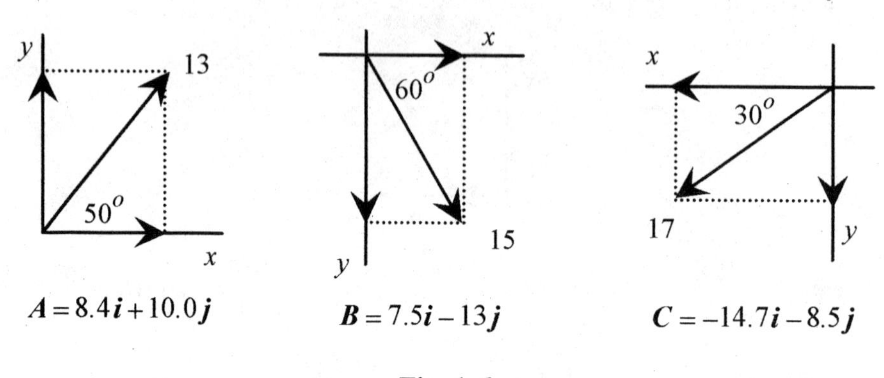

Fig. 1-6

The sum of these vectors $\boldsymbol{R} = \boldsymbol{A} + \boldsymbol{B} + \boldsymbol{C}$ is $\boldsymbol{R} = 1.2\boldsymbol{i} - 11.5\boldsymbol{j}$. This resultant vector is diagrammed in Fig. 1-7 along with the magnitude and angle.

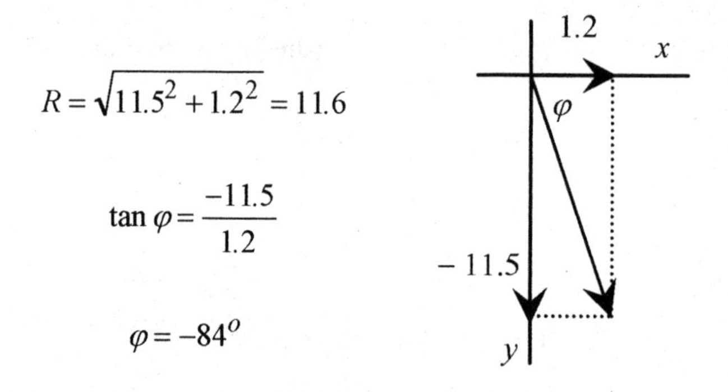

Fig. 1-7

The vector is $\quad \boldsymbol{R} = 11.6\angle -84^o$

The key to getting vector addition problems correct is to use multiple diagrams. Most mistakes in vector problems are sign mistakes. And the way to avoid sign mistakes is to use diagrams extensively.

1-6 Find the resultant of the two forces $F_1 = 800\,\mathrm{N}\angle 47^o$ and $F_2 = 600\,\mathrm{N}\angle 140^o$.

Solution: Place the force vectors with components on the same diagram (Fig. 1-8).

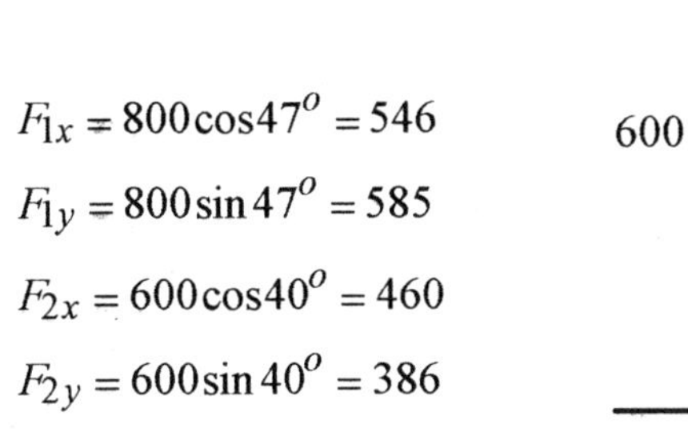

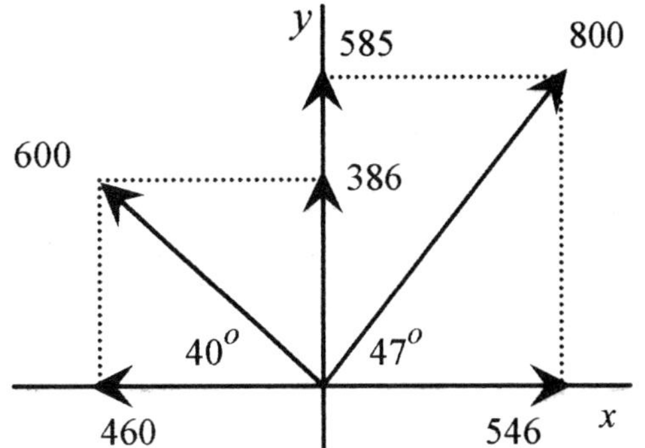

Fig. 1-8

The components of the resultant can be diagrammed directly. If the diagrams are done in this manner it is not necessary to use large angles and misread calculator readouts. Note that component forces in the y direction are both positive, while F_{1x} and F_{2x} are in opposite directions. Taking forces acting in the $+x$ and $+y$ directions as positive, and forces acting in the $-x$ and $-y$ directions as negative, the result of the addition of these two forces is 86 in the $+x$ direction and 971 in the $+y$ direction. Using the diagram it is not necessary to keep track of the plus and minus signs. The diagram makes clear how the components should be added to produce the correct resultant. The resultant written in magnitude and angle form is $\boldsymbol{F} = 975\,\mathrm{N}\angle 85^o$.

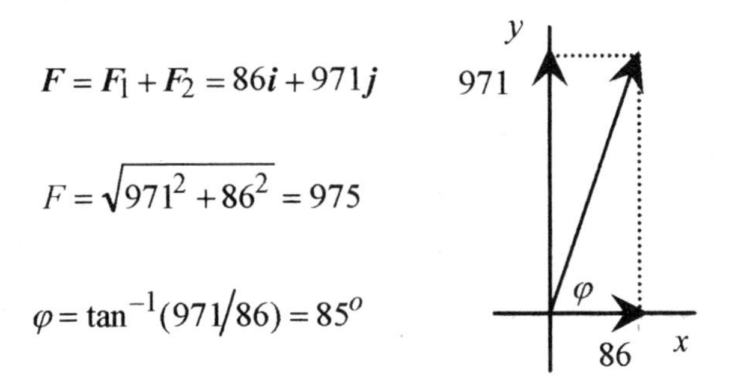

Fig. 1-9

There are two different types of products of vectors. One results in a scalar, and the other results in a vector. The next two sections discuss these products. Depending on the order of topics in your course, you may want to put off reading these sections until they come up in mechanics.

Scalar or Dot Product

The dot product $(\boldsymbol{A}\cdot\boldsymbol{B})$ produces a scalar. There are two definitions of the **dot product**. The most easily visualized is

$$A \cdot B = AB \cos\theta \tag{1-1}$$

where θ is the angle between A and B. This definition can be viewed as the projection of A on B or the component of A in the direction of B times the magnitude of B.

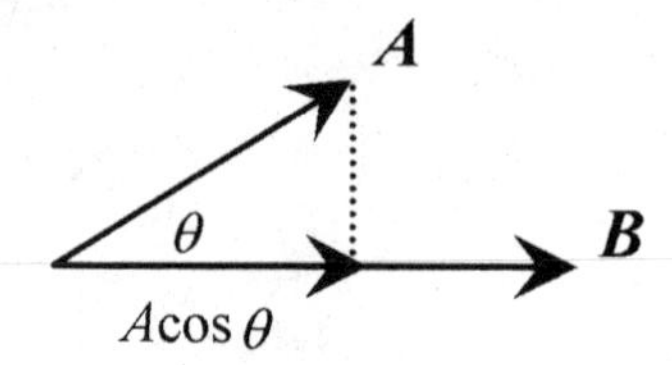

Fig. 1-10

The second definition fits with the unit vector notation

$$A \cdot B = a_x b_x + a_y b_y \tag{1-2}$$

Following $A \cdot B = AB\cos\theta$, $i \cdot i = 1 \cdot 1\cos 0^o = 1$, and $i \cdot j = 1 \cdot 1\cos 90^o = 0$.

Following $A \cdot B = a_x b_x + a_y b_y$, $i \cdot i = 1 \cdot 1 = 1$, and $i \cdot j = 0$.

1-7 Form the dot product of $A = 23\angle 37^o$ and $B = 14\angle -35^o$.

Solution: Using the first definition $A \cdot B = A \cdot B\cos\theta = 23 \cdot 14\cos 72^o = 100$.

The second definition of the dot product requires the components of the vectors. The components of A (from Fig. 1-1) are $a_x = 18.4$ and $a_y = 13.8$. The components of B are in Fig. 1-11.

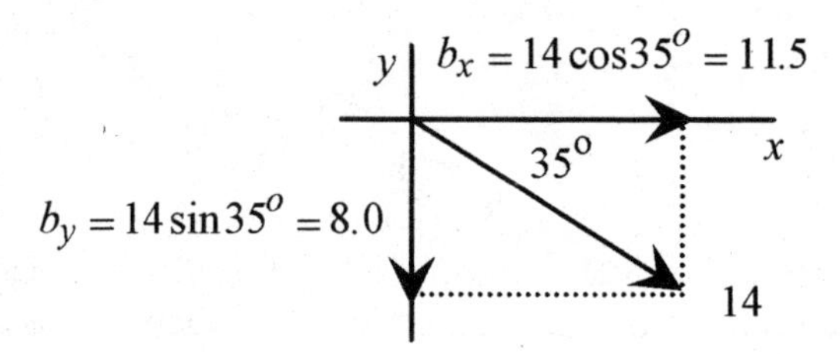

Fig. 1-11

The dot product is

$$A \cdot B = a_x b_x + a_y b_y = 18.4 \cdot 11.5 + 13.8(-8.0) = 100$$

More dot product problems will be done in the chapters where dot products are used in the calculation of work. Work, as defined in mechanics, is the product of the component of an applied force in the direction of a displacement and that displacement. $W = F \cdot s$

Vector or Cross Product

The cross product ($\boldsymbol{A} \times \boldsymbol{B}$) produces a vector. As with the dot product there are two definitions of the cross product. The simplest definition to understand is that the **cross product** of A and $\boldsymbol{B}$ produces a vector of magnitude $AB \sin\phi$ in a direction normal (perpendicular) to the plane of A and $\boldsymbol{B}$ with ϕ the angle between A and $\boldsymbol{B}$. The specific direction is obtained by rotating $\boldsymbol{A}$ into $\boldsymbol{B}$ (crossing $\boldsymbol{A}$ into $\boldsymbol{B}$ or $\boldsymbol{A}$ cross $\boldsymbol{B}$) again using the fingers of the right hand naturally curling (closing) from $\boldsymbol{A}$ to $\boldsymbol{B}$ with the thumb pointing in the direction of the new (product) vector. This is the same procedure as for finding the z direction in an x-y-z right-handed coordinate system. A right-handed coordinate system with the three unit vectors is shown in Fig. 1-12.

Practice visualizing $\boldsymbol{i} \times \boldsymbol{j}$ to produce $\boldsymbol{k}$ and $\boldsymbol{j} \times \boldsymbol{k}$ to produce $\boldsymbol{i}$. The angle between the unit vectors is 90°, and their magnitude is 1; so the resultant vector has magnitude 1 and is in the direction given by this "vector crossed into another vector" procedure. Practice pointing your fingers in the direction of the first vector, curling them into the second vector with your thumb pointing in the direction of the result of this "cross" product until you can quickly see that $\boldsymbol{i} \times \boldsymbol{k} = -\boldsymbol{j}$ and $\boldsymbol{k} \times \boldsymbol{i} = +\boldsymbol{j}$.

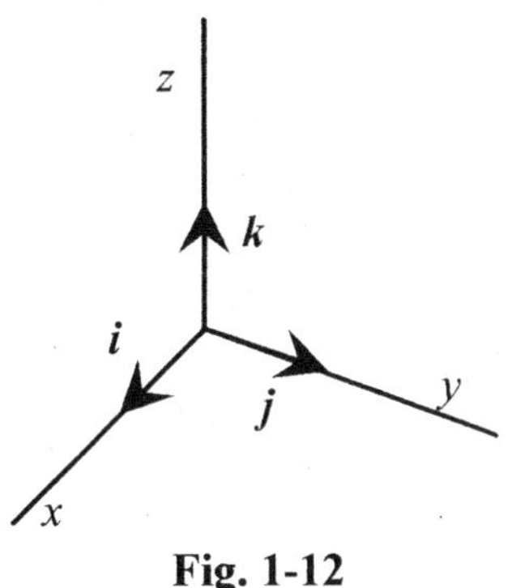

Fig. 1-12

The definition of torque in mechanics is lever arm times force times the sine of the angle between them. Another way of saying this is that torque is lever arm times the component of the force at right angles to the lever arm. In mathematical terms

$$\mathcal{T} = \boldsymbol{r} \times \boldsymbol{F}$$

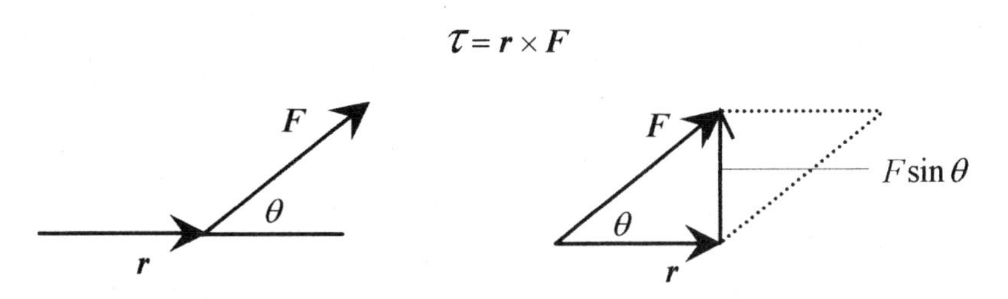

Fig. 1-13

with the magnitude of the torque given by $T = rF \sin\theta$ and the direction of the vector given by the $\boldsymbol{r} \times \boldsymbol{F}$ rule. The $F \sin\theta$ term can be viewed as the component of F perpendicular to r.

In Fig. 1-13 the vectors are arranged as the sides of a parallelogram with $F\sin\theta$ the height of the parallelogram. The product $rF\sin\theta$ is the magnitude of $\boldsymbol{r}\times\boldsymbol{F}$ and is also the area of the parallelogram (a base times the height) formed by $\boldsymbol{r}$ and $\boldsymbol{F}$.

The second definition of the cross product is mathematically easier but harder to relate to physical problems. This definition is expressed as a determinant.

$$\boldsymbol{A}\times\boldsymbol{B}=\begin{vmatrix}\boldsymbol{i} & \boldsymbol{j} & \boldsymbol{k}\\ a_x & a_y & a_z\\ b_x & b_y & b_z\end{vmatrix} \tag{1-3}$$

$$\boldsymbol{A}\times\boldsymbol{B}=\boldsymbol{i}\left[a_yb_z-a_zb_y\right]+\boldsymbol{j}\left[a_zb_x-a_xb_z\right]+\boldsymbol{k}\left[a_xb_y-a_yb_x\right] \tag{1-4}$$

1-8 Form the cross product of $\boldsymbol{A}=23\angle 37^o$ and $\boldsymbol{C}=47\angle 193^o$.

Solution: The vectors and their components are shown in Fig. 1-14. Crossing $\boldsymbol{A}$ into $\boldsymbol{C}$ defines the angle as 156° so that $AC\sin 156^o=440$ with the direction out of the paper as give by the right-hand rule.

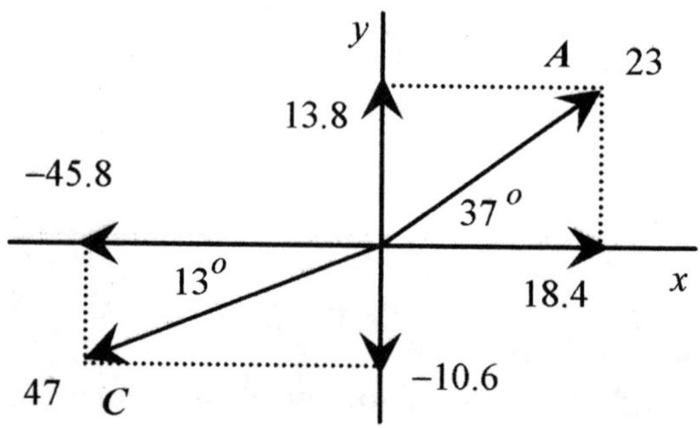

Fig. 1-14

Doing the same problem with determinants

$$\boldsymbol{A}\times\boldsymbol{C}=\begin{vmatrix}\boldsymbol{i} & \boldsymbol{j} & \boldsymbol{k}\\ 18.4 & 13.8 & 0\\ -45.8 & -10.6 & 0\end{vmatrix}$$

$$\boldsymbol{A}\times\boldsymbol{C}=\boldsymbol{k}\left[18.4(-10.6)-13.8(-45.8)\right]=437\boldsymbol{k}$$

The numeric difference, 440 versus 437, is due to rounding the components to three significant figures.

In simple problems it is easier to find the cross product from the geometric definition. In more complicated problems, or where it is difficult to be sure of the direction of the vector product, the determinant form is more convenient.

CHAPTER 2

MOTION IN ONE DIMENSION

The motion of a particle (ballistic missiles, golf balls, and gas molecules are all examples of particles) is described by giving position, velocity, and acceleration usually as a function of time. For convenience in getting started we confine the discussion to one dimension. We also need to differentiate between distance and displacement, and speed and velocity.

If a dog fetching a stick runs in a straight line (the $+x$ direction) $30\,\text{m}$ to pick up the stick and returns (the $-x$ direction) $26\,\text{m}$ then the total distance traveled is $56\,\text{m}$ but the displacement is $+4\,\text{m}$. Distance generally means total distance traveled while **displacement** is the actual difference between end point and beginning.

If the dog were to execute the fetching in $8\,\text{s}$ then the speed of the dog would be the total distance traveled divided by the time or $7\,\text{m/s}$ while the **velocity,** defined as the displacement divided by the time, would be $0.5\,\text{m/s}$.

In a word equation form this is

$$\text{velocity} = \frac{\text{displacement}}{\text{time interval}} = \frac{\Delta x}{\Delta t}$$

$$\text{acceleration} = \frac{\text{difference in velocity}}{\text{time interval}} = \frac{\Delta v}{\Delta t}$$

The "Δ" notation is read as "a change in" or "a small change in."

Let's look at some simple relationships between position and velocity. Remember we are confined to one dimension. The simplest relation is $x = \text{constant}$. As shown in Fig. 2-1, the particle doesn't move. There is no velocity or acceleration because x does not change with time.

The first complication is $x = t$ (Fig. 2-2). The particle moves equal increments in x in equal increments of time. That is, the change in x from $t = 1$ to $t = 2$ is the same (same change) as when t goes from $t = 2$ to $t = 3$. The velocity then is a constant. And by definition the acceleration is zero.

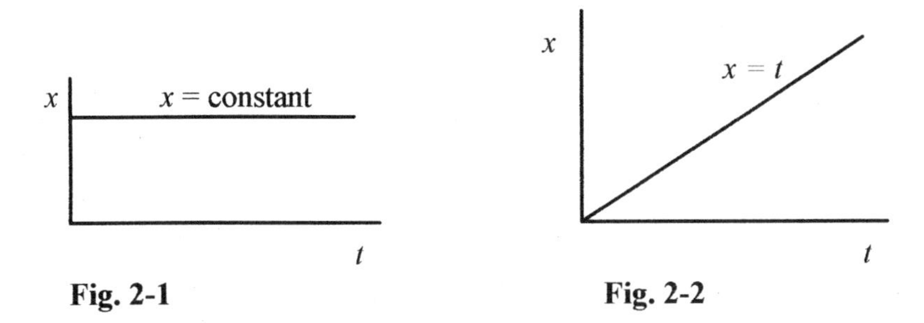

Fig. 2-1 **Fig. 2-2**

The next complication is $x = t^2$. This graph is shown in Fig. 2-3. In this quadratic relationship, the simple definition of velocity begins to break down. Between $t = 0$ and $t = 1$ the velocity is $(1-0)/(1-0) = 1$. Between $t = 1$ and $t = 2$ the velocity is $(4-1)/(2-1) = 3$ and on and on with the velocity changing depending on the time interval chosen. Clearly the acceleration is not a constant. This velocity calculation fits our present definition and is equivalent to drawing a straight line between points on the smooth curve of x versus t.

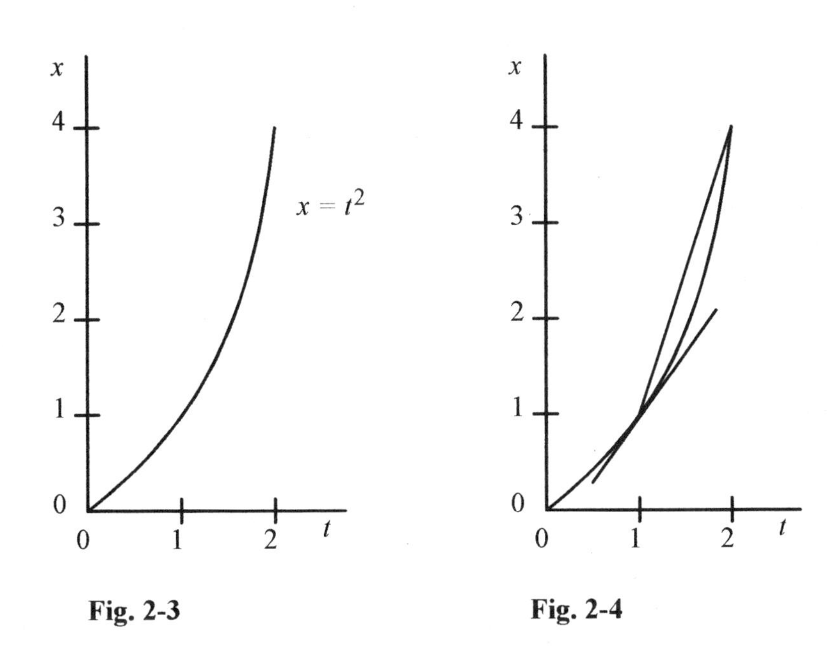

Fig. 2-3 **Fig. 2-4**

Calculating velocity this way presents a problem because depending on the interval we will get different answers for the velocity. To find the velocity at $t = 1$, we found x at $t = 1$ and x at $t = 2$ and performed the velocity calculation. This is not a good approximation to the velocity at $t = 1$ because the average is between $t = 1$ and $t = 2$. A better approximation would be to take values of x between $t = 1$ and $t = 1.1$. Even better would be to take values between $t = 0.99$ and $t = 1.01$. And for even better approximations just shorten the time interval centered about $t = 1$. As the interval gets smaller we will get a better and better measure of velocity. In Fig. 2-4 the slope of the straight line from $x = 1$ to $x = 4$ represents the average velocity between $x = 1$ and $x = 4$. The slope of the straight line tangent to the curve at $x = 1$ represents the velocity at $x = 1$.

Instantaneous Velocity and Acceleration

A more versatile definition of velocity is $\Delta x / \Delta t$, where the interval Δt is very small and centered about the time where the velocity is desired. This approach leads to a general method for obtaining an expression for velocity that can be evaluated at any point rather than going through the numeric calculation whenever we want a velocity.

(C) In words, this definition is stated as the **instantaneous velocity** which is the value of $\Delta x / \Delta t$ as Δt approaches zero. In equation form this is written as

$$v = \lim_{\Delta t \to 0} \frac{\Delta x}{\Delta t} = \frac{dx}{dt}$$

This definition of the **derivative** as the limit of $\Delta x / \Delta t$ as Δt approaches zero is the slope of the tangent to the curve evaluated at the point in question. Thus if we want to find the velocity of any particle traveling according to a polynomial relation between x and t all we need is a general technique for finding the slope of the tangent to the polynomial at any point.

The **instantaneous acceleration** is defined as the value of $\Delta v/\Delta t$ as Δt approaches zero. In equation form this is

$$a = \lim_{\Delta t \to 0} \frac{\Delta v}{\Delta t} = \frac{dv}{dt}$$

The instantaneous acceleration is the slope of the tangent to the curve of v versus t.

(C) The general expression for the slope of any polynomial is discussed in the Introduction, Mathematical Background. For a polynomial of the form $x = ct^n$ the expression for the slope at any point is cnt^{n-1}. Stating this in calculus terms; for any function $x = ct^n$, the derivative of the function is cnt^{n-1}. This can be verified in the case of the parabola by taking successively smaller intervals of Δt and Δx at any point t to verify that the slope at any point is $2t$.

Kinematic Equations of Motion

The kinematic equations of motion are derived under the assumption of constant acceleration. While this may at first seem to be a restriction, there are a large number of problems where the acceleration is a constant. The simplest and most obvious are falling body problems, that is, problems involving bodies falling on (or near) the surface of the earth where the acceleration due to gravity is a constant. Falling body problems are taken up in a separate chapter. In the derivation of the kinematic equations of motion a good image to keep in mind is that of falling bodies.

Starting with the assumption of constant acceleration we can write $a = \frac{v_f - v_o}{t}$,

which can be rearranged to $v_f = v_o + at$.

Usually the *f* subscript is dropped to read $v = v_o + at$.

Now defining the average velocity as $v_{avg} = \frac{v + v_o}{2}$ **(2-1)**

and the displacement as $x = x_o + v_{avg}t$ **(2-2)**

and substituting for v_{avg} with the previous equation $x = x_o + \frac{v + v_o}{2}t$

and further substituting for $v = v_o + at$, we arrive at $x = x_o + v_o t + (1/2)at^2$.

And if $a = \frac{v - v_o}{t}$ is solved for *t* and substituted into $x = x_o + \frac{v + v_o}{2}t$,

then we get $x = x_o + \frac{(v + v_o)}{2}\frac{(v - v_o)}{a}$.

and upon rearranging $v^2 = v_o^2 + 2a(x - x_o)$.

Summarizing, these four kinematic equations of motion are written as

$$v = v_o + at \tag{2-3}$$

$$x - x_o = (1/2)(v + v_o)t \tag{2-4}$$

$$x - x_o = v_o t + (1/2)at^2 \tag{2-5}$$

$$v^2 = v_o^2 + 2a(x - x_o) \tag{2-6}$$

The first three equations relate displacement, velocity, and acceleration in terms of time while the fourth equation does not contain the time.

Now let's apply these four equations to some typical problems. Remember that the kinematic equations of motion allow us to describe the position, velocity, and acceleration of a mass point.

2-1 A train starts from rest (at position zero) and moves with constant acceleration. On first observation the velocity is 20 m/s and 80 s later the velocity is 60 m/s. At 80 s, calculate the position, average velocity, and the constant acceleration over the interval.

Solution: Diagram the problem.

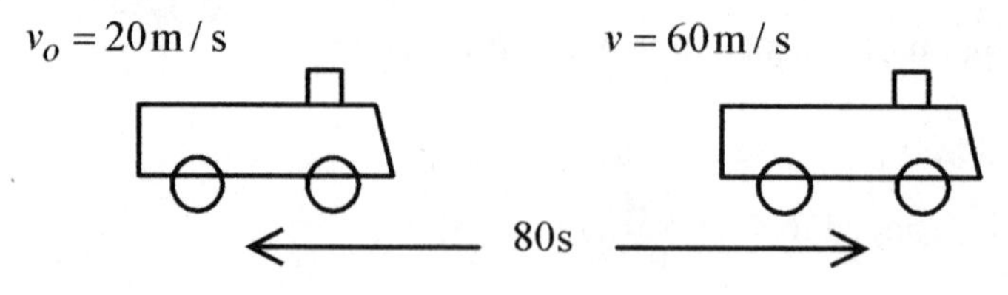

Fig. 2-5

Calculate the acceleration: $a = \frac{v - v_o}{t} = \frac{60\,\text{m/s} - 20\,\text{m/s}}{80\,\text{s}} = 0.50\,\text{m/s}^2$

Calculate the distance traveled over this 80s:

$$x = v_o t + (1/2)at^2 = (20\,\text{m/s})80\,\text{s} + (1/2)(0.50\,\text{m/s}^2)6400\,\text{s}^2 = 3200\,\text{m}$$

The average velocity is

$$v_{\text{avg}} = \frac{v + v_o}{2} = \frac{20\,\text{m/s} + 60\,\text{m/s}}{2} = 40\,\text{m/s}$$

If the acceleration is constant then the average velocity is the average of $20\,\text{m/s}$ and $60\,\text{m/s}$, or $40\,\text{m/s}$, and at an average velocity of $40\,\text{m/s}$ and $80\,\text{s}$, the distance traveled is $3200\,\text{m}$.

2-2 For the situation of problem 2-1, calculate the position of the train at 20s.

Solution: $x = v_o t + (1/2)at^2 = (20\,\text{m/s})20\,\text{s} + (1/2)(0.50\,\text{m/s}^2)(20\,\text{s})^2 = 500\,\text{m}$

2-3 For the situation of problem 2-1 find the time required for the train to reach 100m

Solution: $x - x_o = v_o t + (1/2)at^2$ so $100\,\text{m} = (20\,\text{m/s})t + (1/2)(0.50\,\text{m/s}^2)t^2$

This is a quadratic equation in t. Without units the equation is $t^2 + 80t - 400 = 0$

and has solutions $t = \frac{-80 \pm \sqrt{6400 - 4(1)(400)}}{2(1)} = \frac{-80 \pm 89}{2} = 4.5, -85$

The negative answer is inappropriate for this problem so take $t = 4.5\,\text{s}$.

2-4 For the situation of problem 2-1 find the velocity of the train at 120m.

Solution:

$$v^2 = v_o^2 + 2ax = (20\,\text{m/s})^2 + 2(0.50\,\text{m/s}^2)120\,\text{m} = 400\,\text{m}^2/\text{s}^2 + 120\,\text{m}^2/\text{s}^2 = 520\,\text{m}^2/\text{s}^2$$

$$v = 23\,\text{m/s}$$

2-5 Two vehicles are at position $x = 0$ at $t = 0$. Vehicle 1 is moving at constant velocity of $30\,\text{m/s}$. Vehicle 2, starting from rest, has acceleration of $10\,\text{m/s}^2$. A typical question of this situation is "Where do they pass?"

Solution: First diagram the situation.

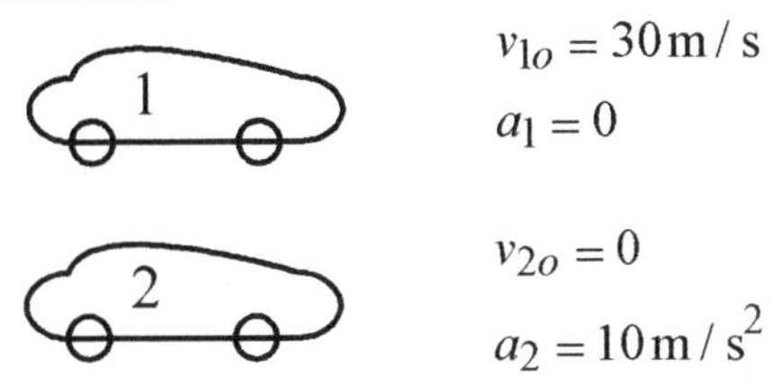

Fig. 2-6

The question "Where do they pass?" translated into algebra means "What is the value of x when they pass?" This can be determined by writing equations for the position of each vehicle and equating $x_1 = v_{1o}t = (30\,\text{m/s})t$ and $x_2 = (1/2)a_2t^2 = (5\,\text{m/s}^2)t^2$

Setting $x_1 = x_2$ gives the time when they pass as $30t = 5t^2$ or $5t(t-6) = 0$; so the vehicles pass at $t = 0$ and $t = 6$. Putting $t = 6\,\text{s}$ in either equation for x (x_1 or x_2) gives $180\,\text{m}$ as the distance.

2-6 For the situation in problem 2-5, when do the vehicles have the same velocity?

Solution: In algebra this means to set the equations for velocity equal ($v_1 = v_2$) and solve for the time. Remember that three of the four equations of motion are functions of time so most questions are answered by first calculating the time for a certain condition to occur.

$$30\,\text{m/s} = (10\,\text{m/s}^2)t \qquad \text{or} \qquad t = 3.0\,\text{s}$$

Now that we know when, we can calculate where they have the same velocity. Use either equation for position and $t = 3.0\,\text{s}$

$$x_1\big|_{t=3.0} = (30\,\text{m/s})3.0\,\text{s} = 90\,\text{m}$$

2-7 For the situation of problem 2-5, what is the position, velocity, and acceleration of each vehicle when vehicle 2 has traveled twice the distance of vehicle 1?

Solution: The time when this occurs is when $x_2 = 2x_1$: $5t^2 = 60t$ or $5t(t-12) = 0$

This gives times of $t = 0$ and $t = 12$. The time $t = 0$ is correct, though uninteresting. The time $t = 12$ is the physically interesting answer.

At $t = 12$: $\quad x_1\big|_{t=12} = (30\,\text{m/s})12\,\text{s} = 360\,\text{m}$

From the original statement of the problem, $v_1 = v_{1o} = 30\,\text{m/s}$ and $a_1 = 0$.

Now solve for the remaining variables for the second vehicle by substitution

$$x_2\big|_{t=12} = (5\,\text{m/s}^2)(12\,\text{s})^2 = 720\,\text{m}$$

$$v_2\big|_{t=12} = 10\,\text{m/s}^2\,12\,\text{s} = 120\,\text{m/s}$$

From the original statement of the problem $a_2 = 10\,\text{m/s}$.

2-8 Two trains are traveling along a straight track, one behind the other. The first train is traveling at $12\,\text{m/s}$. The second train, approaching from the rear, is traveling at $20\,\text{m/s}$. When the second train is $200\,\text{m}$ behind the first, the operator applies the brakes producing a constant deceleration of $0.20\,\text{m/s}^2$. Will the trains collide, and if so where and when?

Solution: First, diagram the situation. Our strategy will be to write down the equations for each train using equations (2-1) through (2-6) and the information provided in the problem. Take $t = 0$ when the brakes are applied and the first train is $200\,\text{m}$ ahead of the second. This makes the position of the first train $200\,\text{m}$ at $t = 0$ (x_{1o} in equation 2-2).

$$x_1 = 200\,\text{m} + (12\,\text{m/s})t \qquad x_2 = (20\,\text{m/s})t - (0.10\,\text{m/s}^2)t^2$$

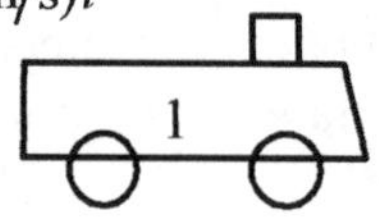

$$v_1 = 12\,\text{m/s} \qquad v_2 = 20\,\text{m/s} - (0.20\,\text{m/s}^2)t$$

$$a_1 = 0 \qquad a_2 = -0.20\,\text{m/s}^2$$

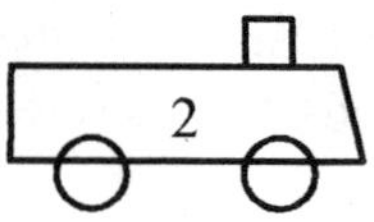

Fig. 2-7

The question as to whether the trains collide means: "Is there a real (time) solution to the equation resulting from setting $x_1 = x_2$.

$$200\,\text{m} + (12\,\text{m/s})t = (20\,\text{m/s})t - (0.10\,\text{m/s}^2)t^2$$

If there are no real solutions to this equation then the trains do not collide. Drop the units and write $0.10t^2 - 8t + 200 = 0$ which is solved by quadratic formula

$$t = \frac{+8 \pm \sqrt{64 - 4(0.10)200}}{2(0.10)} = \frac{8 \pm \sqrt{-16}}{0.20}$$

Since there are no real solutions to this equation there are no times when the trains collide.

2-9 Change problem 2-8 by giving the second train an initial velocity of $25\,\mathrm{m/s}$. This will give a real time for the collision. Find the collision time.

Solution: The situation is now as shown in the Fig. 2-8.

Again setting $x_1 = x_2$ and dropping the units produces $200 + 12t = 25t - 0.10t^2$ or $0.10t^2 - 13t + 200 = 0$ with solutions

$$t = \frac{+13 \pm \sqrt{169 - 4(0.10)200}}{2(0.10)} = \frac{13 \pm \sqrt{89}}{0.20} = \frac{13 \pm 8.3}{0.20} = 17.8,\,112$$

$x_1 = 200\,\mathrm{m} + (12\,\mathrm{m/s})t$

$v_1 = 12\,\mathrm{m/s}$

$a_1 = 0$

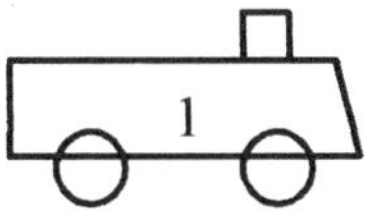

$x_2 = (25\,\mathrm{m/s})t - (0.10\,\mathrm{m/s^2})t^2$

$v_2 = 25\,\mathrm{m/s} - (0.20\,\mathrm{m/s^2})t$

$a_2 = -0.20\,\mathrm{m/s^2}$

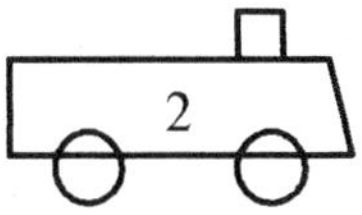

Fig. 2-8

The two times correspond to when $x_1 = x_2$. The earliest time is the first coincidence and the end of the (physical) problem. The position at this time can be obtained from either expression for x.

$x_1\big|_{t=17.8} = 200\,\mathrm{m} + (12\,\mathrm{m/s})17.8\,\mathrm{s} = 414\,\mathrm{m}$ Verify this distance by using x_2.

The velocity of the second train at collision is

$$v_2\big|_{t=17.8} = 25\,\mathrm{m/s} - (0.20\,\mathrm{m/s^2})17.8\,\mathrm{s} = 21.4\,\mathrm{m/s}$$

The relative velocity (between the two trains) is $v = 1.4\,\mathrm{m/s}$.

The two times are the result of the quadratic in t. The two solutions occur when the curves cross. The equation for $x_1 = 200 + 12t$ is a straight line of slope 12 starting at 200. The equation for $x_2 = 25t - 0.10t^2$ is a parabola that opens down. Figure 2-9 (not to scale) shows the two curves.

While the "mathematics" produces two times, the reality of the problem dictates the earlier time as the one for the "collision."

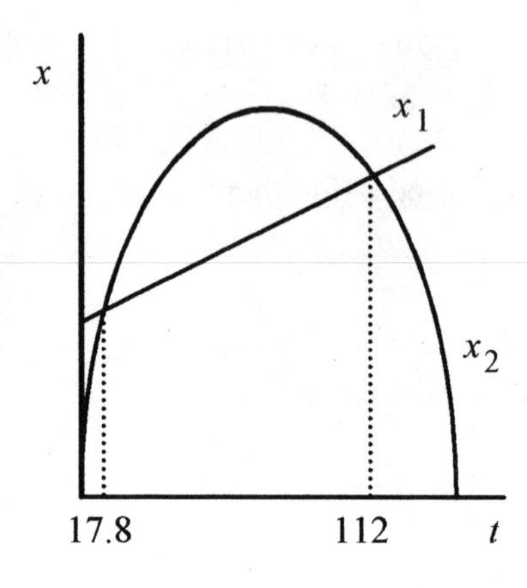

Fig. 2-9

CHAPTER 3

FALLING BODY PROBLEMS

The kinematic equations of motion for constant acceleration, equation 2-1 and equations 2-3 through 2-6 of the previous chapter, can be applied to a large collection of problems known as "falling body problems," problems where the constant acceleration is the acceleration due to gravity on the surface of the earth. These equations from the previous chapter are rewritten here for your convenience.

$$v_{avg} = \frac{v + v_o}{2} \qquad \textbf{(3-1)}$$

$$v = v_o + at \qquad \textbf{(3-2)}$$

$$x - x_o = (1/2)(v + v_o)t \qquad \textbf{(3-3)}$$

$$x - x_o = v_o t + (1/2)at^2 \qquad \textbf{(3-4)}$$

$$v^2 = v_o^2 + 2a(x - x_o) \qquad \textbf{(3-5)}$$

3-1 Consider a ball dropped from the top of a 40 m tall building. Calculate everything possible.

Solution: "Calculate everything possible" is an unusual request. Usually early in your study of falling body problems, there is a problem that asks for something that does not at all seem like it has anything to do with the information given. It's almost like asking "What color is the building?" When this happens, and it probably will, the way to do the problem is to calculate what you can and let the information you generate lead you through the problem. Let's apply this approach to this problem. First, diagram the problem as shown in Fig. 3-1. Place the origin at the top of the building with displacement, velocity, and acceleration (g) all positive down.

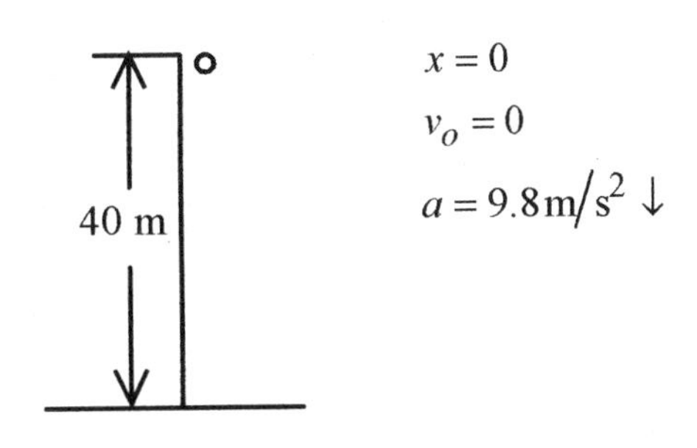

Fig. 3-1

Since most of the kinematic equations contain the time, this is usually one of the first things to calculate. Use equation 3-4 to find the time for the ball to strike the ground.

$$x - x_o = v_o t + (1/2)at^2 \quad \text{or} \quad 40\,\text{m} = (1/2)(9.8\,\text{m/s}^2)t^2 \quad \text{and} \quad t = 2.9\,\text{s}$$

Knowing the time we can calculate the velocity from equation 3-2.

$$v = v_o + gt = (9.8\,\text{m/s}^2)2.9\,\text{s} = 28\,\text{m/s}$$

Alternatively we can use equation 3-4.

$$v^2 = v_o^2 + 2a(x - x_o) = 2(9.8\,\text{m/s}^2)40\,\text{m} = 784\,\text{m}^2/\text{s}^2 \quad \text{or} \quad v = 28\,\text{m/s}$$

NOTE: This last equation is a better one to use because it relies on original data rather than calculated data. If there had been an error in the time calculation it would have been repeated in the $v = v_o + gt$ equation. Also round off errors are eliminated by using equations that rely on original data.

3-2 Now add a complication to problem 3-1 by throwing the ball down with an initial velocity of $8.0\,\text{m/s}$. Find the time for the ball to reach the ground and the velocity on impact.

Solution: Again diagram the problem as in Fig. 3-2. Note that in this problem the displacement and velocity are positive down.

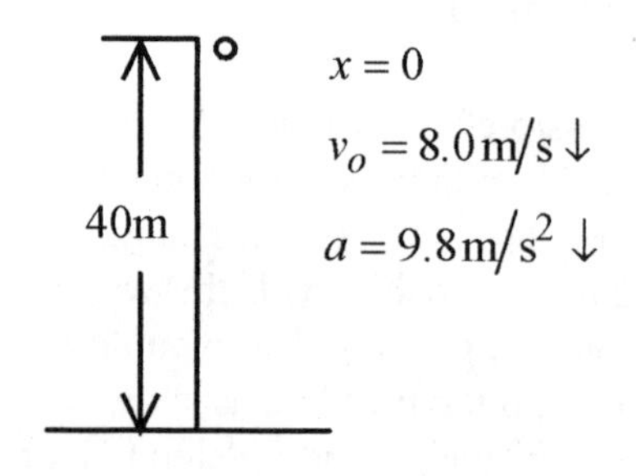

Fig. 3-2

The time of flight is from $40\,\text{m} = (8.0\,\text{m/s})t + (1/2)(9.8\,\text{m/s}^2)t^2$,

which rearranged and without units is $4.9t^2 + 8.0t - 40 = 0$,

with solution by formula of $t = \dfrac{-8.0 \pm \sqrt{64 - 4(4.9)(-40)}}{2(4.9)} = 2.2, -3.8$.

The positive time is the obvious choice.

The velocity at the ground level is from

$$v^2 = \left(8.0\,\mathrm{m/s^2}\right)^2 + 2(9.8\,\mathrm{m/s^2})40\,\mathrm{m} = 848\,\mathrm{m^2/s^2} \qquad v = 29\,\mathrm{m/s}$$

3-3 Add a different complication to problem 3-1. Throw the ball up from the top of the building with a velocity of $8.0\,\mathrm{m/s}$. Find the time for the ball to reach the ground and the velocity on impact

Solution: Diagram the problem as in Fig. 3-3. Take x the displacement as positive down and the velocity as negative up. It is important to remember that the sign of the velocity is opposite that of the displacement. It doesn't matter whether the velocity is negative and the distance down positive or vice versa. It does matter that the signs be opposite! Getting a sign wrong is the source of many, possibly most, errors in falling body problems.

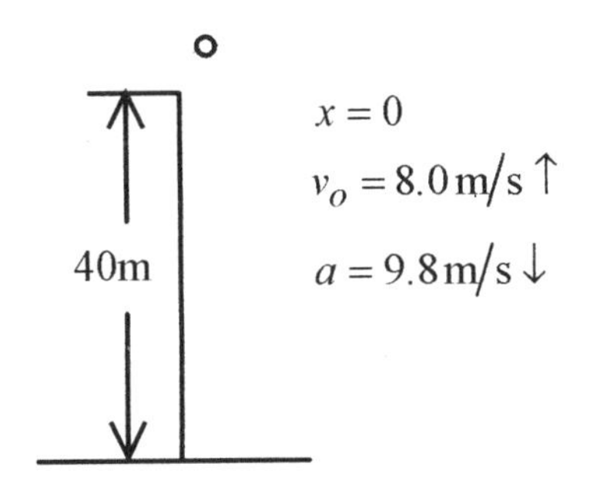

Fig. 3-3

Calculate the time of flight and note the signs: $40\,\mathrm{m} = (-8.0\,\mathrm{m/s})t + (4.9\,\mathrm{m/s^2})t^2$.

Rearranging and without units, this equation is $4.9t^2 - 8.0t - 40 = 0$,

with solution $t = \dfrac{8.0 \pm \sqrt{64 - 4(4.9)(-40)}}{9.8} = 3.8, -2.2$.

The positive time is the obvious choice. Note the numbers used for the solutions to the quadratics in time for this and problem 3-2.

The velocity when the ball strikes the ground is

$$v^2 = (-8.0\,\mathrm{m/s})^2 + 2(9.8\,\mathrm{m/s^2})40\,\mathrm{m} = 848\,\mathrm{m^2/s^2} \quad \text{or} \quad v = 29\,\mathrm{m/s}$$

Notice that whether the v term is a positive or negative number, the result is the same. If the ball is thrown up with a certain velocity or down with the same velocity, the velocity at impact is the same. This is to be expected from the symmetry of the equations. If the ball is thrown up with a certain velocity, then on the way down it passes the same level (from which it was thrown) with (numerically) that same velocity.

3-4 For the situation of problem 3-3 calculate the maximum height above the top of the building and the time for the ball to reach maximum height.

Solution: The time for the ball to reach maximum height is from equation 3-2. Note that at maximum height the velocity must be zero. Again watch the signs closely. It doesn't matter how you choose the signs, but the acceleration has to be opposite the velocity. The equations

$$0 = 8.0\,\text{m/s} - (9.8\,\text{m/s}^2)t \quad \text{and} \quad 0 = -8.0\,\text{m/s} + (9.8\,\text{m/s}^2)t$$

yield the same result, $t = 0.82\,\text{s}$. Because of symmetry, it takes the ball the same amount of time to reach maximum height as it does for the ball to return to the original level.

Calculate the height above the top of the building from equation 3-5.

$$0^2 = (8.0\,\text{m/s})^2 - 2(9.8\,\text{m/s}^2)(x - x_o) \quad \text{or} \quad x = 3.3\,\text{m}$$

A thorough understanding of these four problems will keep you from making sign mistakes in problems like these.

3-5 A bottle of champagne is dropped by a balloonist. The balloon is rising at a constant velocity of $3.0\,\text{m/s}$. It takes $8.0\,\text{s}$ for the bottle of champagne to reach the ground. Find the height of the balloon when the bottle was dropped, the height of the balloon when the bottle reached the ground, and the velocity with which the bottle strikes the ground.

Solution: Diagram the situation as shown in Fig. 3-4 being especially careful about the relative orientation (algebraic signs) of displacement, velocity, and acceleration. There are several possibilities as regards the origin and direction of the coordinate system. Take the origin at the height of the balloon when the bottle is dropped; the position of the balloon at $t = 0$. Take the displacement as positive down. The main reason for taking the displacement as positive down is that the acceleration is down and the initial velocity up making two positives and one negative. As time goes on, however, displacement, velocity, and acceleration will be positive. This choice seems to make for fewer minus signs and less chance for error with an algebraic sign.

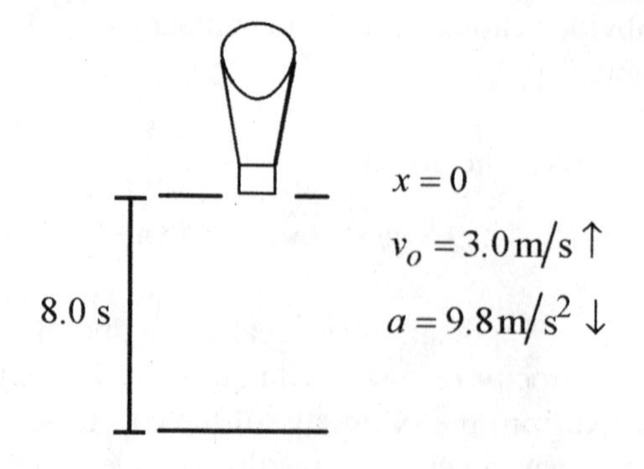

Fig. 3-4

Write the equation for the height of the balloon (when the bottle was dropped) counting time from when the bottle was dropped

$$x = (-3.0\,\text{m/s})t + (4.9\,\text{m/s}^2)t^2 = (-3.0\,\text{m/s})8.0\,\text{s} + (4.9\,\text{m/s}^2)64\,\text{s}^2 = 290\,\text{m}$$

The height of the balloon when the bottle reached the ground would be the height when the bottle was dropped plus the amount the balloon rose in the $8.0\,\text{s}$ it took the bottle to reach the ground, or

$$290\,\text{m} + (3.0\,\text{m/s})8.0\,\text{s} = 314\,\text{m}$$

The velocity on impact $v = v_o + at = -3.0\,\text{m/s} + (9.8\,\text{m/s}^2)8.0\,\text{s} = 75\,\text{m/s}$.

This velocity could also be calculated using

$$v_x^2 = v_o^2 + 2a(x - x_o) = 9.0\,\text{m}^2/\text{s}^2 + (19.6\,\text{m/s}^2)290\,\text{m} \qquad \text{or} \qquad v = 75\,\text{m/s}$$

3-6 A parachutist descending at a constant rate of $2.0\,\text{m/s}$ drops a smoke canister at a height of $300\,\text{m}$. Find the time for the smoke canister to reach the ground and its velocity when it strikes the ground. Then find the time for the parachutist to reach the ground, the position of the parachutist when the smoke canister strikes the ground, and an expression for the distance between the smoke canister and the parachutist.

Solution: Diagram the system as shown in Fig. 3-5 taking displacement, velocity, and acceleration as positive down with the origin at the point where the canister is dropped.

The time for the smoke canister to reach the ground is from equation 3-4.

$$x = v_o t + (1/2)at^2 \qquad \text{or} \qquad 300\,\text{m} = (2.0\,\text{m/s})t + (4.9\,\text{m/s}^2)t^2$$

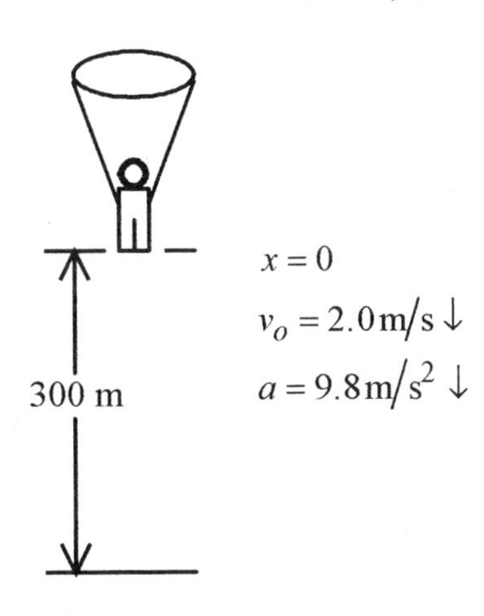

Fig. 3-5

Without units the equation is $4.9t^2 + 2.0t - 300 = 0$ with solutions

$$t = \frac{-2 \pm \sqrt{4 - 4(4.9)(-300)}}{2 \cdot 4.9} = \frac{-2 \pm 76.7}{9.8} = 7.6, -8.0$$

The time for the canister to reach the ground is $7.6\,\mathrm{s}$.

The velocity when it strikes the ground is

$$v^2 = (2.0\,\mathrm{m/s})^2 + 2(9.8\,\mathrm{m/s^2})300\,\mathrm{m} = 5884\,\mathrm{m^2/s^2} \quad \text{or} \quad v = 54\,\mathrm{m/s}$$

The time for the parachutist to reach the ground is from equation 3-2.

$$300\,\mathrm{m} = (2.0\,\mathrm{m/s})t \quad \text{or} \quad t = 150\,\mathrm{s}$$

When the canister strikes the ground the parachutist has dropped $(2.0\,\mathrm{m/s})7.6\,\mathrm{s} = 15\,\mathrm{m}$ and is $285\,\mathrm{m}$ above the ground.

The expression for the distance between the canister and the parachutist is

$$x_c - x_p = (2.0\,\mathrm{m/s})t + (4.9\,\mathrm{m/s^2})t^2 - (2.0\,\mathrm{m/s})t = (4.9\,\mathrm{m/s^2})t^2$$

3-7 A coconut is dropped from a height of $60\,\mathrm{m}$. One second later a second coconut is thrown down with an initial velocity. Both coconuts reach the ground at the same time. What was the initial velocity of the second coconut?

Solution: In problems where there is a time delay it is usually best to calculate the position, velocity, and acceleration of the first particle at the time when the second particle starts to move. In the train problem in the previous chapter there was a position difference between the two trains at $t = 0$ that was easily translated into the equations. It is possible to do time delay problems with a time differential in one set of equations. The difficulty with this approach is that it is easy to get an algebraic sign wrong. If you say that the time for the second particle is the time for the first plus the difference between them, then it is essential that the algebraic sign of the difference be correct. It is much easier, especially when you are learning how to do problems with a time delay, to do them in this slower, but inherently more accurate, method. First, calculate the state of the first particle when the second one begins moving. Then, write the two sets of equations describing the motion with this instant as $t = 0$.

Calculate the position and velocity of the first coconut at the end of one second, the time when the second one starts.

$$x = (1/2)at^2 = (1/2)(9.8\,\mathrm{m/s^2})(1.0\,\mathrm{s})^2 = 4.9\,\mathrm{m} \qquad v = at = 9.8\,\mathrm{m/s^2}\,1.0\,\mathrm{s} = 9.8\,\mathrm{m/s}$$

Since everything (position, velocity, and acceleration) is positive down, orient the coordinate system for positive down with the origin at the top. Now diagram the problem as in Fig. 3-6. At the instant the second coconut is thrown down, the first coconut has position $4.9\,\text{m}$, velocity $9.8\,\text{m/s}$ and acceleration $9.8\,\text{m/s}^2$.

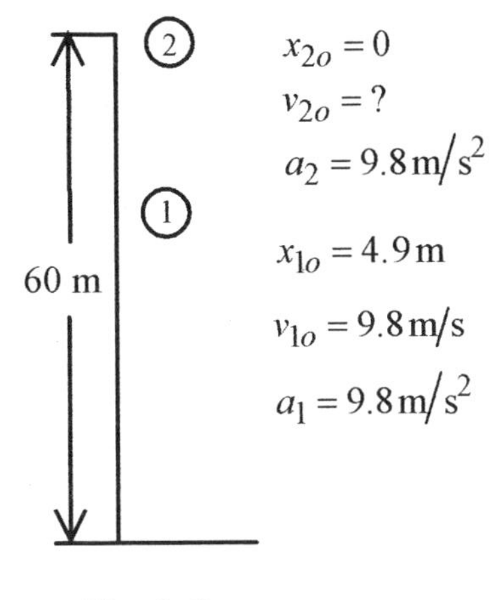

Fig. 3-6

Since both coconuts strike the ground at the same time use the conditions of the first coconut to find the total time.

First calculate the velocity at impact of the first coconut.

$$v_1^2 = v_{1o}^2 + 2a(x_1 - x_{1o}) = (9.8\,\text{m/s})^2 + 2(9.8\,\text{m/s}^2)(60 - 4.9)\,\text{m} = 1174\,\text{m}^2/\text{s}^2$$

or

$$v_1 = 34.3\,\text{m/s}$$

The time for the second coconut to reach the ground is the same as the time for the first coconut to go from $4.9\,\text{m}$ to $60\,\text{m}$ or the time for the first coconut to go from $9.8\,\text{m/s}$ to $34.3\,\text{m/s}$.

This comes from $v = v_o + at$, where v_o is v_{1o}, the velocity of the first coconut when the second one is thrown down, and v is the velocity of the first coconut at the ground.

$$34.3\,\text{m/s} = 9.8\,\text{m/s} + (9.8\,\text{m/s}^2)t \quad \text{or} \quad t = 2.5\,\text{s}$$

Now that we have the time for the second coconut to travel the $60\,\text{m}$ we can find its initial velocity from $x - x_o = v_{2o}t + (1/2)at^2$ where t is the total time for the second coconut.

$$60\,\text{m} = v_{2o}(2.5\,\text{s}) + (4.9\,\text{m/s}^2)(2.5\,\text{s})^2 \quad \text{or} \quad v_{2o} = 11.5\,\text{m/s}$$

Review this problem until all the different times and velocities are clear in your mind. Set up the problem and do it yourself without reference and you will know you understand it.

3-8 A boat is passing under a bridge. The deck of the boat is 15m below the bridge. A small package is to be dropped from the bridge onto the deck of the boat when the boat is 25m from just below the drop point. What (boat) speed is necessary to have the package land in the boat?

Solution: Calculate the time for the package to fall the 15.0m using $x - x_o = v_o t + (1/2)at^2$.

$$15\text{m} = (4.9\,\text{m/s}^2)t^2 \quad \text{or} \quad t = 1.7\,\text{s}$$

The boat must move at $25\text{m}/1.7\text{s} = 14\,\text{m/s}$

3-9 You are observing steel balls falling at a constant velocity in a liquid-filled tank. The window you are using is one meter high and the bottom of the window is twelve meters from the bottom of the tank. You observe a ball falling past the window taking 3.0s to pass the window. Calculate the time required to reach the bottom of the tank, after the ball has reached the bottom of the window.

Solution: The situation is diagrammed in Fig. 3-7.

The observed velocity is $1.0\text{m}/3.0\text{s}$ so $12\,\text{m} = (1.0\,\text{m}/3.0\,\text{s})t$ or $t = 36\text{s}$

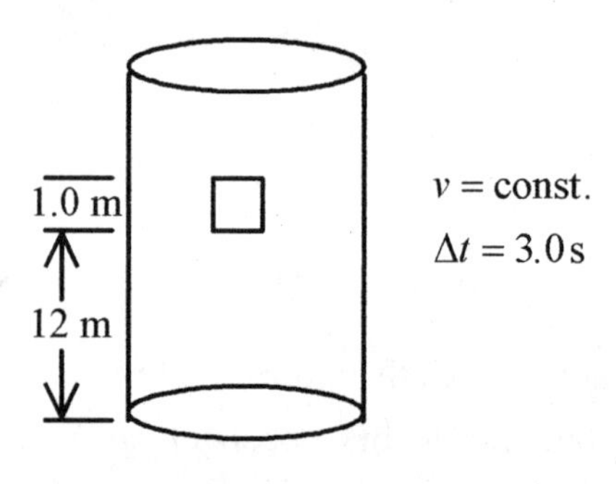

Fig. 3-7

3-10 In a situation similar to problem 3-9, the tank is filled with a different liquid causing the acceleration in the tank to be $6.0\,\text{m/s}^2$ and the time to traverse the window, 0.40s. Calculate the height of the liquid above the window, the time to reach the bottom of the tank, and the velocity of the ball when it reaches the bottom of the tank.

Solution: Diagram the problem as in Fig. 3-8.

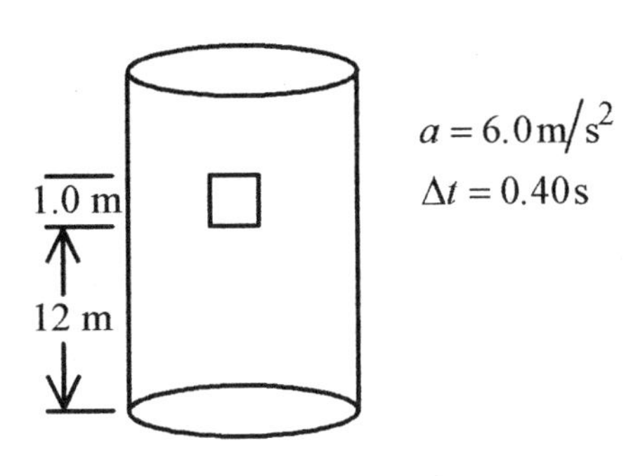

Fig. 3-8

From the data about the window, calculate the velocity of the ball at the top of the window.

$$x - x_o = v_t t + (1/2)at^2 \qquad 1.0\,\mathrm{m} = v_t 0.40\,\mathrm{s} + (3.0\,\mathrm{m/s^2})(0.40\,\mathrm{s})^2 \qquad \text{or} \qquad v_t = 1.3\,\mathrm{m/s}$$

The velocity at the bottom of the window is from

$$v_b^2 = v_t^2 + 2a(x - x_o) = (1.3\,\mathrm{m/s})^2 + 2(6.0\,\mathrm{m/s^2})1.0\,\mathrm{m} \qquad \text{or} \qquad v_b = 3.7\,\mathrm{m/s}$$

The time to reach the bottom of the tank is from $x - x_o = v_b t + (1/2)at^2$

$$12\,\mathrm{m} = (3.7\,\mathrm{m/s})t + (3.0\,\mathrm{m/s^2})t^2$$

and eliminating the units $3.0t^2 + 3.7t - 12 = 0$ so

$$t = \frac{-3.7 \pm \sqrt{(3.7)^2 - 4(3.0)(-12)}}{2(3)} = \frac{-3.7 \pm 12.6}{6} = 1.5, -2.7$$

This (positive) time is for the ball to travel from the bottom of the window to the bottom of the tank.

Assuming the ball started at zero velocity the distance from the top of the liquid to the top of the window, x_t, will come from $v_t^2 = v_o^2 + 2ax_t$.

$$(1.3\,\mathrm{m/s})^2 = 2(6.0\,\mathrm{m/s})x_t \qquad \text{or} \qquad x_t = 0.14\,\mathrm{m}$$

The velocity with which the ball strikes the bottom of the tank is

$$v = v_b + at = (3.7\,\mathrm{m/s}) + (6.0\,\mathrm{m/s^2})1.5\,\mathrm{s} = 12.7\,\mathrm{m/s}$$

3-11 A ball is observed to pass a $1.4\,\mathrm{m}$ tall window going up and later going down. The total observation time is $0.40\,\mathrm{s}$ ($0.20\,\mathrm{s}$ going up and $0.20\,\mathrm{s}$ going down). How high does the ball rise above the window?

Solution: Diagram the problem as in Fig. 3-9. This is another example of a question that seems totally unrelated to the information in the problem. When you don't know where to start, and you do not see a clear path to the desired answer, simply start where you can, calculating what you can and hopefully learning enough to answer the specific question.

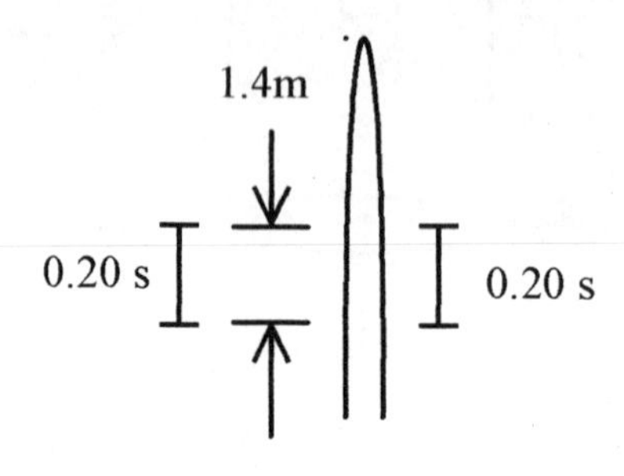

Fig. 3-9

One of the first things we can calculate in this problem is the average velocity at the middle of the window.

$$v_{avg} = (1.4\,\mathrm{m}/0.20\,\mathrm{s}) = 7.0\,\mathrm{m/s}$$

This average velocity is the velocity of the ball on the way up (and on the way down) at the middle of the window. Add this feature to the problem. With this information we can use $v^2 = v_o^2 + 2ax_t$ to find the distance the ball rises above the midpoint of the window.

$$(7.0\,\mathrm{m/s})^2 = 2(9.8\,\mathrm{m/s^2})x_t \quad \text{or} \quad x = 2.5\,\mathrm{m}$$

So the ball rises $2.5 - 0.7 = 1.8\,\mathrm{m}$ above the top of the window.

Second Solution: View the ball as decelerating as it goes up past the window and find v_b at the bottom of the window from $x - x_o = v_o t + (1/2)at^2$.

$$1.4\,\mathrm{m} = v_b(0.20\,\mathrm{s}) - (4.9\,\mathrm{m/s^2})(0.20\,\mathrm{s})^2 \quad \text{or} \quad v_b = 8.0\,\mathrm{m/s}$$

Velocity and displacement are taken as positive so acceleration is negative.

Now the distance to maximum height (velocity zero) is

$$(8.0\,\mathrm{m/s})^2 = 2(9.8\,\mathrm{m/s^2})x_m \quad \text{or} \quad x_m = 3.2\,\mathrm{m}$$

Again the maximum height of the ball above the top of the window is $3.2\,\mathrm{m} - 1.4\,\mathrm{m} = 1.8\,\mathrm{m}$.

CHAPTER 4

PROJECTILE MOTION

In order to understand projectile motion you have to look at the motion in two directions with one direction oriented in the direction of constant acceleration and the other direction at a right angle to it so as to form an x-y coordinate system. In most problems, where an initial velocity and angle with the horizontal is given, the velocity is written in component form.

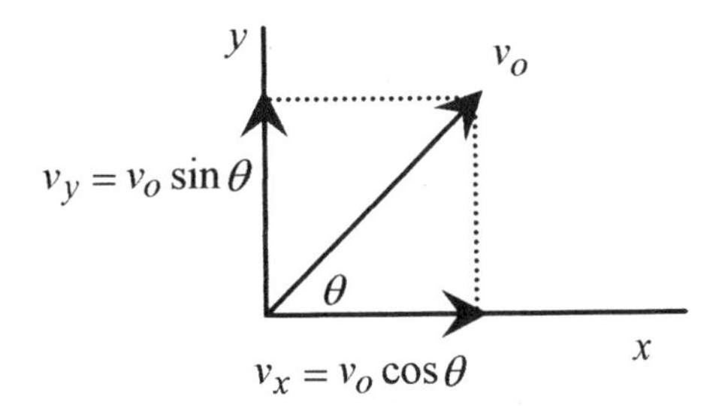

Fig. 4-1

Note that the motion is effectively separated into horizontal and vertical components (one in the direction of the constant acceleration and one at a right angle to the acceleration).

The motion is a parabola in this coordinate system. The symmetry of parabolas (in this case parabolic motion) is helpful in understanding the motion. (See the Introduction, Mathematical Background, for a discussion of the properties of parabolas.) In order to better understand how the motion is separated and how the parabolic property helps us to understand projectile problems we will do a simple problem without numbers just to get a feel for the features of the motion. If you are familiar with the properties of parabolas you may want to skip this discussion and go directly to the worked problems.

In Fig. 4-1 the velocity of the projectile is written in component form. The acceleration in the horizontal direction is zero and in the vertical direction is due to gravity. With the velocity components, and the acceleration in the direction of an axis, we can write the six equations describing acceleration, velocity, and position in the x and y directions.

These equations are based on the kinematic equations of motion for constant acceleration: $a = \text{const.}$, $v = v_o + at$, and $s = s_o + v_o t + (1/2)at^2$.

The six equations are:

$$a_x = 0 \qquad a_y = -g$$

$$v_x = v_o \cos\theta \qquad v_y = v_o \sin\theta - gt$$

$$x = v_o(\cos\theta)t \qquad y = v_o(\sin\theta)t - (1/2)gt^2$$

The equations for x and y can be looked upon as parametric equations in time. Parametric equations such as $x = f(t)$ and $y = f(t)$ are equations that can be combined to produce $y = f(x)$ or $x = f(y)$. To find the position of the particle in x-y, or y as a function of x, solve one equation for t and substitute into the other. In this case solve $x = \cdots$ for t, as this is the simplest choice.

$$t = \frac{x}{v_o \cos\theta} \qquad \text{Substitute into} \quad y = \cdots \quad \text{to obtain}$$

$$y = (\tan\theta)x - \frac{g}{2v_o^2 \cos^2\theta}x^2$$

This is of the form $y = -ax^2 + bx = x(-ax + b)$, which is a parabola that opens down and intercepts the y-axis (makes $y = 0$) at $x = 0$ and $x = b/a$.

The range, or value of x when $y = 0$, can be determined from the factored form of this equation

$$y = x\left[\tan\theta - \frac{g}{2v_o^2 \cos^2\theta}x\right]$$

which tells us that $y = 0$ at $x = 0$ and

$$x = \frac{2v_o^2}{g}\cos^2\theta \tan\theta = \frac{2v_o^2}{g}\sin\theta\cos\theta$$

or using the trigonometric identity $2\sin\theta\cos\theta = \sin 2\theta$

$$x = \frac{v_o^2}{g}\sin 2\theta$$

The maximum range occurs for an angle of 45°, corresponding to $\sin 2\theta = 1$.

The main point of this discussion is that the motion is a parabola, and the properties of parabolas can be used in solving problems in projectile motion. That the maximum range occurs at 45° is not surprising, and is generally not of interest in problems. The expression for maximum range is only of passing interest, since the range is one of the easier things to calculate in any problem.

Do not consume precious memory space memorizing formulas for the range or time of flight. If you work the problems by first writing down the six equations describing the motion, then the time of flight, range, and many other things are easily calculated.

In order to set up a problem in projectile motion, first orient one axis of a right angle (x-y) coordinate system in the direction of constant acceleration. Remember to place the origin of the coordinate system and the positive direction for x and y for (your) convenience in solving the problems. Then write equations for acceleration, velocity, and position for each direction. Along with the symmetry of the motion, these equations can be used to find the characteristics (position, velocity, and acceleration) of the projectile at any point in space or time. The procedure is illustrated with a simple problem.

4-1 A soccer player kicks a ball at an initial velocity of $18\,\mathrm{m/s}$ at an angle of 36° to the horizontal. Find the time of flight, range, maximum height, and velocity components at $t = 0$, mid-range, and at impact.

Solution: The acceleration is down so set up the coordinate system with x horizontal (along the ground) and y vertical, and place v_o on the graph. Write the velocity in component form, and calculate v_{xo} and v_{yo}.

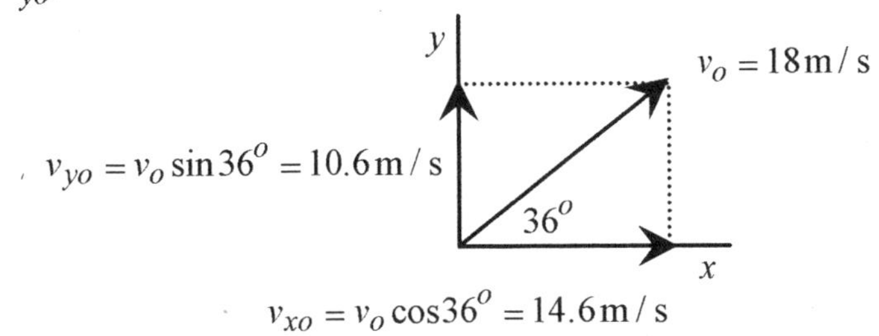

Fig. 4-2

Now, <u>and this is the key to getting the problem right</u>, write down the six equations governing the motion starting with the accelerations (the accelerations are the easiest to write).

$$a_x = 0 \qquad\qquad a_y = -9.8\,\mathrm{m/s^2}$$

$$v_x = 14.6\,\mathrm{m/s} \qquad\qquad v_y = 10.6\,\mathrm{m/s} - (9.8\,\mathrm{m/s^2})t$$

$$x = (14.6\,\mathrm{m/s})t \qquad\qquad y = (10.6\,\mathrm{m/s})t - (4.9\,\mathrm{m/s^2})t^2$$

Now look at the motion, which is parabolic in x-y, keeping in mind the properties of parabolas.

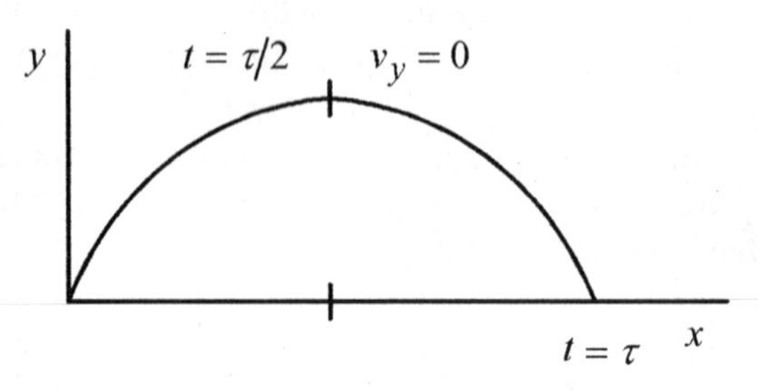

Fig. 4-3

The ball is on the ground at $t=0$ and $t=\tau$, the time of flight. To find these times, set $y=0$ and obtain (without units)

$$10.6t-4.9t^2=0 \qquad \text{or} \qquad t(10.6-4.9t)=0$$

which gives $t=0$ and $t=10.6/4.9=2.16\text{s}$, the time of flight.

The range is the value of x at 2.16s: $x|_{t=2.16}=(14.6\text{m/s})2.16\text{s}=31.5\text{m}$

The maximum height, and $v_y=0$, occur at 1.08s (one-half the time of flight). Therefore the maximum height is y at $t=1.08\text{s}$

$$y_{\max}=y|_{t=1.08}=(10.6\text{m/s})1.08s-(4.9\text{m/s}^2)(1.08\text{s})^2=5.73\text{m}$$

The velocity components are:

at	$t=0$:	$v_x=14.6\text{m/s}$,	$v_y=10.6\text{m/s}$
at	$t=1.08\text{s}$	$v_x=14.6\text{m/s}$,	$v_y=0$
at	$t=2.16\text{s}$	$v_x=14.6\text{m/s}$,	$v_y=-10.6\text{m/s}$

The procedure for doing the problem is to write the initial velocity in component form, write the six equations for acceleration, velocity, and position, and then perform the mathematical operations answering questions about the problem.

4-2 For the situation of problem 4-1 suppose the field is covered with fog down to 3.0m above the ground. What are the times for the ball's entry into and exit out of the fog?

Solution: The times are found from the equation for y as a function of time with y set equal to 3.0m.

$$3.0\,\text{m} = (10.6\,\text{m/s})t - (4.9\,\text{m/s}^2)t^2$$

This is a quadratic and removing the units it reads $4.9t^2 - 10.6t + 3.0 = 0$

and is solved with the quadratic formula.

$$t = \frac{10.6 \pm \sqrt{10.6^2 - 4(4.9)3.0}}{2(4.9)} = \frac{10.6 \pm 7.3}{9.8} = 0.34, 1.8$$

As an exercise find the x positions for these times.

4-3 An airplane traveling at $100\,\text{m/s}$ drops a bomb from a height of $1500\,\text{m}$. Find the time of flight, distance traveled, and the velocity components as the bomb strikes the ground.

Solution: Start by placing the origin of the coordinate system at the point where the bomb is released. And take the direction the bomb falls due to gravity as positive x and the horizontal position as y. This is different from the conventional orientation, but it is convenient in this problem because the six equations all come out positive!

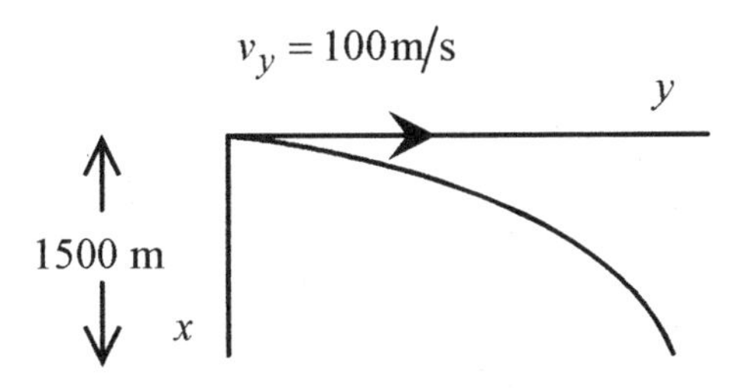

Fig. 4-4

Now write down the six equations governing the motion.

$$a_x = 9.8\,\text{m/s}^2 \qquad a_y = 0$$

$$v_x = (9.8\,\text{m/s}^2)t \qquad v_y = v_o = 100\,\text{m/s}$$

$$x = (4.9\,\text{m/s}^2)t \qquad y = (100\,\text{m/s})t$$

With these six equations we can answer all the questions in the problem. We need only to "translate" the word questions into algebra questions.

a) The time for the bomb to reach the earth means: Find the time when $x = 1500\,\text{m}$?

$$1500\,\text{m} = (4.9\,\text{m/s}^2)t^2 \qquad \text{yields} \qquad \tau = 17\,\text{s}$$

b) How far does the bomb travel horizontally means: Find the value of y when $t = 17\,\text{s}$?

$$y|_{t=17} = (100\,\text{m/s})17\,\text{s} = 1700\,\text{m}$$

c) Find the velocity components at impact means: Find v_x and v_y at $t = 17\,\text{s}$?

$$v_x|_{t=17} = (9.8\,\text{m/s}^2)17\,\text{s} = 167\,\text{m/s}$$

$$v_y|_{t=17} = 100\,\text{m/s}$$

d) Where is the airplane when the bomb strikes the earth means: What is y at $t = 17\,\text{s}$? Remember that the plane and the bomb have the same velocity in the horizontal direction.

$$y|_{\text{plane}} = (100\,\text{m/s})17\,\text{s} = 1700\,\text{m}$$

4-4 A baseball is hit at a 45° angle and a height of $0.90\,\text{m}$. The ball travels a total distance of $120\,\text{m}$. What is the initial velocity of the ball? What is the height of the ball above a $3.0\,\text{m}$ fence $100\,\text{m}$ from where the ball is hit?

Solution: This problem is unique in that it requests the initial velocity, a number usually given in the problem. Also note that the angle is 45°, the angle for maximum range. The other interesting feature of the problem is the question concerning the height at some specific point down range. Set up the problem in the conventional way assigning v_o to the initial velocity.

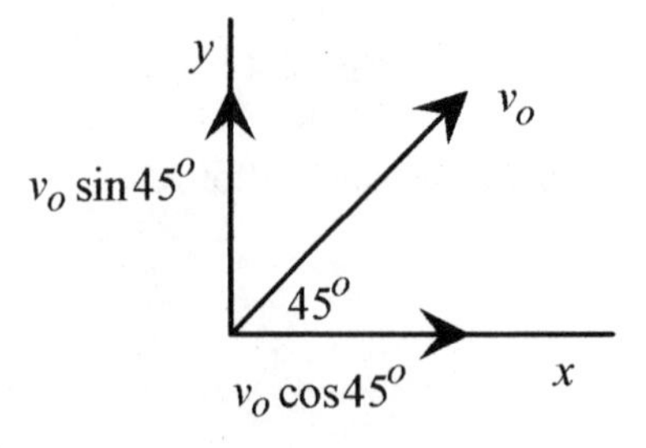

Fig. 4-5

The next question is how to handle the ball being hit at $0.90\,\text{m}$ above the ground. Since v_o is specified at the point where the ball is hit, and this is where we start counting time, then the origin for the coordinate system should be put here. We just need to keep in mind that this origin is $0.90\,\text{m}$ above the ground level.

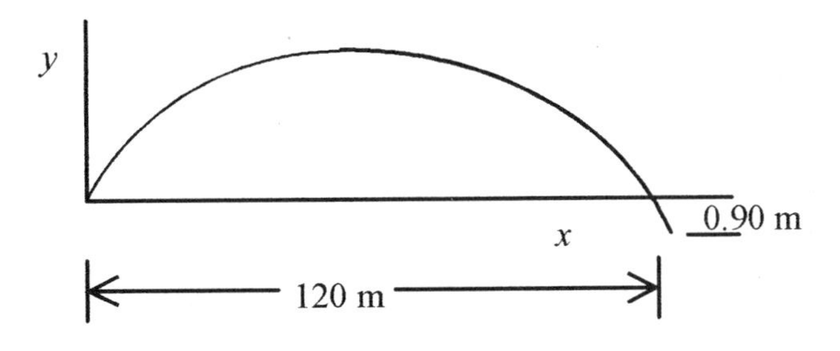

Fig. 4-6

Now write the six equations.

$$a_x = 0 \qquad a_y = -9.8\,\mathrm{m/s^2}$$

$$v_x = v_o \cos 45^o \qquad v_y = v_o \sin 45^o - 9.8\,\mathrm{m/s^2}\, t$$

$$x = v_o \cos 45^o t \qquad y = v_o \sin 45^o t - 4.9\,\mathrm{m/s^2}\, t^2$$

The first question reduces to finding v_o for the $120\mathrm{m}$ hit. Algebraically this means that when $y = -0.9\,\mathrm{m}$, $x = 120\mathrm{m}$, so write the two equations (for x and y) with these conditions

$$120\mathrm{m} = v_o \cos 45^o t \qquad -0.9\,\mathrm{m} = v_o \sin 45^o t - (4.9\,\mathrm{m/s^2})t^2$$

Note that $v_o \cos 45^o t = v_o \sin 45^o t = 120\mathrm{m}$ so write

$$0 = 0.90\,\mathrm{m} + 120\mathrm{m} - (4.9\,\mathrm{m/s^2})t^2 \quad \text{or} \quad t^2 = \frac{120.9}{4.9}\mathrm{s^2} \quad \text{and} \quad t = 4.97\,\mathrm{s}$$

This is the time of flight, so put this time into the equation for x.

$$120\mathrm{m} = v_o \cos 45^o (4.97\,\mathrm{s}), \text{ to find } v_o \text{ as } v_o = \frac{120\mathrm{m}}{\cos 45^o \cdot 4.97\,\mathrm{s}} = 34.1\,\mathrm{m/s}$$

The next part of the problem asks for the height above the fence at $100\mathrm{m}$. Again we need to find the time for x to be $100\mathrm{m}$ down range and substitute this value of t into the y equation

$$100\mathrm{m} = (34.1\,\mathrm{m/s})\cos 45^o t \quad \text{or} \quad t = \frac{100\mathrm{s}}{34.1 \cos 45^o} = 4.15\,\mathrm{s}$$

This is the time for the ball to go $100\mathrm{m}$. The height of the ball at this time is

$$y|_{t=4.15} = (34.1\,\mathrm{m/s})\cos 45^o (4.15\mathrm{s}) - (4.9\,\mathrm{m/s^2})(4.15\mathrm{s})^2 = 15.6\,\mathrm{m}$$

Remember that this is 15.6 m above the zero of the coordinate system which is 0.90 m above the ground. So the ball is 16.5 m above the ground at this point and for a 3.0 m high fence is 13.5 m above the fence.

4-5 A physicist turned motorcycle stunt rider will jump a 20 m wide row of cars. The launch ramp is 30° and 9.0 m high. The land ramp is also 30° and is 6.0 m high. Find the minimum speed for the launch.

Solution:

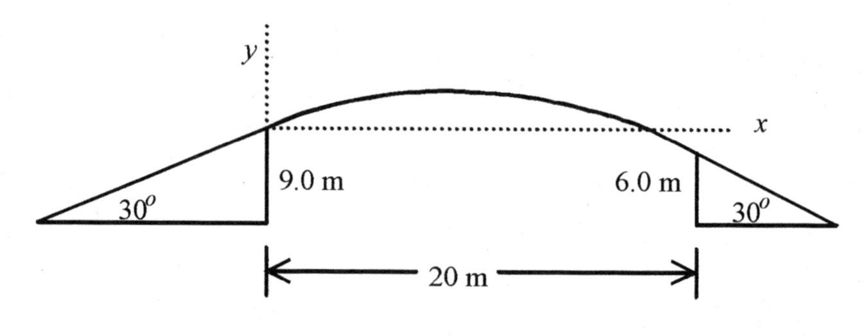

Fig. 4-7

First write down the six equations of motion.

$$a_x = 0 \qquad a_y = -9.8\,\mathrm{m/s^2}$$

$$v_x = v_o \cos 30^o \qquad v_y = v_o \sin 30^o - (9.8\,\mathrm{m/s^2})t$$

$$x = v_o \cos 30^o t \qquad y = v_o \sin 30^o t - (4.9\,\mathrm{m/s^2})t^2$$

The minimum v_o is dictated by the condition that the rider be at $x = 100\,\mathrm{m}$ and $y = -3.0\,\mathrm{m}$. So put these conditions into the equations for x and y.

$$100\,\mathrm{m} = v_o \cos 30^o t \qquad -3.0\,\mathrm{m} = v_o \sin 30^o t - (4.9\,\mathrm{m/s^2})t^2$$

At this point we can solve for v_o by substituting $t = \dfrac{100\,\mathrm{m}}{v_o \cos 30^o}$ from the first equation into the second equation.

$$-3.0\,\mathrm{m} = (100\,\mathrm{m}) \tan 30^o - \frac{4.9 \times 10^4\,\mathrm{m^3}}{v_o^2 \cos^2 30^o\,\mathrm{s^2}} \quad \text{yields} \quad v_o = 32.8\,\mathrm{m/s}$$

This is the minimum velocity to make the jump.

As a follow on to this problem calculate the velocity to just hit the end of the ramp. This velocity gives an upper limit to the jump velocity.

Go back through the problems and notice the common approach:

1) orient a coordinate system,
2) write the six equations of motion, and
3) translate the word questions into algebra.

In these problems we have not always used each of the six equations to answer the question. Few problems require the detailed description of the motion available in the six equations. If, however, each problem is described by these six equations you will always have all the information to answer any possible question about the motion.

CHAPTER 5

FORCES (including friction)

In the development of mechanics, the first thing to learn is the interrelation of position, velocity, and acceleration (for constant acceleration). These interrelations are described with the four kinematic equations of motion and discussed in earlier chapters.

In this chapter we relate constant force to constant acceleration and then to all of kinematics. Operationally, force can be defined as what makes masses accelerate. Actually it is the unbalanced force on a mass that makes it accelerate. Simply: **Unbalanced forces make masses accelerate.**

Force, acceleration, and mass are related by **Newton's second law** as $F = ma$.

5-1 A 40-N force is applied to a 20kg block resting on a horizontal frictionless table. Find the acceleration?

Solution: The acceleration is $a = \dfrac{F}{m} = \dfrac{40\,\text{N}}{20\,\text{kg}} = 2.0\,\text{m/s}$

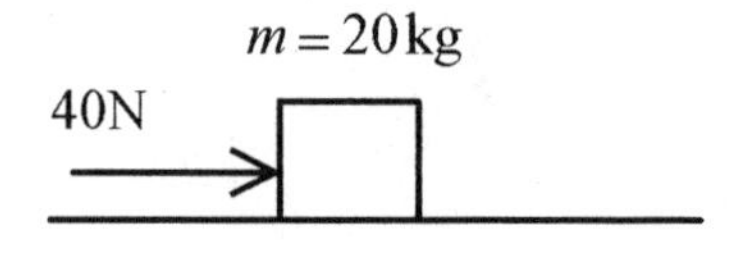

Fig. 5-1

Once the acceleration is known, the kinematic equations of motion can be applied to find position, velocity, and acceleration as a function of time.

5-2 Now place the 20kg mass (with no external force applied) on a frictionless 50° incline plane. What is the acceleration of the mass?

Solution: The force acting on the mass is due to gravity so set up a vector diagram (some authors call this a force diagram or free body diagram) starting with this 196N force acting down. The mass is constrained to move down the plane, so the force we want is the component of this gravitational force acting down the plane. In Fig. 5-2 the gravitational force is shown with components down the plane, F_p, and normal, F_n, to the plane. Notice the geometry of the situation where the 50° angle of the plane is the same as the angle between the gravitational force and the normal (force).

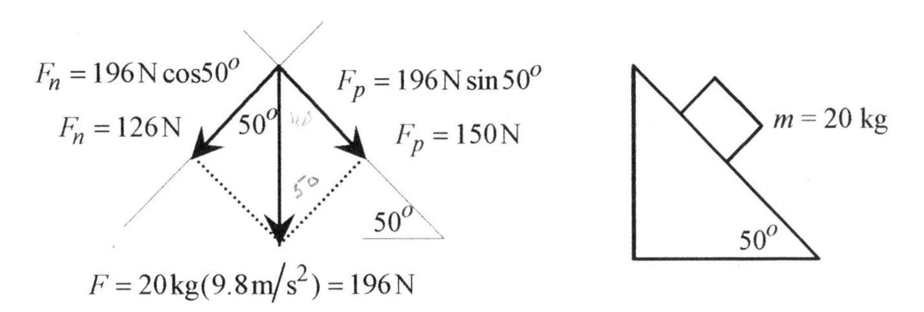

Fig. 5-2

The $150\,\text{N}$ force acting down the plane causes the $20\,\text{kg}$ mass to accelerate at

$$a = \frac{150\,\text{N}}{20\,\text{kg}} = 7.5\,\text{m/s}$$

5-3 Place a $10\,\text{kg}$ mass on a frictionless 35° incline plane, and attach a second $20\,\text{kg}$ mass via a cord to hang vertically as shown in Fig. 5-3. Calculate the acceleration of the system.

Solution:

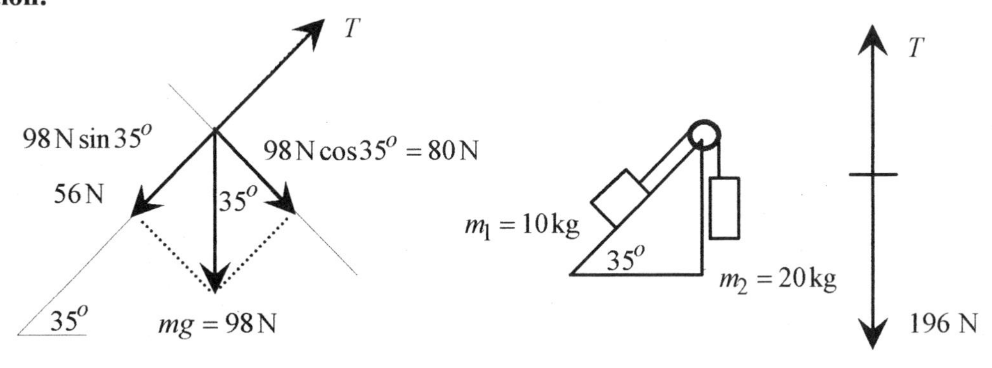

Fig. 5-3

This problem introduces the concept of the tension in the connecting cord. The most convenient way to visualize this tension is that if the cord were cut and a force meter inserted it would read a certain tension (force). This tension acts, as shown, on each mass. Notice the 35° angle of the plane and the 35° angle between the normal and the direction of *mg*.

We'll do the problem first with the tension and later without the tension.

To write equations relating the unbalanced force to the acceleration, we need to make an assumption as to which way the masses move. Assume that the system (of masses) moves to the right or clockwise.

Start by writing equations relating the unbalanced force on each mass. On m_1 the unbalanced force is $T - 56\,\text{N}$, and this force makes the $10\,\text{kg}$ accelerate.

$$T - 56\,\text{N} = 10\,\text{kg} \cdot a_1$$

Likewise on m_2, the unbalanced force to make the mass accelerate is

$$196\,\text{N} - T = 20\,\text{kg} \cdot a_2$$

The assumption that the system accelerates clockwise dictates the directions of the forces in these two equations. Look again at these equations and notice that if we had assumed the acceleration were counterclockwise then the signs would be different. In frictionless problems it is all right to have a negative acceleration. The negative sign means that you guessed wrong in assuming the acceleration direction. <u>This is not true with friction problems</u>.

Add the two equations and take the accelerations the same. The accelerations are the same because the masses are linked together with the cord.

$$140\,\text{N} = 30\,\text{kg} \cdot a \quad \text{or} \quad a = 4.7\,\text{m/s}^2$$

Now the tension in the cord is easily calculated from the equation for m_1.

$$T = 56\,\text{N} + 10\,\text{kg} \cdot 4.7\,\text{m}/^{2} = 103\,\text{N}$$

Another way to view this problem is to say that the unbalanced force on m_1 and m_2 is $(196 - 56)\,\text{N} = 140\,\text{N}$. This force makes the total mass of $30\,\text{kg}$ accelerate, so

$$140\,\text{N} = 30\,\text{kg} \cdot a \quad \text{or} \quad a = 4.7\,\text{m/s}^2$$

This is a very quick way to find the acceleration of the system, but it is inconvenient to find the tension in the cord.

5-4 The next complication in force problems with incline planes is a double-incline plane as shown in Fig. 5-4.

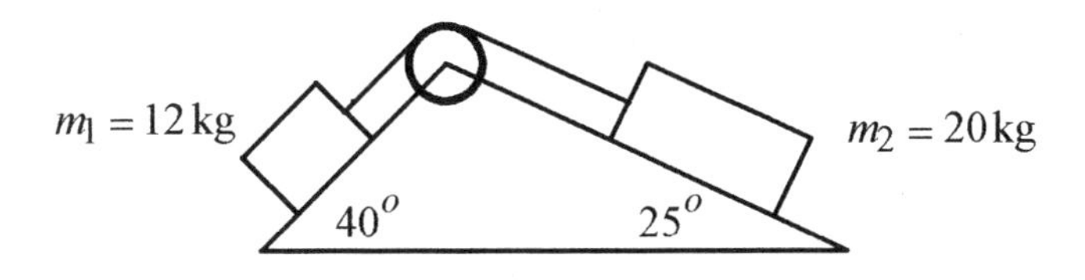

Fig. 5-4

Solution: The vector diagram for these masses is shown in Fig. 5-5. Assume the system moves clockwise.

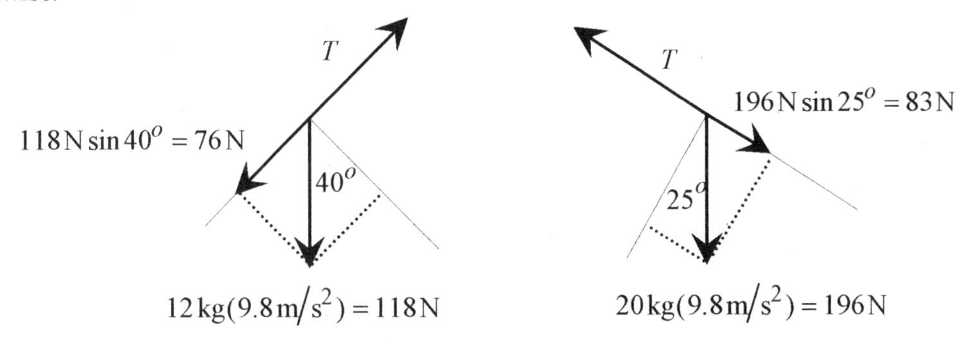

Fig. 5-5

The equation for m_1 is $\quad T - 76\,\text{N} = 12\,\text{kg}\cdot a$

The equation for m_2 is $\quad 83\,\text{N} - T = 20\,\text{kg}\cdot a$

Adding the two equations $\quad 7.0\,\text{N} = 32\,\text{kg}\cdot a \quad$ or $\quad a = 0.22\,\text{m/s}^2$

The tension in the cord is $T = 76\,\text{N} + 12\,\text{kg}\cdot 0.22\,\text{m/s}^2 = 79\,\text{N}$

5-5 Another popular problem is the Atwood machine. A simple example is shown in Fig. 5-6. Find the acceleration of the system.

Solution: Assume the system moves counterclockwise.

The equation for m_1 is $\quad 98\,\text{N} - T = 10\,\text{kg}\cdot a$

The equation for m_2 is $\quad T - 147\,\text{N} = 15\,\text{kg}\cdot a$

Add the equations $-49\,\text{N} = 25\,\text{kg} \cdot a$ or $a = -2.0\,\text{m/s}^2$

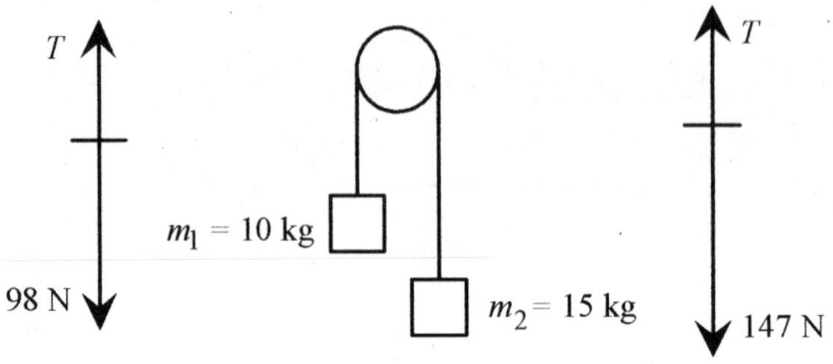

Fig. 5-6

The system accelerates clockwise! The tension in the cord is from the equation for m_1.

$$T = 98\,\text{N} + (10\,\text{kg})2.0\,\text{m/s}^2 = 118\,\text{N}$$

Friction

A good first problem in friction is the arrangement shown in Fig. 5-7 where the problem is first done without friction and then with friction.

For the purposes of doing problems there are two important properties of frictional forces:

- **they oppose the motion, and**
- **they are less than or equal to a constant (the coefficient of friction) times the normal force, the force at the frictioning surface. ($f \le \mu N$)**

5-6 Two masses are arranged with one, of $50\,\text{kg}$, on a frictionless table and the other, of $30\,\text{kg}$, attached by a cord and hanging, over a frictionless pulley, off the edge of the table. Find the acceleration of the system.

Solution: Assume the system moves clockwise.

The equation for m_1 is $T = 50\,\text{kg} \cdot a$

The equation for m_2 is $294\,\text{N} - T = 30\,\text{kg} \cdot a$

Adding the two equations $294\,\text{N} = 80\,\text{kg} \cdot a$ or $a = 3.7\,\text{m/s}^2$

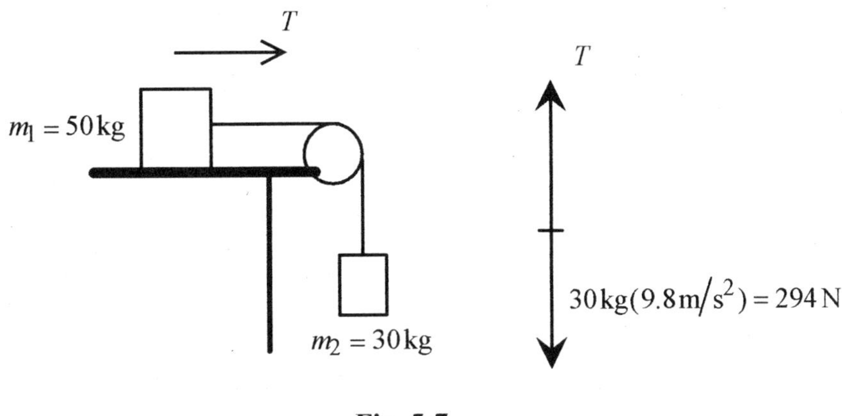

Fig. 5-7

5-7 For the situation of problem 5-6 add a coefficient of friction $\mu = 0.20$ between the block and table.

Solution: This makes the vector diagram look like Fig. 5-8. Notice that the only addition is the frictional retarding force $\mu N = 98\,N$.

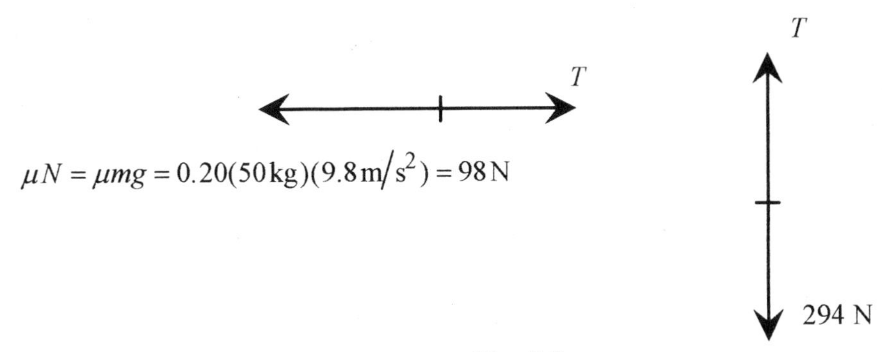

Fig. 5-8

The equation for m_1 is $T - 98\,N = 50\,kg \cdot a$

The equation for m_2 is $294\,N - T = 30\,kg \cdot a$

Add the equations to obtain $196\,N = 80\,kg \cdot a$ or $a = 2.4\,m/s^2$

5-8 For the situation of problem 5-7 increase the coefficient of friction to 0.80. What is the acceleration of the system?

Solution: This makes the vector diagram look like Fig. 5-9.

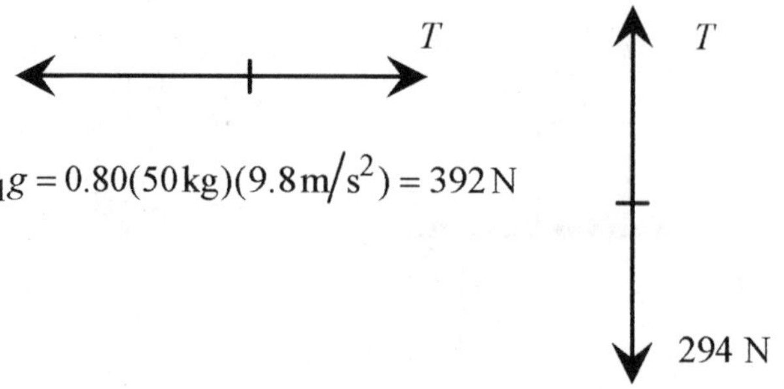

T T

$$f = \mu N = \mu m_1 g = 0.80(50\,\text{kg})(9.8\,\text{m/s}^2) = 392\,\text{N}$$

294 N

Fig. 5-9

The equation for m_1 is $T - 392\,\text{N} = 50\,\text{kg} \cdot a$

The equation for m_2 is $294 - T = 30\,\text{kg} \cdot a$

Adding the two equations $-98\,\text{N} = 80\,\text{kg} \cdot a$ or $a = -1.2\,\text{m/s}^2$

Clearly the system does not accelerate counterclockwise! This is an illustration of the second property of frictional forces: the frictional force is less than or equal to μN. In this case the frictional force just balances the tension in the cord. The frictional force, as calculated, can be up to $392\,\text{N}$. In order for m_2 not to move, the tension in the cord must be $294\,\text{N}$. (If m_2 is not moving, the forces must be in balance.) Therefore the forces on m_1 must be $294\,\text{N}$ due to the tension in the cord and $294\,\text{N}$ due to friction.

Another important point with frictional forces is the normal force as illustrated by this next problem.

5-9 A $50\,\text{kg}$ sled is pulled along a level surface at constant velocity by a constant force of $200\,\text{N}$ at an angle of 30°. What is the coefficient of friction between the sled and the surface?

Solution: The first step in this problem is to find the components of the applied force and the *mg* of the sled. These are shown on the vector diagram. Notice that the normal force, the actual force between the sled and the friction surface is *mg* minus the component of the applied force lifting up on the sled. Sometimes it is easy to miss this vertical component of the applied force. This is the major mistake in this type of problem. To help visualize this vertical component of force try placing the origin for the vectors in the middle of the sled rather than at one end where a rope would be attached.

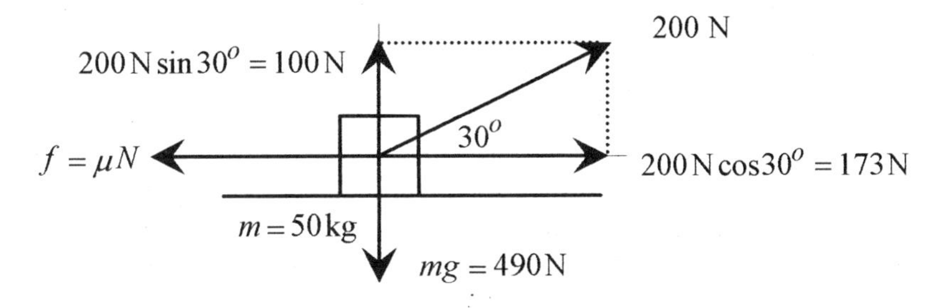

Fig. 5-10

The normal force is $mg(=490\,\mathrm{N})$ minus the vertical component of the applied force, $100\,\mathrm{N}$, or $390\,\mathrm{N}$. Since the sled is moving at constant velocity the forces must be in equilibrium. The horizontal component of the applied force must equal the frictional retarding force, the coefficient of friction times the normal force.

$$\mu \cdot 390\,\mathrm{N} = 173\,\mathrm{N} \qquad \text{so} \qquad \mu = 0.44$$

5-10 A 65°-incline plane has a mass of $20\,\mathrm{kg}$ on the plane where the coefficient of friction is 0.40 and a mass of $50\,\mathrm{kg}$ hanging free. Calculate the acceleration of the system.

Solution: Set up the vector diagram assuming motion counterclockwise. In the diagram the frictional force is calculated as a maximum of $33\,\mathrm{N}$.

The equation for m_2 is $\qquad 490\,\mathrm{N} - T = 50\,\mathrm{kg} \cdot a$

The equation for m_1 is $\qquad T - 178\,\mathrm{N} - 33\,\mathrm{N} = 20\,\mathrm{kg} \cdot a$

Adding the equations $297\,\mathrm{N} = 70\,\mathrm{kg} \cdot a \quad$ or $\quad a = 4.2\,\mathrm{m/s^2}$

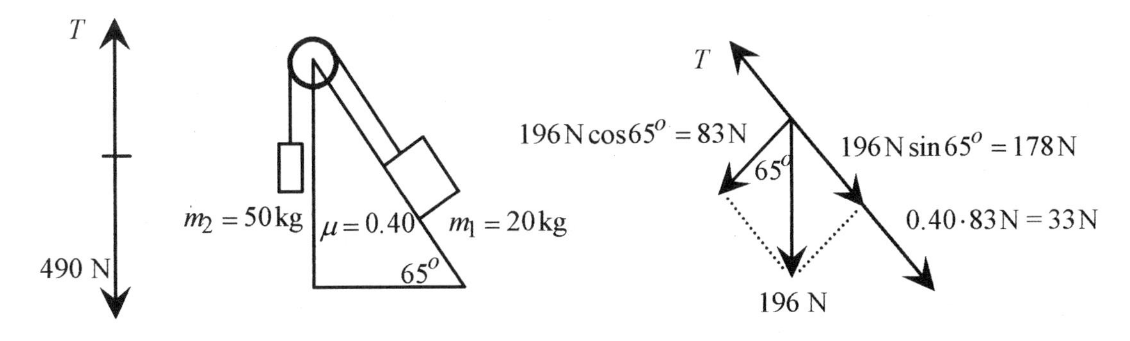

Fig. 5-11

In doing friction problems like this, you have to be careful with the frictional force. If the forces up and down the plane are nearly balanced, it is important to remember that the frictional force is not $33\,\mathrm{N}$ but that it can be up to $33\,\mathrm{N}$. In a problem similar to this, it could easily happen that the blocks would not move.

5-11 Consider a double-incline plane with angles, masses, and coefficients of friction as shown in Fig. 5-12. Calculate the acceleration of the system.

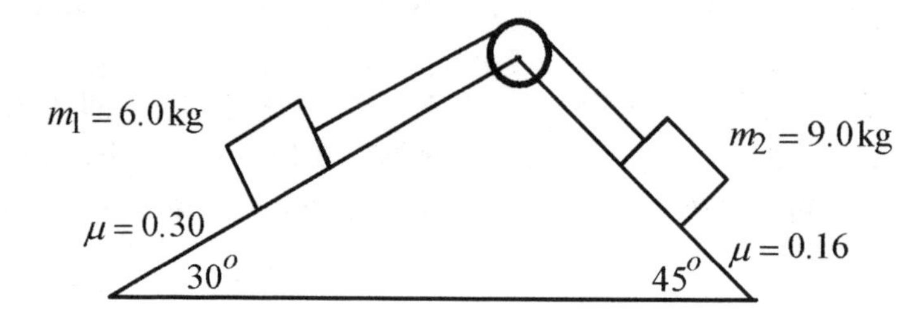

Fig. 5-12

Solution: Assume the system moves counterclockwise and set up the vector diagram.

The equation for m_1 is $\quad (30 - 15 - T)\,\mathrm{N} = 6.0\,\mathrm{kg} \cdot \mathrm{a}$

The equation for m_2 is $\quad (T - 62 - 10)\,\mathrm{N} = 9.0\,\mathrm{kg} \cdot \mathrm{a}$

Adding the equations $\quad -57\,\mathrm{N} = 15\,\mathrm{kg} \cdot \mathrm{a}$

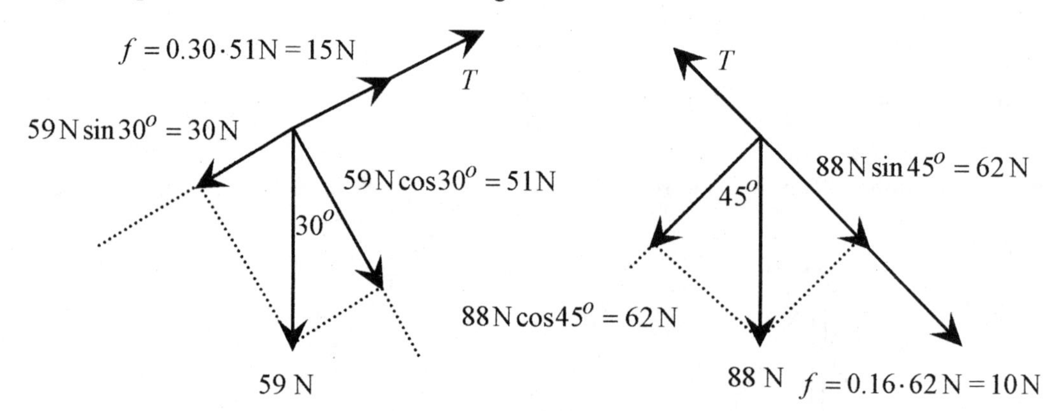

Fig. 5-13

This is a negative acceleration. We cannot simply reverse the sign of a to find the correct answer since if the system were moving the other way the frictional forces would have to be reversed. Further analysis requires us to consider another vector diagram assuming the system moves in a clockwise direction.

The numbers have all been calculated so the vector diagram can be written down easily and is shown in Fig. 5-14

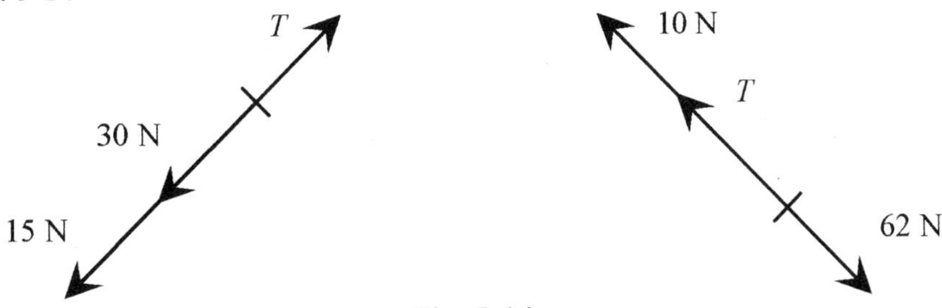

Fig. 5-14

The new equations are: $(T - 30 - 15)\,\text{N} = 6.0\,\text{kg} \cdot a$

$$(62 - T - 10)\,\text{N} = 9.0\,\text{kg} \cdot a$$

Adding the equations yields $7.0\,\text{N} = 15\,\text{kg} \cdot a$ or $a = 0.54\,\text{m/s}^2$

The tension in the cord is from the first equation: $T = 45\,\text{N} + 6.0\,\text{kg}(0.54\,\text{m/s}^2) = 48\,\text{N}$

If the acceleration of the system were negative for both the clockwise and counterclockwise calculations then we would conclude that the system would not move.

The final complication in these incline plane problems is the introduction of three masses and two connecting cords.

5-12 Consider a flat surface with a coefficient of friction of 0.20 with a 2.0 kg mass and a 3.0 kg mass connected together with a 6.0 kg mass along a slant as shown in Fig. 5-15. The coefficient of friction on the slant is zero. Calculate the acceleration of the system.

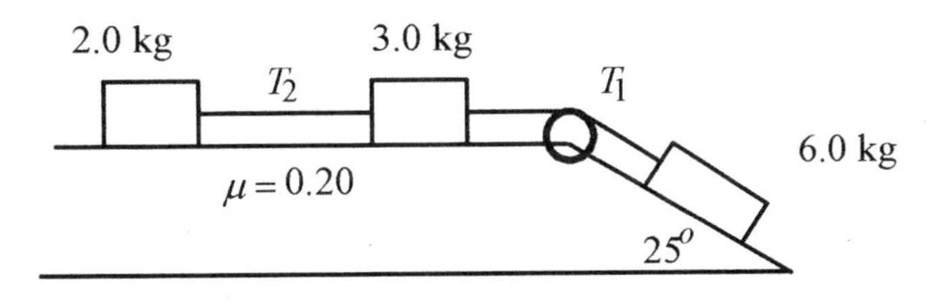

Fig. 5-15

Solution: In this problem note that the tensions are different leading to three unknowns, T_1, T_2, and a. The solution will require three equations from three vector diagrams as shown in Fig. 5-16. Assume the acceleration of the system is clockwise.

The frictional retarding force on the 2.0 kg block is $f = \mu N = 0.20(19.6\,\text{N}) = 3.92\,\text{N}$.

The frictional retarding force on the 3.0 kg block is $f = \mu N = 0.20(29.4\,\text{N}) = 5.88\,\text{N}$.

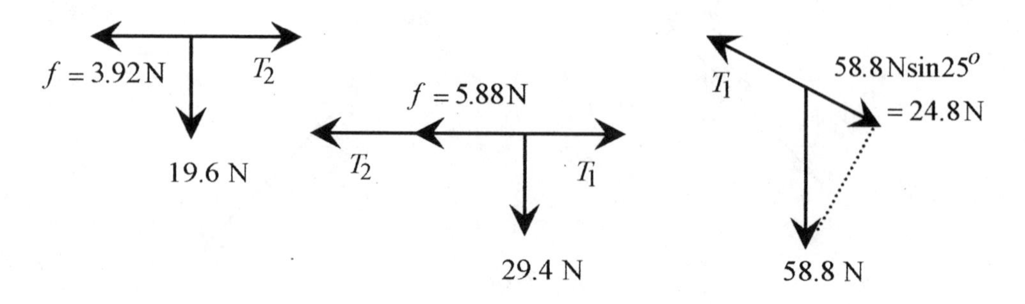

Fig. 5-16

The equations for the masses are:

$$T_2 - 3.92\,\text{N} = 2.0\,\text{kg}\cdot a$$
$$T_1 - T_2 - 5.88\,\text{N} = 3.0\,\text{kg}\cdot a$$
$$24.8\,\text{N} - T_1 = 6.0\,\text{kg}\cdot a$$

The last two equations added together are: $18.9\,\text{N} - T_2 = 9.0\,\text{kg}\cdot a$

and adding this to the first equation yields $15.0\,\text{N} = 11\,\text{kg}\cdot a$ or $a = 1.36\,\text{m/s}^2$

The tensions come from the first and third equations for the masses.

$$T_2 = 3.92\,\text{N} + 2.0\,\text{kg}(1.36\,\text{m/s}^2) = 6.64\,\text{N}$$
$$T_1 = 24.8\,\text{N} - 6.0\,\text{kg}(1.36\,\text{m/s}^2) = 16.6\,\text{N}$$

Circular Motion

A particle moving at constant speed in a circle is **uniform circular motion**. There is an acceleration and a force associated with such a particle. Figure 5-17 shows the velocity vectors associated with a particle in uniform circular motion.

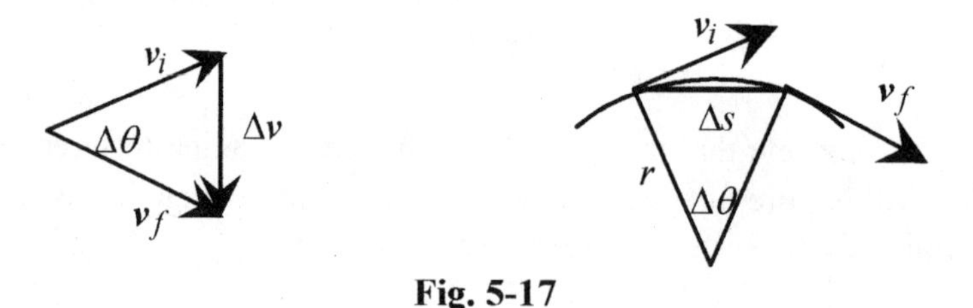

Fig. 5-17

While the length of the velocity vector remains the same, the direction changes continually. The diagram on the left of Fig. 5-17 shows v_i and v_f, with the tail of the vectors at the same point, and Δv. Looking at the limiting case where $\Delta\theta \to 0$ the ratio

$$\frac{|\Delta v|}{v} = \frac{\Delta s}{r}$$

is valid because of similar triangles. With this statement write

$$\frac{|\Delta v|}{\Delta t} = \frac{v}{r}\frac{\Delta s}{\Delta t}$$

which in the limiting situation when $\Delta\theta \to 0$ or $\Delta t \to 0$ (same criterion) is

$$\lim_{\Delta t \to 0} \frac{\Delta v}{\Delta t} = a_{\text{rad}} = \frac{v^2}{r}$$

The "rad" subscript is a reminder that the acceleration points radially inward. The direction of a_{rad} is also clear from the limiting situation. As $\Delta t \to 0$ the angle between v and $\Delta v \to 90^o$, so the acceleration is directed radially inward.

5-13 A certain car driven in a circle can exert a maximum side force of $0.95g$. What is the maximum speed for this car driven in a circle of 160m radius?

Solution: An $0.95g$ means a side force or acceleration directed at right angles to the direction of travel of $0.95(9.8\,\text{m/s}^2) = 9.3\,\text{m/s}^2$. Using $a_{\text{rad}} = v^2/r$

$$v = \sqrt{a_{\text{rad}} r} = \sqrt{9.3\,\text{m/s}^2 (160\,\text{m})} = 38.6\,\text{m/s}$$

At a speed greater than $38.6\,\text{m/s}$ the car will "slide out of the circle."

The acceleration directed toward the center of the circle for a mass in uniforn circular motion is called the centripetal acceleration. The force associated with moving a mass in a circle is ma_{rad} and is called the centripetal force.

5-14 What is the (centripetal) force produced by the tires acting on the pavement for the car of problem 5-13 if the car is 1200kg of mass?

Solution: $F = ma_{rad} = 1200\text{kg}(9.3\text{m/s}^2) = 11200\text{N}$

5-15 An 0.60kg rubber stopper is whirled in a horizontal circle of 0.80m radius at a rate of 3.0 revolutions per second. What is the tension in the string?

Solution: Three revolutions per second means three $2\pi r$'s (circumferences) per second so

$$v = \frac{3 \cdot 2\pi(0.80\text{m})}{1.0\text{s}} = 15\text{m/s}$$

The tension in the string, which is providing the centripetal force, is

$$F = m\frac{v^2}{r} = 0.60\text{kg}\frac{(15\text{m/s})^2}{0.80\text{m}} = 170\text{N}$$

5-16 A 1-oz gold coin is placed on a turntable turning at 33-1/3 RPM (revolutions per minute). What is the coefficient of friction between the coin and the turntable if the maximum radius, before the coin slips, is 0.14m?

Solution: The frictional force between the coin and turntable provides the center-directed force to keep the coin on the turntable (see Fig. 5-18). This center-directed force must equal mv^2/r.

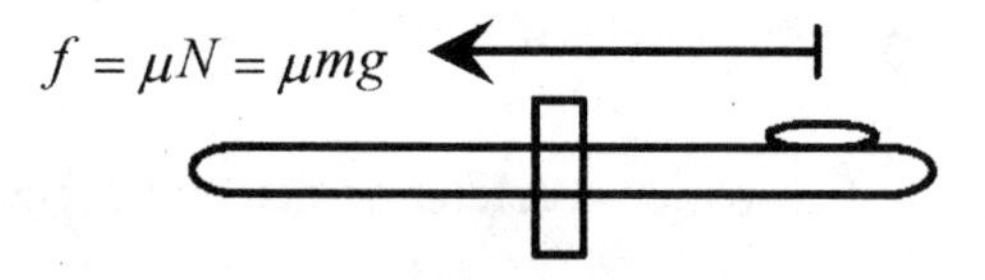

Fig. 5-18

This is a force-balance problem. In equation form: frictional force = centripetal force, or

$$\mu mg = mv^2/r \quad \text{or} \quad \mu = v^2/rg$$

Therefore,

$$\mu = \frac{v^2}{rg} = \frac{(0.49\text{m/s})^2}{0.14\text{m}(9.8\text{m/s}^2)} = 0.18$$

5-17 A conical pendulum is a mass on the end of a cord, where the mass moves at constant speed in a circle with the cord tracing out a cone. A conical pendulum of length 1.2 m moves in a circle of radius $0.20\,\text{m}$. What is the period of the pendulum?

Solution: Note that the mass is not given. Start with a vector diagram of the forces on the mass as shown in Fig. 5-19.

The mg must equal the vertical component of the tension in the string $T\cos\theta = mg$.

The horizontal component is due to the centripetal force so $T\sin\theta = mv^2/r$.

Dividing the second equation by the first eliminates T and m or $\tan\theta = v^2/rg$.

The angle is from $\sin\theta = r/L$ making $\tan\theta = 0.17$ and

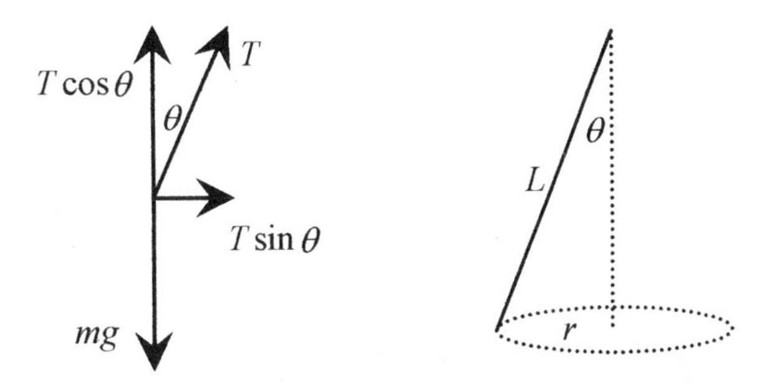

Fig. 5-19

$$v = \sqrt{rg\tan\theta} = \sqrt{0.20\text{m}(9.8\text{m/s}^2)0.18} = 0.59\text{m/s}$$

The period is from $2\pi r/T = v$ or $T = \dfrac{2\pi r}{v} = \dfrac{2\pi(0.20\text{m})}{0.59\text{m/s}} = 2.1\text{s}$

5-18 What is the speed, at which no side force due to friction is required, for a car traveling on a 20° banked road?

Solution: Figure 5-20 shows a profile of the car on the banking and the vector diagram.

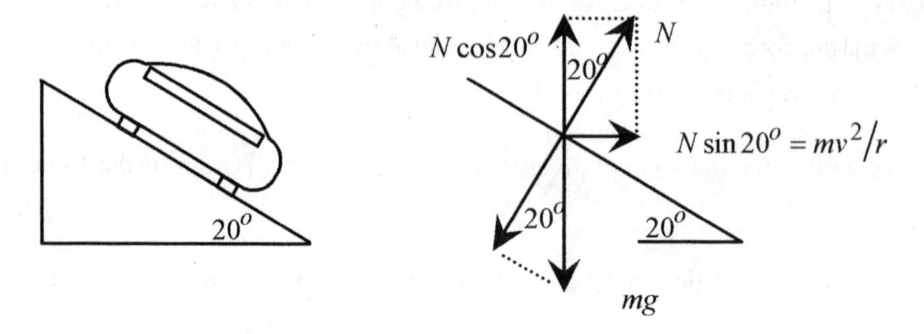

Fig. 5-20

Start with mg acting down. The normal force (between the car and road) is normal to the surface with the vertical component equal to mg and the horizontal component equal to mv^2/r.

$$N\cos 20^o = mg \qquad N\sin 20^o = mv^2/r$$

And again,

$$\frac{N\sin 20^o}{N\cos 20^o} = \tan 20^o = \frac{v^2}{rg}$$

so,

$$v = \sqrt{rg\tan 20^o} = \sqrt{240\,\mathrm{m}(9.8\,\mathrm{m/s^2})\tan 20^o} = 29\,\mathrm{m/s}$$

Notice that in this problem the mass of the car does not enter into the calculation.

Go back over problems 5-13 through 5-18 and notice how each problem could be written asking for a different variable. For example in problem 5-13 give the side force and the velocity and ask for the radius. In problem 5-16 give the speed and the coefficient of friction and ask for the radius. Changing the problems like this is an excellent way to generate practice (for the test) problems.

CHAPTER 6

APPARENT WEIGHT

Apparent weight problems usually are encountered in the discussion of force. Though relatively easy, they cause problems for some people. This short chapter should give you enough background and examples to show you how to do apparent weight problems.

One of the first points to get straight is the difference between **mass** and **weight**. Mass is that property that makes different objects accelerate differently when equal forces are applied. Operationally mass is the m in the equation $F = ma$. Mass is measured in kilograms or slugs. Weight in equation form is

$$W = mg \tag{6-1}$$

and is the force something exerts. Weight is measured in Newtons or pounds. Popular usage scrambles weight and mass. Gold and silver are usually sold by the gram or kilogram while nails and chicken feed are sold by the pound.

The first step in keeping this all straight is to remember that weight is force. A scale that measures weight in pounds or Newtons is just a force meter. Tension in a rope is a force.

A person of $80\,\text{kg}$ mass standing on the surface of the earth is subject to the force of gravity that acts between the $80\,\text{kg}$ person and the mass of the earth. This force is expressed as an acceleration due to gravity. On the earth, this acceleration is $9.8\,\text{m/s}^2$. The force on the person, also called weight is

$$F \quad \text{or} \quad W = 80\,\text{kg} \cdot 9.8\,\text{m/s}^2 = 784\,\text{N}$$

A force meter placed between the person and the earth would read $784\,\text{N}$.

To understand apparent weight take a "thought trip" (based on experience) on an elevator. Imagine a force meter between you and the floor of the elevator. As you step on the elevator the force meter reads the same as when you are standing on the ground. As the elevator accelerates upward, the force meter registers higher, reading maximum at maximum acceleration. As the acceleration decreases and the elevator assumes a constant velocity (upward) the force meter reads the same as when the elevator was not moving. This constant velocity condition is equivalent to forces in equilibrium, so that the only thing contributing to the reading of your force meter is the mg due to the earth. As the elevator slows, by decelerating, the force meter reads less than the force for constant velocity. As the elevator comes to rest at a higher level, the force meter again reads the same as when you were at ground level.

Let's go back over this elevator ride and put in some numbers. At zero or constant velocity the force meter reads $784\,\mathrm{N}$. If the elevator were accelerating upward at $2.0\,\mathrm{m/s^2}$, then the force meter would read

$$F \text{ or } W = 784\,\mathrm{N} + 2.0\,\mathrm{m/s^2}\,(80\,\mathrm{kg}) = 944\,\mathrm{N}$$

If the elevator were accelerating downward (or decelerating near the top of its trip up) at $3.0\,\mathrm{m/s^2}$, then the force meter would read

$$F \text{ or } W = 784\,\mathrm{N} - 3.0\,\mathrm{m/s^2}\,(80\,\mathrm{kg}) = 544\,\mathrm{N}$$

If the elevator were in free fall, then the force meter would read zero. The acceleration acting on the elevator is the same as on the person.

6-1 A $12\,\mathrm{kg}$ flower pot is hanging by a cord from the roof of an elevator. What is the tension in the cord when the elevator is stationary and when it is accelerating upward at $3.0\,\mathrm{m/s^2}$?

Solution: When the elevator is stationary the tension in the cord is

$$T = 12\,\mathrm{kg}(9.8\,\mathrm{m/s^2}) = 118\,\mathrm{N}$$

When the elevator is accelerating upward the tension in the cord is increased by an amount equal to ma

$$T = 118\,\mathrm{N} + 12\,\mathrm{kg}\cdot 3.0\,\mathrm{m/s^2} = 154\,\mathrm{N}$$

Second Solution: Now take another, more analytical, look at the problem. When the elevator is at rest the tension in the cord equals mg, so writing a force statement, $T - mg = ma$ where a is zero, produces $T = mg$.

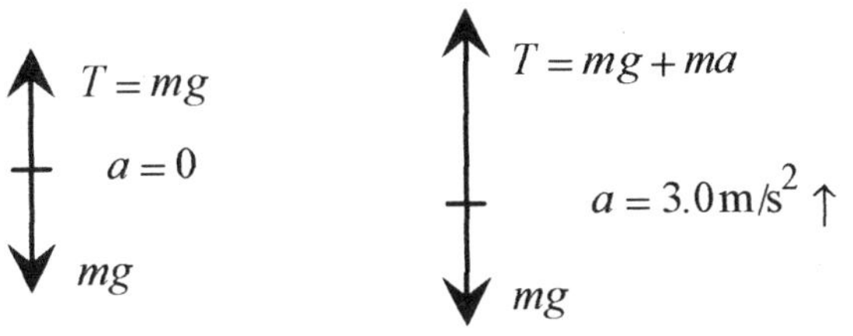

$T = mg$ $a = 0$ mg

$T = mg + ma$ $a = 3.0\,\text{m/s}^2 \uparrow$ mg

Fig. 6-1

When the elevator is accelerating upward at $3.0\,\text{m/s}^2$ the $T - mg = ma$ equation has an acceleration. The hard part of doing the problem this way is keeping the algebraic sign of the acceleration correct. Looking at Fig. 6-1, T and a have the same sign so write

$$T - 12\,\text{kg}(9.8\,\text{m/s}^2) = 12\,\text{kg}(3.0\,\text{m/s}^2) \quad \text{or} \quad T = 154\,\text{N}$$

If you use this vector approach to doing apparent weight problems be especially careful of the signs. However you do the problems, go through a little thought experiment to make sure you have the signs right.

Another difficulty with the vector approach to apparent weight has to do with the interpretation of the vector diagram (on the right in Fig. 6-1). If you are an observer outside the elevator you observe an acceleration of the flower pot, as the analysis indicates. If, however, you are riding in the elevator you observe no acceleration of the flower pot because you are riding in the accelerating reference frame.

6-2 A rope (fastened at the top) is hanging over a cliff. What is the tension in the rope with a $70\,\text{kg}$ mountain climber sliding down the rope at a constant acceleration of $6.0\,\text{m/s}^2$.

Solution: Performing a little "thought experiment," the tension in the rope is less than if he were at zero velocity or sliding at constant velocity. In this instance the $6.0\,\text{m/s}^2$ must be subtracted from the $9.8\,\text{m/s}^2$.

$$T = (9.8 - 6.0)\,\text{m/s}^2\,(70\,\text{kg}) = 266\,\text{N}$$

When in doubt as to whether to add or subtract the acceleration from g, look to the extreme situation. In this problem the person sliding down the rope at constant velocity would produce $(9.8\,\text{m/s}^2)70\,\text{kg} = 686\,\text{N}$ of tension in the rope. If the person were in "free fall," that is accelerating at $9.8\,\text{m/s}^2$, he would produce no tension in the rope. Thus, any acceleration down the rope should be subtracted from the $9.8\,\text{m/s}^2$.

6-3 What is the tension in a rope with the $70\,\text{kg}$ mountain climber of problem 6-3 accelerating (climbing) up the rope at a constant acceleration of $1.0\,\text{m/s}^2$.

Solution: In this case the tension in the rope is increased by ma.

$$T = (9.8 + 1.0)\,\text{m/s}^2\,(70\,\text{kg}) = 756\,\text{N}$$

6-4 A $100\,\text{kg}$ astronaut produces a force (weight) on the surface of the earth of $9800\,\text{N}$. What force (weight) would the astronaut produce on the surface of the moon where the "g" is about one-sixth of the g on earth?

Solution: The force or weight would be $W_{\text{moon}} = 100\,\text{kg}(9.8\,\text{m/s}^2)\dfrac{1}{6} = 1630\,\text{N}$

6-5 A $16\,\text{kg}$ monkey wishes to raise a $20\,\text{kg}$ mass by climbing (accelerating) up a rope that passes over a pulley attached to the mass. How much must the monkey accelerate up the rope in order to raise the mass?

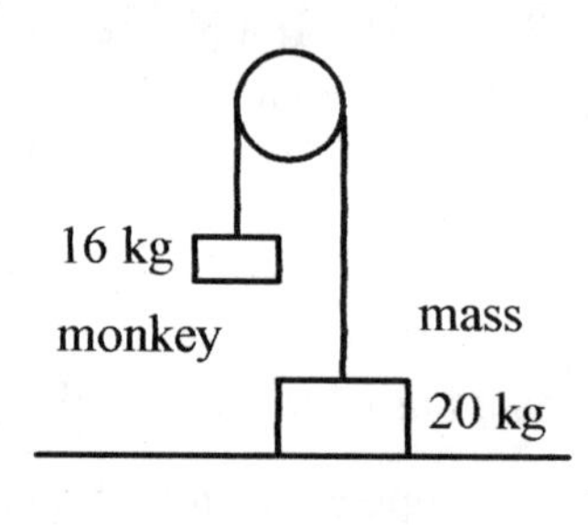

Fig. 6-2

Solution: The mass produces a force or tension in the rope of

$$T_{\text{mass}} = 20\,\text{kg}(9.8\,\text{m/s}^2) = 196\,\text{N}$$

The mass of the monkey hanging on the rope produces a force of

$$T_{\text{monkey}} = 16\,\text{kg}(9.8\,\text{m/s}^2) = 157\,\text{N}$$

To just balance the mass the monkey must accelerate up the rope to produce a force of $(196 - 157)\,\text{N} = 39\,\text{N}$.

$$39\,\text{N} = (16\,\text{kg})a \quad \text{or} \quad a = 2.4\,\text{m/s}^2$$

At this acceleration, the tension in the rope is $196\,\text{N}$.

To raise the mass, the monkey must accelerate at a rate greater than $2.4\ \text{m/s}^2$.

Any acceleration greater than $2.4\,\text{m/s}^2$ will increase the tension in the rope by an amount equal to the "additional" acceleration times the mass of the monkey.

CHAPTER 7

WORK AND THE DEFINITE INTEGRAL

Work is defined as the force, in the direction of a displacement, times that displacement. Consider the application of a constant force, F, applied over a distance, x, as depicted in Fig. 7-1.

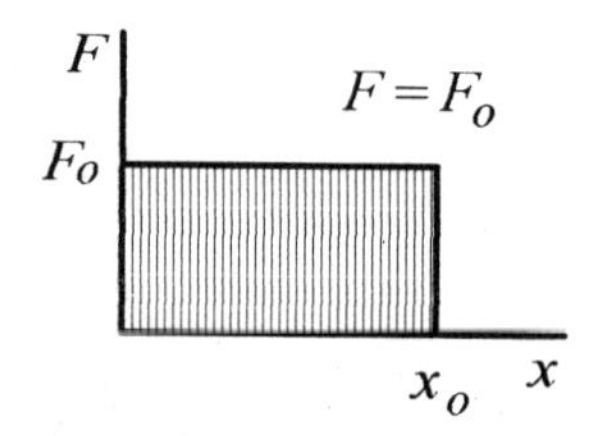

Fig. 7-1

According to the definition, the work is

$$W = F_o x_o \quad \textbf{(7-1)}$$

which is seen as the shaded rectangular area in Fig. 7-1. In calculus terminology the work is the area under the curve. The area under this curve is the area of the rectangle, $F_o x_o$. The unit of work is the Joule. $(\mathrm{J} = \mathrm{N} \cdot \mathrm{m})$

7-1 Calculate the work performed by a constant force of $60\,\mathrm{N}$ acting in the direction of a displacement of $3.0\,\mathrm{m}$.

Solution:

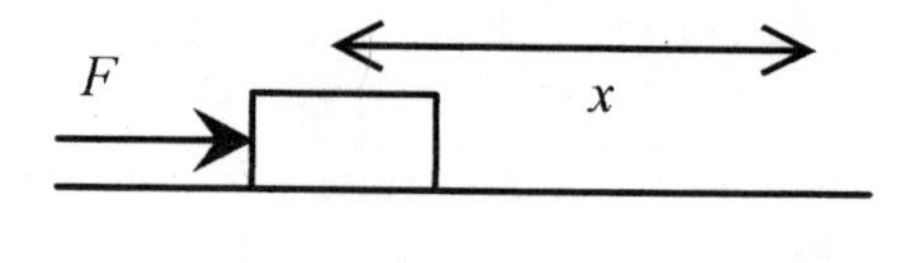

Fig. 7-2

The force is acting in the direction of the displacement, so the work is the product of force times displacement.

$$W = F \cdot x = 60\,\mathrm{N} \cdot 3.0\,\mathrm{m} = 180\,\mathrm{J}$$

In problem 7-1 the force is in the direction of the displacement, and the work is simply the product of force and displacement. In this next problem the force is not in the direction of the displacement. The simplest way to obtain the component of the force in the direction of the displacement is to form the dot product of force and displacement. Using s as a general symbol for displacement, **work** is defined as the dot product of the force vector times the displacement vector.

$$W = \mathbf{F} \cdot \mathbf{s} \qquad \textbf{(7-2)}$$

Work is a scalar, the dot product of force and displacement. See the Introduction, Mathematical Background, for a discussion of the dot product.

7-2 Calculate the work performed by a constant force of $40\,\mathrm{N}$ pushing a block a distance of $3.0\,\mathrm{m}$ along a horizontal surface at an angle of 20° with the horizontal as shown in Fig. 7-3.

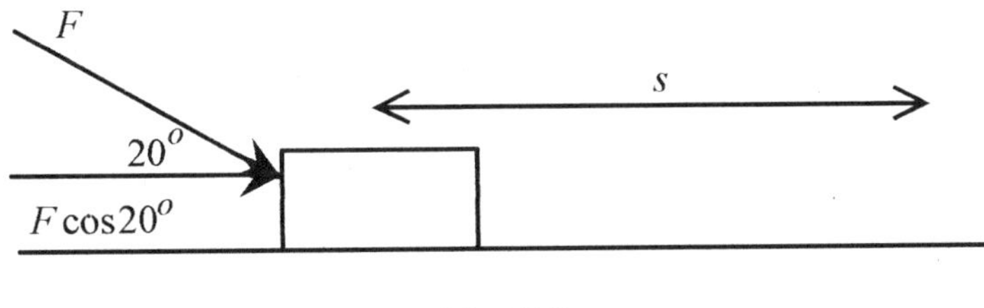

Fig. 7-3

Solution: The work performed is calculated with equation 7-2.

$$W = \mathbf{F} \cdot \mathbf{s} = 40\,\mathrm{N}\cos 20^o \cdot 3.0\,\mathrm{m} = 113\,\mathrm{J}$$

7-3 Calculate the work done by the force of gravity acting on a $70\,\mathrm{kg}$ student sliding down a 30° inclined slide a slant distance of $10\,\mathrm{m}$.

Solution: The first thing to calculate is the force due to gravity, *mg*, which is $70\,\mathrm{kg} \cdot 9.8\,\mathrm{m/s^2} = 686\,\mathrm{N}$. Place this vector on Fig. 7-4 pointing down.

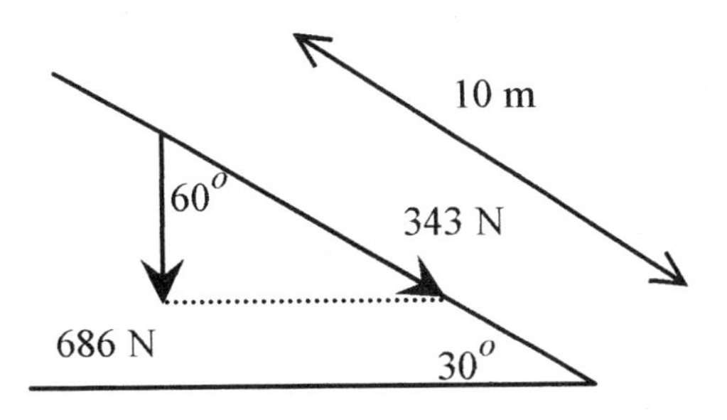

Fig. 7-4

Calculate the force in the direction of the incline $F = 686\,\text{N}\cos 60^o = 343\,\text{N}$

The work performed by the gravitational force is $W = \boldsymbol{F} \cdot \boldsymbol{s} = 343\,\text{N} \cdot 10\,\text{m} = 3430\,\text{J}$

Now consider variable forces and look first at a force that is proportional to x, the displacement, or $F = kx$ where k is a constant. This form of force is encountered in springs where one unit of force compresses (or elongates) the spring a certain distance and the next unit of force doubles the compression, and so on. The graph of force versus displacement is shown in Fig. 7-5.

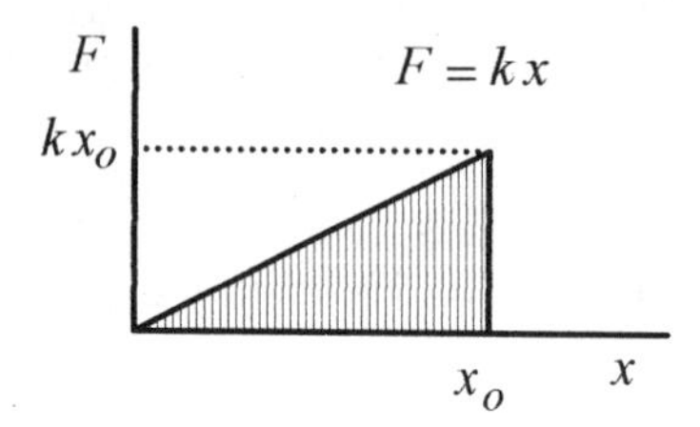

Fig. 7-5

The area under any curve of force versus distance can be approximated by taking a sum of forces times increments of distances over the curve. In this instance take each increment as Δx and the associated forces as $F_1, F_2, F_3...$ up to N forces and increments as in Fig. 7-6. In mathematical language this sum of the forces would be written as

$$\sum_{n=1}^{n=N} F\Delta x \tag{7-3}$$

where F represents the N discrete forces measured at intervals of Δx from 0 to x_o. This sum is a collection of rectangles of height $F_1, F_2, ...F_N$ and width Δx.

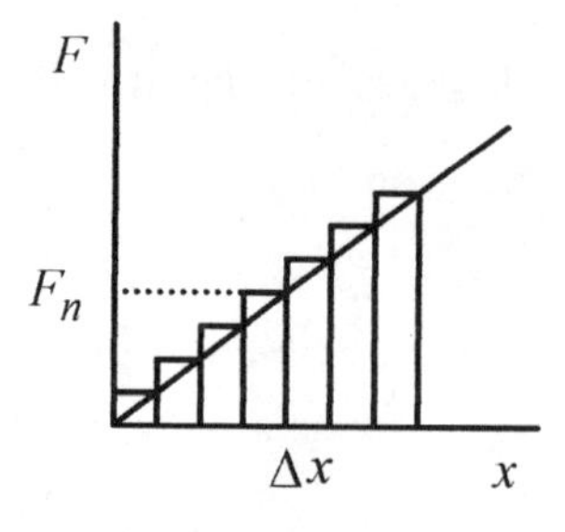

Fig. 7-6

Better approximations are obtained by increasing the number of intervals (decreasing the width of each one). In the limiting case the area would be written as

$$\lim_{\Delta x \to 0} = \sum_{n=1}^{\infty} F\Delta x \tag{7-4}$$

which is a statement that says as $\Delta x \to 0$ the rectangles become progressively narrower.

Geometrically this limit of the sum represents the "area under the curve" between $x = 0$ and $x = x_o$ and physically represents the work performed by this linear force over x. For a linear spring then the work performed is the shaded triangular area in Fig. 7-5 or

$$\frac{kx^2}{2} \tag{7-5}$$

7-4 Calculate the work performed in compressing a spring with a force constant of $200\,\mathrm{N/m}$ the first $3.0\,\mathrm{cm}$.

Solution: The work performed is calculated with equation 7-5 as

$$W = \frac{kx^2}{2} = \frac{200\,\mathrm{N}}{\mathrm{m}} \cdot \frac{(3.0 \times 10^{-2}\,\mathrm{m})^2}{2} = 9.0 \times 10^{-2}\,\mathrm{N \cdot m} = 9.0 \times 10^{-2}\,\mathrm{J}$$

7-5 Calculate the work performed in compressing the spring of problem 7-4 the next $3.0\,\mathrm{cm}$.

Solution: The general expression for the work performed is $kx^2/2$. The work to compress the spring from $3.0\,\mathrm{cm}$ to $6.0\,\mathrm{cm}$ is the work to compress it $6.0\,\mathrm{cm}$ less the work necessary to compress it $3.0\,\mathrm{cm}$.

$$W_{3 \to 6} = \frac{200\,\mathrm{N}}{\mathrm{m}} \cdot \frac{1}{2}\left[(6.0 \times 10^{-2}\,\mathrm{m})^2 - (3.0 \times 10^{-2}\,\mathrm{m})^2\right] = 0.27\,\mathrm{J}$$

(C) This process of taking an infinite number of sums of infinitesimally small increments as described in equation 7-4 and Fig. 7-6 defines an integral.

$$\lim_{\Delta x \to 0} \sum_{n=1}^{\infty} F\Delta x = \int_0^{x_o} F dx \tag{7-6}$$

For the linear or Hooke's law spring where $F = kx$ the integral becomes

$$\int_0^{x_o} kx\,dx = \frac{kx_o^2}{2} \tag{7-7}$$

which corresponds to the area under the curve of kx in the interval from 0 to x_o.

The area is the area of the triangle and is equal to $kx_o^2/2$. We will not develop the general technique for obtaining integrals of power law functions. We will, however, present a justification for the general formula for power law integrals. Consider the following table.

Force	"Work" Integral	"Work" Area
constant F_o	$\int_0^{x_o} F_o dx$	$F_o x_o$
linear kx	$\int_0^{x_o} kx\,dx$	$\frac{kx_o^2}{2}$
quadratic kx^2	$\int_0^{x_o} kx^2 dx$	$\frac{kx_o^3}{3}$
power kx^n	$\int_0^{x_o} kx^n dx$	$\frac{kx_o^{n+1}}{n+1}$

The first two entries in the table are based on the area of a rectangle and a triangle, easy examples done above. The third entry, for the quadratic, follows the pattern of the previous two and can be verified by "numerically" integrating the quadratic, that is, by taking small increments and calculating the associated force and multiplying to find the area of each little approximating rectangle. See the section on integrals in the Introduction, Mathematical Background, for more on this technique. This process also can be done for higher power curves to come to the conclusion shown in the last entry in the table, the one for the general power law curve.

We come now to another definition of **work** as the integral of the dot product of force and differential distance.

$$W = \int \mathbf{F} \cdot d\mathbf{s} \tag{7-8}$$

Most of the situations encountered in the elementary course are for constant or linear forces. The linear force is encountered in springs. Linear springs are characterized by the constant, k. The k in the equation $F = kx$ has the units N/m. The work performed in compressing a "kx" spring is $kx^2/2$. As a final problem consider the work performed in compressing a "quadratic" spring.

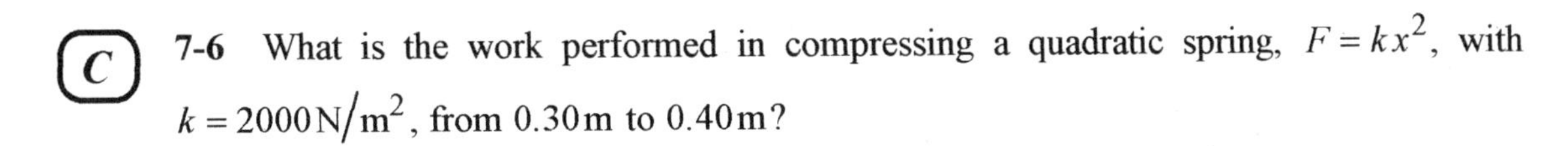

C **7-6** What is the work performed in compressing a quadratic spring, $F = kx^2$, with $k = 2000\,\mathrm{N/m^2}$, from $0.30\,\mathrm{m}$ to $0.40\,\mathrm{m}$?

Solution: The force is in the direction of the displacement so equation 7-8 reduces to

$$W = \int \mathbf{F} \cdot d\mathbf{s} = \int_{0.3}^{0.4} kx^2 dx = \left.\frac{kx^3}{3}\right|_{0.3}^{0.4} = \frac{2000\,\mathrm{N}}{\mathrm{m}^2}\frac{1}{3}\left[(0.4\,\mathrm{m})^2 - (0.3\,\mathrm{m})^2\right] = 47\,\mathrm{N \cdot m} = 47\,\mathrm{J}$$

Go back through this chapter and notice the development of the definition of work. The first definition was the product of the force in the direction of displacement times displacement (equation 7-1). In order to handle forces not in the direction of the displacement the dot product was introduced (equation7-2). Finally, for nonconstant forces the integral was introduced giving the last definition (equation 7-8).

CHAPTER 8

WORK-ENERGY PROBLEMS

Work-energy analysis is a very powerful tool for solving problems. Force analysis works well in problems where there is a constant force or where there is a balance of forces. But when the force acts over a distance, especially when that force is not constant, then work-energy analysis is the appropriate analytic tool. To handle the transformation of force over a distance (or work performed) to energy, we first need to define the concepts of kinetic and potential energy.

Kinetic Energy

A constant force applied to a mass over a distance produces an acceleration according to $F = ma$. This acceleration over a distance changes the velocity of the mass in accord with the kinematic equation $v^2 = v_o^2 + 2a(x - x_o)$. If a in this equation is replaced with F/m then $v^2 = v_o^2 + 2(F/m)(x - x_o)$. If the constant force over a distance is identified as the work then

$$F(x - x_o) = W = mv^2/2 - mv_o^2/2$$

This simple derivation suggests a more formal look at the work performed on a mass by a force.

(C) The work performed on a mass by a force is defined as the integral of the force times the distance over a specific distance. Mathematically this is $\int \mathbf{F} \cdot d\mathbf{s}$. If the force is replaced by ma and the distance taken in one dimension, for convenience, then the work integral becomes

$$W = \int_{x_1}^{x_2} ma\,dx \qquad \textbf{(8-1)}$$

The acceleration can be written as a chain derivative (see the Introduction, Mathematical Background).

$$a = \frac{dv}{dt} = \frac{dv}{dx}\frac{dx}{dt} \qquad \text{so} \qquad adx = \frac{dv}{dx}\frac{dx}{dt}dx = \frac{dx}{dt}dv = vdv$$

which makes the integral read

$$W = \int_{v_1}^{v_2} mv\,dv = \left.\frac{mv^2}{2}\right|_{v_1}^{v_2} = \frac{mv_2^2}{2} - \frac{mv_1^2}{2} \qquad \textbf{(8-2)}$$

When a force is applied to a mass over a distance and that mass accelerates, then there is a difference in velocity of the mass between before and after the application of the force. The work performed is manifest in this velocity and the $mv^2/2$ form is called kinetic energy. Work is transformed into kinetic energy.

Gravitational Potential Energy

Gravitational potential energy comes from the acceleration due to gravity and the accompanying force on a mass on the surface of the earth. If a mass is raised from a height y_1 to a height y_2, directly opposite the acceleration due to gravity, then the work performed appears as gravitational potential energy.

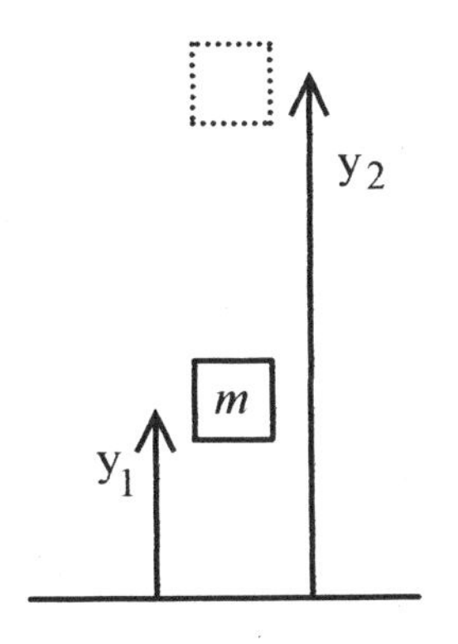

Fig. 8-1

$$W = \int F \cdot ds = \int_{y_1}^{y_2} mg\,dy = mgy_2 - mgy_1 \qquad \textbf{(8-3)}$$

This *mgy* term is the gravitational potential. In mechanical systems the total energy of the system is the sum of the kinetic plus potential energies.

In a mechanical system on the surface of the earth, work performed on the system (mass) can appear as kinetic or potential energy.

Work Performed = Change in Kinetic Energy + Change in Potential Energy

A convenient way of understanding this is to say that work "goes into" kinetic or potential energy. Goes into may be shortened to the one very descriptive word "**Goesinto**." As the problems get more complicated you will find this "**Goesinto**" concept more and more useful.

Look first at a block sliding down a frictionless incline plane as shown in Fig. 8-2. The velocity at the bottom of the plane is related to the height, h, from which the block "fell".

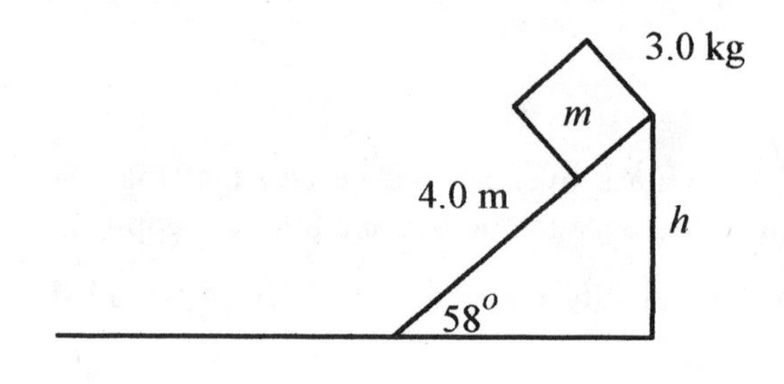

Fig. 8-2

The kinetic energy at the bottom of the plane is $mv^2/2$, and this must equal mgh, the potential energy at the top of the plane. The gravitational potential energy **Goesinto** kinetic energy.

$$\frac{mv^2}{2} = mgh \qquad \text{or} \qquad v = \sqrt{2gh}$$

For the geometric situation shown in Fig. 8-2, h is $3.4\,\text{m}$ so

$$v = \sqrt{2(9.8\,\text{m/s}^2)3.4\,\text{m}} = 8.2\,\text{m/s}$$

If the block were to continue to slide horizontally along a frictionless surface it would maintain this $8.2\,\text{m/s}$ velocity.

Frictional Forces

Frictional forces are proportional to the normal force (the force between the two surfaces) and a constant characteristic of the interface (the surfaces involved). They also act to oppose the motion. Frictional forces acting over a distance result in energy lost due to friction. In the example shown in Fig. 8-2, a coefficient of friction between the block and a flat surface at the bottom of the plane would result in energy lost to friction and the block eventually coming to rest. In this problem, potential energy at the top of the incline plane **Goesinto** kinetic energy at the bottom of the plane, and this energy **Goesinto** work to overcome friction.

Give the flat surface (Fig. 8-2) a coefficient of 0.20 and the frictional retarding force is

$$f = \mu mg = 0.20(3.0\,\text{kg})9.8\,\text{m/s}^2 = 5.9\,\text{N}$$

To find the distance the block slides, we can go back to the kinetic energy at the bottom of the plane or the potential energy at the top of the plane. Since there are no losses these numbers must be the same. Take the potential energy at the top of the plane because it is easier to calculate and is original data in the problem. This energy must equal the work due to friction fL or

$$mgh = fL = \mu mgL$$

$$3.0\,\mathrm{kg}(9.8\,\mathrm{m/s^2})3.4\,\mathrm{m} = 100\,\mathrm{J} = 0.20(3.0\,\mathrm{kg})(9.8\,\mathrm{m/s^2})L \quad \text{or} \quad L = 17\,\mathrm{m}$$

8-1 Place a $3.0\,\mathrm{kg}$ block at the top of a $3.4\,\mathrm{m}$ high frictionless incline. At the bottom of the incline the block encounters a spring with a constant of $400\,\mathrm{N/m}$ on a horizontal surface. No energy is lost to friction. How far is the spring compressed?

Solution: The energy at the top of the plane, which is the same as the energy at the bottom of the plane, **Goesinto** compressing the spring. The energy at the top of the plane is, from the example problem, $100\,\mathrm{J}$ so

$$100\,\mathrm{J} = \frac{kx^2}{2} \quad \text{or} \quad x = \left[\frac{200\,\mathrm{J}\cdot\mathrm{m}}{400\,\mathrm{N}}\right]^{1/2} = 0.71\,\mathrm{m}$$

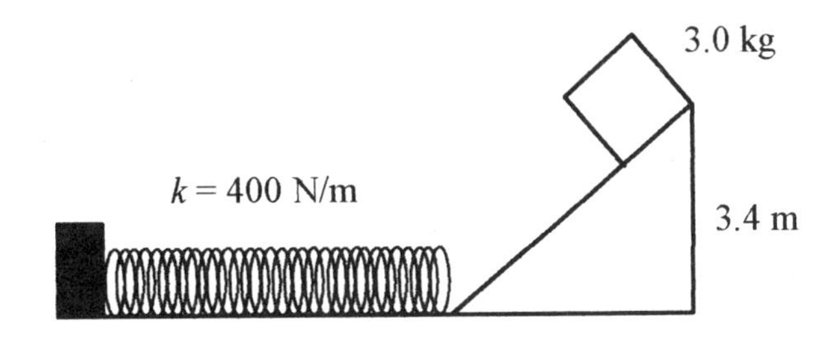

Fig. 8-3

8-2 Now complicate problem 8-1 by adding a coefficient of friction of 0.20 for the horizontal surface. How far does the block slide while compressing the spring?

Solution: Now the potential energy **Goesinto** compressing the spring <u>and</u> overcoming friction. This statement is very helpful in writing the equation. In words, the potential energy, mgh, equals the energy to compress the spring, $kx^2/2$ plus the work to overcome friction, μmgx, or

$$mgh = \frac{kx^2}{2} + \mu mgx$$

or

$$100\,\mathrm{J} = (200\,\mathrm{N/m})x^2 + 0.20(3.0\,\mathrm{kg})(9.8\,\mathrm{m/s^2})x$$

Eliminating the units, which are correct, the equation becomes

$$200x^2 + 5.9x - 100 = 0$$

with solutions

$$x = \frac{-5.9 \pm \sqrt{5.9^2 - 4(200)(-100)}}{2(200)} = \frac{-5.9 \pm 283}{400} = 0.70\,\text{m}$$

The negative root is inappropriate for this problem. The block compresses the spring $0.70\,\text{m}$ while sliding on this frictional surface.

It is instructive to do a problem first with force analysis and then with work-energy analysis. Consider the case of a block sliding down an incline plane with a coefficient of friction. As will become evident, it is possible to do this problem with force analysis techniques. Work-energy analysis is, however, conceptually and computationally easier. The force analysis follows the procedures in the chapter on forces.

8-3 Consider the incline plane with friction shown in Fig. 8-4 and calculate the velocity of the block at the bottom of the plane using force analysis, then work-energy analysis.

Solution: The vector diagram is also shown in Fig. 8-4. The unbalanced force of $20\,\text{N}$ acts on the $5.0\,\text{kg}$ block causing it to accelerate at $4.0\,\text{m/s}^2$ down the plane.

The slant height of the plane is $6.0\,\text{m} / \sin 35^o = 10.5\,\text{m}$.

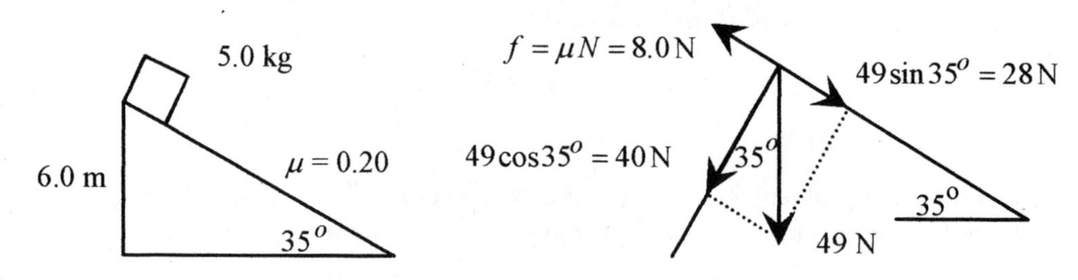

Fig. 8-4

This acceleration over the $10.5\,\text{m}$ results in a velocity

$$v^2 = 2a(x - x_o) = 2(4.0\,\text{m/s}^2)10.5\,\text{m} \quad \text{or} \quad v = 9.2\,\text{m/s}$$

Second Solution: Now do the same problem using work-energy analysis. When the block is at the top of the plane, the energy is

$$mgh = 5.0\,\text{kg}(9.8\,\text{m/s}^2)6.0\,\text{m} = 294\,\text{J}$$

At the bottom of the plane all of this (potential) energy has gone into kinetic energy (velocity of the block) except for the amount used to do work against the frictional force. The work (against the frictional force) is

$$\mu NL = 0.20(40\,\mathrm{N})10.5\,\mathrm{m} = 84\,\mathrm{J}.$$

The energy "left" after the block has slid down the plane is the 294 J at the top of the plane minus the 84 J lost due to friction, and this energy (= 210 J) is manifest in the velocity of the block (kinetic energy). This "energy" is equal to the $mv^2/2$, the kinetic energy of the block at the bottom of the plane.

$$\frac{(5.0\,\mathrm{kg})v^2}{2} = 210\,\mathrm{J} \qquad \text{or} \qquad v = 9.2\,\mathrm{m/s}$$

8-4 A 5.0 kg block sliding at 12 m/s along a frictionless surface encounters a spring of force constant of 500 N/m resting on a surface with coefficient of friction 0.25. The friction surface is only under the spring. What is the maximum compression of the spring?

Solution: Before looking at Fig. 8-5 try drawing the diagram yourself. On exams you sometimes have to work a problem from only a word description. As you are working through these problems practice drawing the diagrams, and do not be hesitant in using several diagrams. Notice the previous problem where the system was depicted with the vector diagram beside it. Sketching the system and all the vectors on a single diagram is unnecessary confusion. Use multiple diagrams and reduce the clutter.

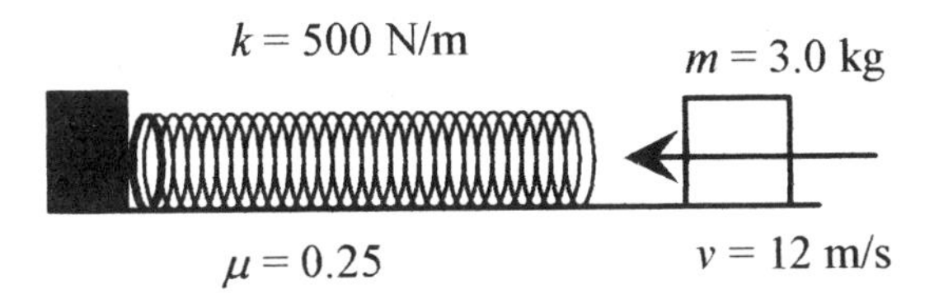

Fig. 8-5

When the block encounters the spring and the friction surface all the energy is kinetic, $mv^2/2$. This energy **Goesinto** compressing the spring and sliding along the friction surface. In equation form

$$\frac{mv^2}{2} = \frac{kx^2}{2} + \mu mgx$$

Be careful in writing this statement. Many people write the kinetic energy correctly, the energy to compress the spring correctly, and then write the <u>force</u> to slide along the friction surface. Be aware of this mistake and check the units to be sure the units on one side of the equation are the same as the units on the other side of the equation. You may have noticed that in all problems the first equation is written with units just as a reminder that checking units will reduce the number of errors. It is very frustrating to make a simple mistake that could have been avoided had units been checked in the problem before proceeding with the numerical calculations.

Putting in numbers

$$\frac{3.0\,\text{kg}(12\,\text{m/s})^2}{2} = \frac{500\,\text{N}}{\text{m}}\frac{x^2}{2} + 0.25(3.0\,\text{kg})(9.8\,\text{m/s}^2)x$$

The units are correct so without units the equation is

$$250x^2 + 7.35x - 216 = 0$$

and

$$x = \frac{-7.35 \pm \sqrt{7.35^2 - 4(250)(-216)}}{2 \cdot 250} = \frac{-7.4 \pm 464.8}{500} = 0.91\,\text{m}$$

This is the amount the spring is compressed as the block comes to a stop.

8-5 For the situation described in problem 8-4, find the velocity with which the block leaves the spring.

Solution: Start this problem with the spring compressed the $0.91\,\text{m}$. At this point the potential energy stored in the spring is used to push the block $0.91\,\text{m}$ back across the friction surface. The energy remaining as the block leaves the spring is the original kinetic energy less the energy expended in two $0.91\,\text{m}$ trips across this friction surface. In word-equation form this is

$$KE_{\text{initial}} - \text{Two}(\mu mgx\text{'s}) = KE_{\text{final}}$$

This is a **Goesinto** problem. Initial *KE* **Goesinto** two trips across the friction surface plus the final *KE*.

The total work performed in the block moving across this surface is

$$2\left[0.25(3.0\,\text{kg})(9.8\,\text{m/s}^2)0.91\,\text{m}\right] = 13.4\,\text{J}$$

The velocity of the block as it flies off the spring is

$$\frac{(3.0\,\text{kg})144\,\text{m}^2/\text{s}^2}{2} - 13.4\,\text{J} = \frac{(3.0\,\text{kg})v_f^2}{2} \quad \text{or} \quad v_f = 11.6\,\text{m/s}$$

8-6 Go back to an incline plane problem (similar to problem 8-3) but with the addition of a spring with a force constant of $300\,\mathrm{N/m}$ placed at the bottom of the incline, as shown in Fig. 8-6. Calculate the compression of the spring when the block slides down the plane.

Solution: An approximate answer can be obtained by setting *mgh,* where *h* is the vertical distance from the block to the spring, equal to $kx^2/2$, the compression of the spring.

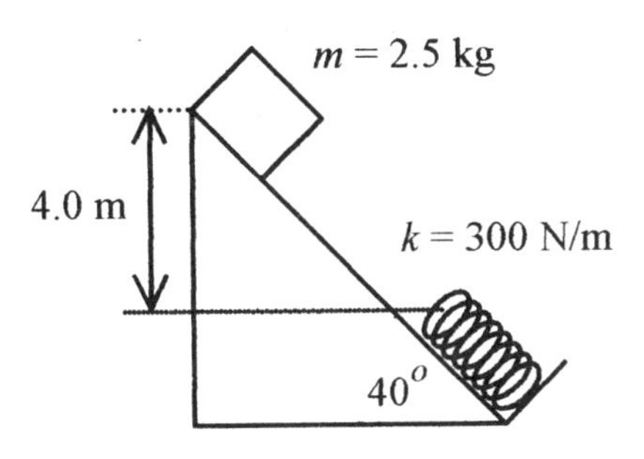

Fig. 8-6

$$2.5\,\mathrm{kg}(9.8\,\mathrm{m/s^2})4.0\,\mathrm{m} = \frac{300\,\mathrm{N}}{\mathrm{m}}\frac{x^2}{2} \quad \text{or} \quad x = 0.81\,\mathrm{m}$$

The compression of the spring gives the block a little more potential energy so a more correct statement would take into account this "additional height," h'.

$$mgh + (mgx)\sin 40^o = \frac{kx^2}{2} \qquad\qquad h' = x(\sin 40^o)$$

$$2.5\,\mathrm{kg}(9.8\,\mathrm{m/s^2})4.0\,\mathrm{m} + 2.5\,\mathrm{kg}(9.8\,\mathrm{m/s^2})x(\sin 40^o) = \frac{300\,\mathrm{N}}{\mathrm{m}}\frac{x^2}{2}$$

Eliminating the units and cleaning up the numbers yields a quadratic

$$150x^2 - 15.7x - 98 = 0$$

with solution

$$x = \frac{15.7 \pm \sqrt{15.7^2 - 4(150)(-98)}}{2(150)} = \frac{15.7 \pm 243}{300} = 0.86\,\mathrm{m}$$

This value is somewhat larger than calculated above as expected, due to the "additional height."

The procedure of finding an approximate solution and then refining it to the exact solution is very helpful in attacking problems. First you get an approximate solution that is close to the exact one, and second you educate yourself as to how to proceed to the exact solution. This procedure is used in the last problem in this chapter.

8-7 Place a mass on a track made up of a flat section, L, with coefficient of friction, μ, and two frictionless semicircular surfaces of radius R. Let the mass start from the top of one of the semicircles and calculate where it comes to rest.

Solution:

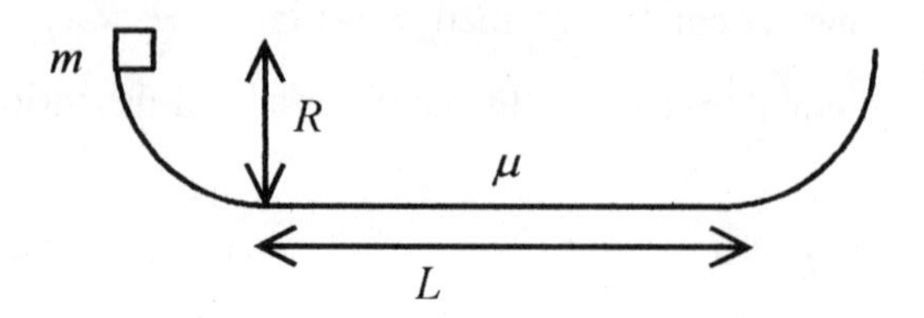

Fig. 8-7

The initial potential energy is mgR. When the mass encounters the friction surface, this (potential) energy is dissipated in doing work to overcome friction. Assuming the energy lost due to friction in one traverse is less than the initial potential energy, the mass will rise to a height (on the opposite semicircle) R' dictated by the energy statement

$$mgR' = mgR - \mu mgL$$

After another traverse of the flat portion of the track the height will be dictated by

$$mgR'' = mgR - 2\mu mgL$$

and so on until all the original potential energy is dissipated.

8-8 For the track shown in problem 8-7 take $\mu = 0.10$, $R = 1.0\,\text{m}$, and $L = 2.0\,\text{m}$ and find where the mass stops.

Solution: The only way the mass loses energy is by sliding along the flat "friction" part of the track. The initial potential energy **Goesinto** frictional work.

$$(mg)1.0\,\text{m} = 0.10(mgx) \qquad \text{or} \qquad x = 10\,\text{m}$$

The mass crosses the friction area 5 times and ends up at the edge of the friction surface opposite from where it started.

8-9 Consider a $10\,\text{gm}$ bullet passing through a $3.0\,\text{kg}$ block resting on a table. The velocity of the bullet is $400\,\text{m/s}$ on entering the block and $250\,\text{m/s}$ on exit. Calculate the energy lost by the bullet in passing through the block.

Solution: The energy is all kinetic, so calculating before and after

$$KE_{\mathrm{i}} = (1/2)(10 \times 10^{-3}\,\mathrm{kg})(400\,\mathrm{m/s})^2 = 800\,\mathrm{J}$$

$$KE_{\mathrm{f}} = (1/2)(10 \times 10^{-3}\,\mathrm{kg})(250\,\mathrm{m/s})^2 = 312\,\mathrm{J}$$

The kinetic energy lost or $\Delta KE = 488\,\mathrm{J}$.

8-10 If 30% of the energy lost in problem 8-9 is available to move the block along a surface with coefficient of friction 0.80, how far will it move?

Solution: Equate the energy available to the work performed

$$0.30\Delta KE = \mu mgL \quad \text{so} \quad 0.30(488\,\mathrm{J}) = 0.80(3.0\,\mathrm{kg})(9.8\,\mathrm{m/s^2})\,L \quad \text{or} \quad L = 6.2\,\mathrm{m}$$

8-11 Consider an elevator weighing $4000\,\mathrm{N}$ held $5.0\,\mathrm{m}$ above a spring with a force constant of $8000\,\mathrm{N/m}$. The elevator falls onto the spring while subject to a frictional retarding force (brake) of $1000\,\mathrm{N}$. Describe the motion of the elevator.

Solution:

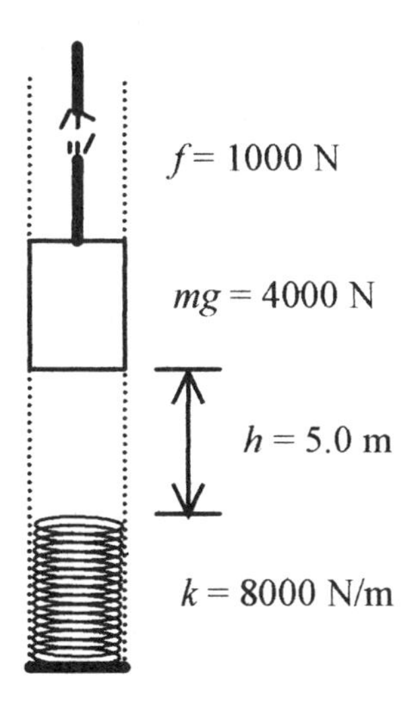

Fig. 8-8

The potential energy of the elevator with respect to the top of the spring is

$$mgh = 4000\,\mathrm{N}(5.0\,\mathrm{m}) = 20000\,\mathrm{J}$$

When the elevator falls, this energy, less the energy lost due to the frictional brake, is available to compress the spring.

$$20000\,\mathrm{J} - 1000\,\mathrm{N}(5.0\,\mathrm{m}) = 15000\,\mathrm{J} = \text{Energy to compress spring}$$

The friction brake stays on while the elevator is in contact with the spring so this energy **Goesinto** $kx^2/2 + fx$. In equation form this is

$$15000\,\text{J} = k(x^2/2) + fx$$

This expression is only approximately correct since if the spring is compressed a distance x, then this mgx provides an additional amount of potential energy. The correct equation that determines the compression of the spring is

$$mg(h+x) = \frac{kx^2}{2} + f(h+x) \quad \text{or} \quad \frac{kx^2}{2} + (f - mg)x + (f - mg)h = 0$$

Putting numbers into the equation

$$(4000\,\text{N/m})x^2 + (1000\,\text{N} - 4000\,\text{N})x + (1000\,\text{N} - 4000\,\text{N})5.0\,\text{m} = 0$$

The units are correct so $4000x^2 - 3000x - 15000 = 0$ or $4x^2 - 3x - 15 = 0$
which has solution

$$x = \frac{3 \pm \sqrt{9 - 4(4)(-15)}}{2(4)} = 2.3\,\text{m}$$

The energy stored in the spring at this (compressed) point is

$$\frac{kx^2}{2} = 8000\,\text{N/m}\frac{(2.3\,\text{m})^2}{2} = 22040\,\text{J}$$

This amount of energy will raise the elevator and do work against friction according to the equation

$$22040\,\text{J} = mgx' + fx' \quad \text{or} \quad 22040\,\text{J} = (5000\,\text{N})x' \quad \text{or} \quad x' = 4.4\,\text{m}$$

This places the elevator $2.1\,\text{m}$ above the spring.

This analysis can be repeated until the elevator comes to a stop. As an exercise perform one more "bounce" of the elevator starting at $2.1\,\text{m}$ above the spring.

The elevator will continue to bounce up and down off the spring until the weight of the elevator is equal to the force of the spring plus the frictional force

$$4000\,\text{N} = 1000\,\text{N} + (8000\,\text{N/m})s \quad \text{or} \quad s = 0.38\,\text{m}$$

so the elevator comes to rest with the spring compressed $0.38\,\text{m}$.

CHAPTER 9

MOMENTUM ANALYSIS

A discussion of the application of the law of conservation of momentum starts with a consideration of the center of mass of a collection of particles. For discrete mass points the **center of mass** is defined as

$$x_{cm} = \frac{\sum_i m_i x_i}{\sum_i m_i} \tag{9-1}$$

and likewise for y and z if the masses are distributed in two or three dimensions.
A more powerful vector notation is

$$\mathbf{r}_{cm} = \frac{1}{M}\sum_i m_i \mathbf{r}_i \tag{9-2}$$

9-1 Calculate the center of mass for a distribution of mass points as shown in Fig. 9-1.

Solution:

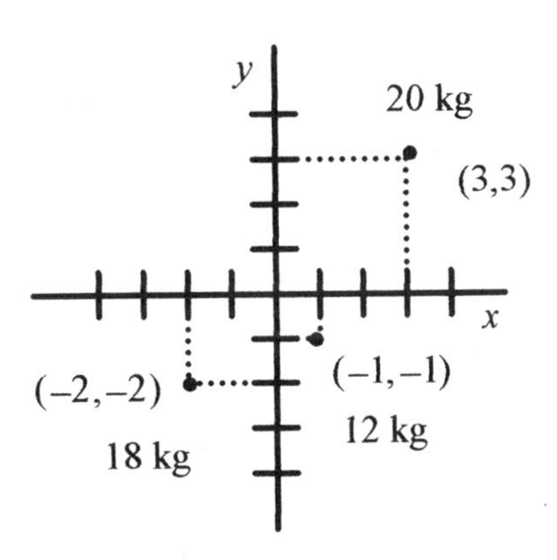

Fig. 9-1

The $\mathbf{r}_{cm}$ form indicates that the calculation is to be done in vector notation, so (The units are often left out of calculations involving unit vectors and added at the end.)

$$r_{cm} = \frac{1}{50}\left[20(3\boldsymbol{i}+3\boldsymbol{j})+18(-2\boldsymbol{i}-2\boldsymbol{j})+12(\boldsymbol{i}-\boldsymbol{j})\right]$$

$$r_{cm} = \frac{1}{50}\left[36\boldsymbol{i}+12\boldsymbol{j})\right] = \frac{36}{50}\boldsymbol{i}+\frac{12}{50}\boldsymbol{j} = 0.72\boldsymbol{i}+0.24\boldsymbol{j}$$

These mass points act as if all of their mass (50 kg) were at the point (0.72,0.24). If these three masses were placed on a plate of negligible mass, the balance point would be at (0.72,0.24).

The law of **conservation of momentum** can be viewed as a consequence of the statement: The total mass of a collection of particles times the acceleration of the center of mass equals the applied, or external, force, that is, the sum of the forces on all the individual component masses. This is of course a vector equation.

$$M\boldsymbol{a}_{cm} = \boldsymbol{F}_{ext} \qquad \textbf{(9-3)}$$

9-2 For the same collection of masses at the same points, as in problem 9-1, add forces as shown in Fig. 9-2 to each mass and find the resultant acceleration of the center of mass.

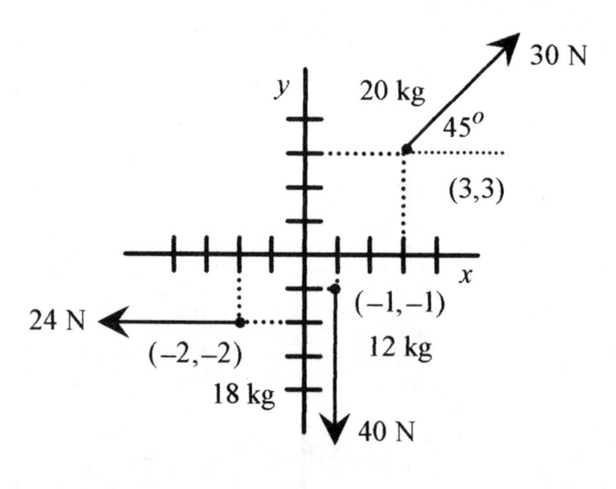

Fig. 9-2

Solution: Remember: the acceleration of the center of mass is the total mass times the vector sum of all these forces. Applying equation 9-3

$$(50\,\text{kg})\boldsymbol{a}_{cm} = \left[30\cos 45^o\boldsymbol{i}+30\sin 45^o\boldsymbol{j}-24\boldsymbol{i}-40\boldsymbol{j}\right]\text{N} = \left[-3\boldsymbol{i}-19\boldsymbol{j}\right]\text{N}$$

so

$$\boldsymbol{a}_{cm} = \left[-\frac{3}{50}\boldsymbol{i}-\frac{19}{50}\boldsymbol{j}\right]\frac{\text{m}}{\text{s}^2}$$

(C) If the previous statement $F_{ext} = ma_{cm}$ is viewed as $F_{ext} = \frac{d}{dt}(mv_{cm})$, then if $F_{ext} = 0$, mv_{cm} must be a constant. (The derivative of a constant is zero; or viewed graphically, if the curve of mv versus time is a constant then the slope is zero.)

Stated another way: For a system with no external forces the sum of the momentum vectors $m_1v_1 + m_2v_2 + \ldots$, which add to mv_{cm}, must all add to zero.

9-3 A $5.0\,\text{g}$ pellet is compressed against a spring in a gun of mass $300\,\text{g}$. The spring is released and the gun allowed to recoil with no friction as the pellet leaves the gun. If the speed of the recoiling gun is $8.0\,\text{m/s}$, what is the speed of the pellet?

Solution: This problem is solved by application of the law of conservation of momentum. This law can be applied because there is no external force. Since there is no external force, all the mv's must add to zero.

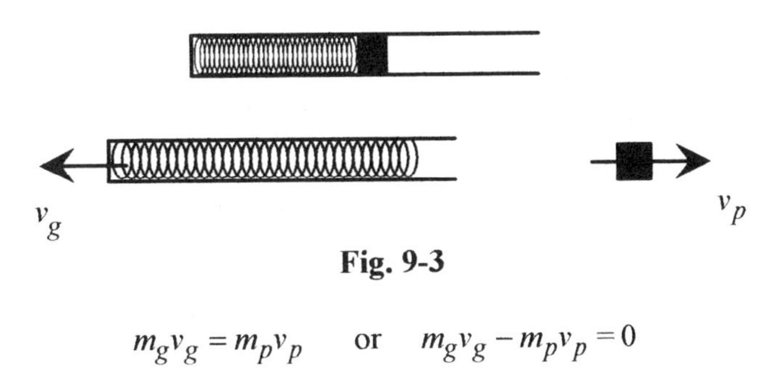

Fig. 9-3

$$m_gv_g = m_pv_p \qquad \text{or} \qquad m_gv_g - m_pv_p = 0$$

The conservation of momentum statement on the left is based on the simple observation "bullet go one way, gun go the other," while the formal statement that the mv's add to zero is on the right. With a well-labeled diagram of the situation, the statement on the left is probably easier to visualize, that is, the momenta are equal and opposite. Putting in the numbers

$$300\,\text{g}(8.0\,\text{m/s}) = 5.0\,\text{g} \cdot v_p \qquad \text{or} \qquad v_p = 480\,\text{m/s}$$

As a check, note that the momentum of the gun $p_g = m_gv_g = 2.4\,\text{kg}\cdot\text{m/s}$ and the momentum of the pellet $p_p = m_pv_p = 2.4\,\text{kg}\cdot\text{m/s}$ are numerically equal and since they are in opposite directions add to zero.

The energy of each is $mv^2/2$ or $p^2/2m$. So for the gun

$$KE_{\text{gun}} = \frac{(2.4\,\text{kg}\cdot\text{m/s})^2}{2\cdot 0.30\,\text{kg}} = 9.6\,\text{J}$$

Performing the same calculation for the pellet

$$KE_{\text{pellet}} = \frac{(2.4\,\text{kg}\cdot\text{m/s})^2}{2\cdot 0.010\,\text{kg}} = 288\,\text{J}$$

The total energy stored in the spring is the sum of these energies or $298\,\text{J}$.

9-4 Make the pellet gun of problem 9-3 fully automatic and capable of firing 10 pellets per second. Calculate the force these pellets make on a target where the pellets do not bounce.

Solution: This problem is solved by calculating the average momentum transferred to the target per unit of time. The momentum of each pellet is $2.4\,\text{kg}\cdot\text{m/s}$. The force on the target is calculated from the simple expression

$$F = \frac{\Delta p}{\Delta t} = \frac{10(2.4\,\text{kg}\cdot\text{m/s})}{1.0\,\text{s}} = 24\,\text{N}$$

The total momentum transferred each second is 10 individual momenta of each pellet.

9-5 A $75\,\text{kg}$ hockey player traveling at $12\,\text{m/s}$ collides with a $90\,\text{kg}$ player traveling, at right angles to the first, at $15\,\text{m/s}$. The players stick together. Find their resultant velocity and direction. Assume the ice surface to be frictionless.

Solution: This problem can be analyzed by conservation of momentum. Calculate the momenta and draw a vector diagram.

$$p_1 = 75\,\text{kg}(12\,\text{m/s}) = 900\,\text{kg}\cdot\text{m/s} \quad \text{and} \quad p_2 = 90\,\text{kg}(15\,\text{m/s}) = 1350\,\text{kg}\cdot\text{m/s}$$

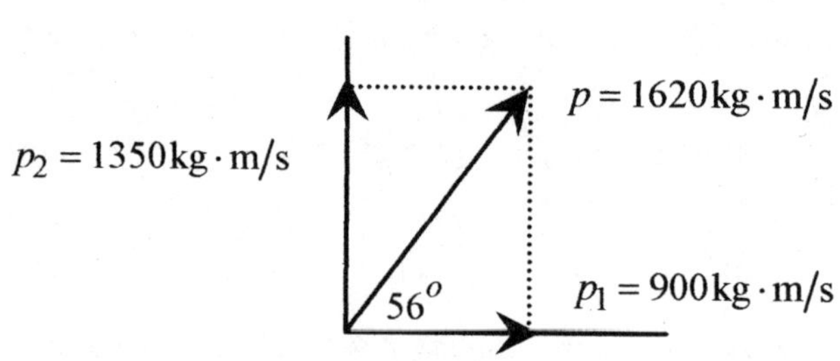

Fig. 9-4

The angle of the two hockey players is from $\tan\theta = \frac{1350}{900} = 1.5$ or $\theta = 56^o$, and the resulting momentum is $p = \sqrt{1350^2 + 900^2}\ \text{kg}\cdot\text{m/s} = 1620\,\text{kg}\cdot\text{m/s}$

The players move off with velocity $v = \dfrac{p}{m_1 + m_2} = 9.83\dfrac{\text{m}}{\text{s}}$ at an angle of 56° to the original direction of the $75\,\text{kg}$ player.

Second Solution: A more formal approach is to write a conservation of momentum statement equating the total (vector) momentum before the collision to the total (vector) momentum after the collision. Take the plus $\boldsymbol{i}$ direction as the initial direction of the first player and the plus $\boldsymbol{j}$ direction as the original direction of the second player. Using the numbers already calculated

$$[900\boldsymbol{i} + 1350\boldsymbol{j}]\text{kg}\cdot\text{m/s} = (165\text{kg})\boldsymbol{v}_f \quad \text{or} \quad \boldsymbol{v}_f = [5.45\boldsymbol{i} + 8.18\boldsymbol{j}]\text{m/s}$$

As an exercise verify the final velocity of $9.83\,\text{m/s}$ at the 56° angle.

9-6 James Bond is skiing along being pursued by Goldfinger, also on skis. Assume no friction. Mr. Bond, at $100\,\text{kg}$, fires backward a $40\,\text{g}$ bullet at $800\,\text{m/s}$. Goldfinger, at $120\,\text{kg}$, fires forward at Bond with a similar weapon. What is the relative velocity change after the exchange of 6 shots each? No bullets hit Bond or Goldfinger.

Solution: The problem is analyzed with conservation of momentum. The $m_b v_b$ of the bullet fired by Bond increases his momentum by $m_B \Delta v_B$. Remember that each bullet Bond fires increase his velocity. Set $m_b v_b = m_B \Delta v_B$ and solve for Δv_B

$$40 \times 10^{-3}\,\text{kg}(800\,\text{m/s}) = (100\,\text{kg})\Delta v_B \quad \text{or} \quad \Delta v_B = 0.32\,\text{m/s}$$

The $40 \times 10^{-3}\,\text{kg}$ bullet is small compared to the $100\,\text{kg}$ of Bond, and it would not affect the calculation. The Δv_B notation is used to indicate that each bullet fired by Bond causes a change in his velocity.

Goldfinger, on the other hand, has his momentum decreased. In his case, $m_b v_b = m_G \Delta v_G$. Putting in the numbers

$$32\,\text{kg}\cdot\text{m/s} = (120\,\text{kg})\Delta v_G \quad \text{or} \quad \Delta v_G = 0.26\,\text{m/s}$$

Bond goes faster and Goldfinger goes slower with the total change in velocity $0.58\,\text{m/s}$ for each pair of shots fired. For six shots this amounts to a difference of $3.48\,\text{m/s}$. If Bond and Goldfinger had been traveling at the same speeds, then after this exchange Bond would have a relative speed advantage of 3.48 m/s.

9-7 A 3000 kg closed boxcar traveling at 3.0 m/s is overtaken by a 1000 kg open boxcar traveling at 5.0 m/s. The cars couple together. Find the resultant speed of the combination.

Solution: The momentum before coupling is the same as the momentum after coupling (no external forces).

$$3000\,\text{kg}(3.0\,\text{m/s}) + 1000\,\text{kg}(5.0\,\text{m/s}) = (4000\,\text{kg})v \quad \text{or} \quad v = 3.5\,\text{m/s}$$

9-8 Continuing problem 9-7, rain falls into the open boxcar so that the mass increases at 1.0 kg/s. What is the velocity of the boxcars at 500 s?

Solution: The total momentum of the boxcars is $4000\,\text{kg}(3.5\,\text{m/s}) = 14000\,\text{kg}\cdot\text{m/s}$. Assume there is no horizontal component of the rain to change the momentum in the direction of motion of the boxcars. The mass increases by $(1.0\,\text{kg/s})3.5\,\text{m/s} = 500\,\text{kg}$. The momentum is a constant, so the new velocity is

$$14000\,\text{kg}\cdot\text{m/s} = (4500\,\text{kg})v_R \quad \text{or} \quad v_R = 3.11\,\text{m/s}$$

(C) **9-9** For the situation described in problems 9-7 and 9-8, what is the rate of change in velocity for the boxcars?

Solution: This is a very interesting calculus problem that involves taking the total derivative. Since there are no external forces, the total change in mv must equal zero so

$$d(mv) = mdv + vdm = 0 \quad \text{or} \quad mdv = -vdm$$

Now write m as a function of time $m = m_o + rt = 4000\,\text{kg} + (1.0\,\text{kg/s})t$.

The derivative of m is $dm = rdt$.

Using the previous two equations and rearranging $\dfrac{dv}{v} = -\dfrac{dm}{m} = -\dfrac{r}{m_o + rt}dt$

Introduce a change of variable $u = m_o + rt$ with $du = rdt$ so $\dfrac{dv}{v} = -\dfrac{du}{u}$.

Integrating, $\ln v = -\ln u + \ln K$. Use $\ln K$ because it is a convenient form for the constant.

Now rearrange: $\ln v + \ln u = \ln K$ and $\ln uv = \ln K$ or $uv = K$

Change the variable back to t so $(m_o + rt)v = K$.

Evaluate the constant from the condition that at

$$t = 0,\ m_o = 14000\,\text{kg}\cdot\text{m/s} \text{ so that } K = 14000\,\text{kg}\cdot\text{m/s}$$

The relation between v and t is $v = \dfrac{14000\,\text{kg}\cdot\text{m/s}}{(4000\,\text{kg} + 1.0\,\text{kg/s}\cdot t)}$

The velocity at $t = 500\,\text{s}$ is $v\big|_{t=500} = \dfrac{14000\,\text{kg}\cdot\text{m/s}}{4500\,\text{kg}} = 3.11\dfrac{\text{m}}{\text{s}}$

Conservation of momentum and a little calculus produce the v versus t relation.

9-10 A 3.0kg cat is in a 24kg boat. The cat is 10m from the shore. The cat walks 3.0m toward the shore. How far is the cat from the shore? Assume no friction between boat and water.

Solution: There are no external forces so the center of mass of the cat-boat system is constant. Knowing that the center of mass doesn't move is all that is necessary to do this problem.

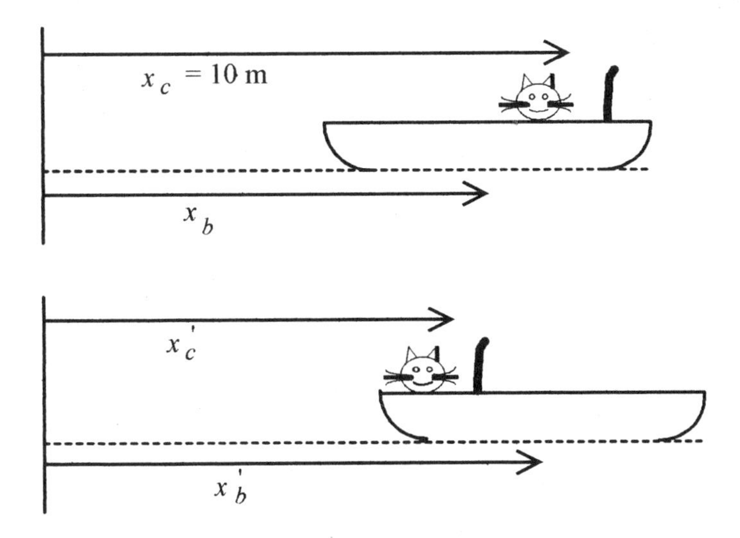

Fig. 9-5

Referring to Fig. 9-5, write the center of mass of the cat-boat system before the cat walks. (M is the mass of the boat and m the cat.) Then write the center of mass of the cat-boat system after the cat walks.

$$x_{cm} = \frac{Mx_b + mx_c}{M + m} \qquad x_{cm} = \frac{Mx_b' + mx_c'}{M + m}$$

Because there are no external forces the centers of mass are the same so

$$Mx_b + mx_c = Mx_b' + mx_c'$$

Watch the algebra and solve for x_c'. Use a word statement before writing the algebra statement. "The final distance of the cat from the shore is equal to the original distance of the cat from the shore plus the displacement of the cat relative to the boat and the displacement of the boat relative to the shore." Read the sentence, look at the diagram, and write

$$x_c' = 10\,\text{m} - 3.0\,\text{m} + (x_b' - x_b)$$

Now substitute from $M(x_b' - x_b) = m(x_c - x_c')$ or $x_b' - x_b = (m/M)(x_c - x_c')$,

$$x_c' = 7 + (1/8)(10 - x_c') \text{ and } 8x_c' = 56 + (10 - x_c') \text{ so } x_c' = 7.33\,\text{m}$$

9-11 Rabbit food in the form of pellets is poured onto a scale pan at the rate of 400 pellets per second. Each pellet has a mass of $20\,\text{g}$ and falls a distance of $2.0\,\text{m}$. Assuming the pellets do not bounce, calculate the scale reading at $7.0\,\text{s}$.

Solution: There are two components of the force. Momentum is transferred to the scale pan giving a constant force. The pellets are accumulating in the scale pan giving a time dependent part of the force.

First calculate the velocity of the pellets as they hit the scale pan.

$$(1/2)mv^2 = mgh \quad \text{or} \quad v = \sqrt{2gh} = \sqrt{2(9.8\,\text{m/s}^2)2.0\,\text{m}} = 6.26\,\text{m/s}$$

The momentum for each pellet is $mv = 0.020\,\text{kg}(6.26\,\text{m/s}) = 0.125\ \text{kg}\cdot\text{m/s}$

The total force due to this momentum transfer is 400 mv's per second

$$F_{\text{momentum}} = \frac{400(0.125\,\text{kg}\cdot\text{m/s})}{1.0\,\text{s}} = 50\,\text{N}$$

The force due to accumulation is the weight of each pellet times the number per second (μ), times the number of seconds.

$$F_{\text{accumulation}} = mg\mu\Delta t = 20\times10^{-3}\,\text{kg}(9.8\,\text{m/s}^2)(400/\text{s})7.0\,\text{s} = 549\,N$$

The scale reading at $7.0\,\text{s}$ is the sum of these two components or $599\,\text{N}$.

CHAPTER 10

COLLISION AND IMPULSE

Momentum provides a new analysis technique. With force analysis and work-energy analysis, application of the appropriate analysis technique to "the problems at the end of the chapter" was reasonably straight forward. At the end of the chapter on forces use force analysis. At the end of the chapter on work-energy use work-energy analysis. However, at the end of the chapter on momentum we need to use momentum and work-energy analysis. Knowing when to use work-energy analysis and when to use momentum analysis adds a whole new dimension to problem solving.

As will become evident in this chapter, there are certain types of collision problems where conservation of momentum and conservation of energy can be applied and certain problems where only conservation of momentum can be applied. This is no where better illustrated than in the ballistic pendulum problem (presented later), where conservation of momentum is applied in one part of the problem and conservation of energy in another part of the problem. Learning where to apply these two conservation laws is difficult. As the problems in this chapter are presented, be aware of which law is being applied and why.

Before doing specific problems take a general look at collisions in one dimension. Throughout the discussion A and B will designate the two particles, and 1 and 2 will designate before and after the collision.

There are two types of collisions:

In **elastic collisions** both momentum and energy are conserved. Examples of elastic collisions are billiard balls or any collision where the participants bounce.

In **inelastic collisions** only momentum is conserved. Examples of inelastic collisions are railroad cars coupling or a steel ball thrown into a piece of clay or any collision where the participants stick together.

It is this sticking or nonsticking that determines if energy is conserved or not. When we say that energy is not conserved in a collision where the participants stick together, we mean only that mechanical energy, $(1/2)mv^2$, is not conserved.

Inelastic Collisions

Consider first inelastic collisions, where the particles stick together. Using the notation described above

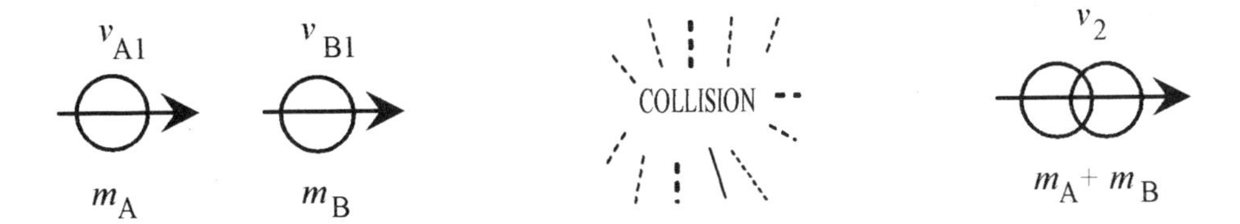

Fig. 10-1

$$v_{A2} = v_{B2} = v_2$$

This equation states that after the collision, the velocity of the A particle is the same as the velocity of the B particle. Applying the law of conservation of momentum and this condition

$$m_A v_{A1} + m_B v_{B1} = (m_A + m_B)v_2$$

The left side of the equation is the momentum before collision, and the right side is the momentum of the two masses stuck together traveling at the same velocity after the collision.

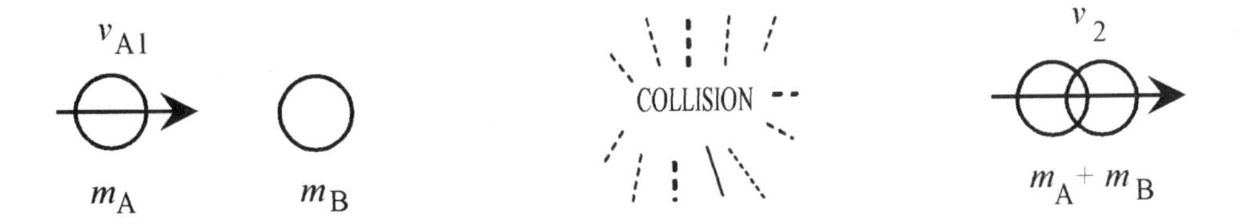

Fig. 10-2

Take the special case where the B particle is initially stationary. While mechanical energy is not conserved in the collision the kinetic energies can be written as

$$K_1 = (1/2)m_A v_{A1}^2 \quad \text{and} \quad K_2 = (1/2)(m_A + m_B)v_2^2$$

K_2 can be rewritten using the conservation of momentum statement

$$m_A v_{A1} = (m_A + m_B)v_2$$

$$K_2 = \frac{1}{2}(m_A + m_B)\left(\frac{m_A v_{A1}}{m_A + m_B}\right)^2 = \frac{1}{2}\frac{m_A^2}{m_A + m_B}v_{A1}^2$$

Comparing these two equations gives a relationship for the energy before and after collision.

$$K_2 = \frac{m_A}{m_A + m_B} K_1$$

10-1 A ballistic pendulum, a device for measuring the speed of a bullet, consists of a block of wood suspended by cords. When the bullet is fired into the block, the block is free to rise. How high does a 5.0 kg block rise when a 12 g bullet traveling at 350 m/s is fired into it?

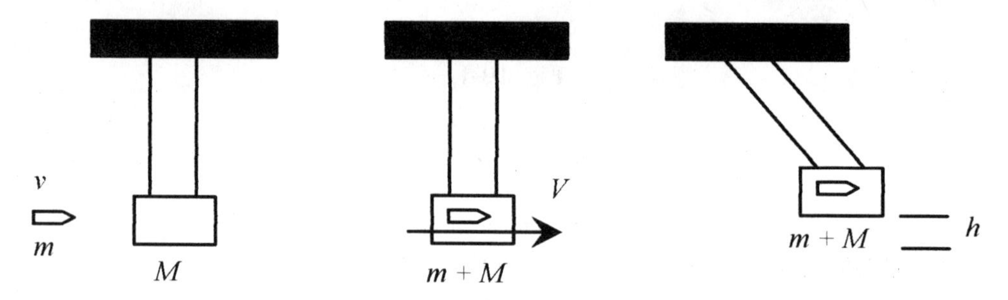

Fig. 10-3

Solution: This is a most interesting and instructive problem. The collision between the bullet and the block is clearly inelastic (the bullet comes to rest in the block). Part of the kinetic energy of the bullet **Goesinto** friction as the bullet burrows its way into the block. Therefore mechanical energy is not conserved.

Because the collision is inelastic, apply conservation of momentum to the collision. Before the collision, all the momentum is in the mv of the bullet. After the collision, the momentum is in the $(m+M)V$ of the block and bullet. We assume that the bullet comes to rest (transfers all its momentum) before there is appreciable motion of the bullet-block combination.

$$mv = (m+M)V$$

After the collision, the rise of the block is determined by energy analysis. The kinetic energy of the block **Goesinto** potential energy.

$$(1/2)(m+M)V^2 = (m+M)gh \quad \text{or} \quad V^2/2 = gh$$

Substituting for V from $mv = (m+M)V$

$$\frac{1}{2}\left(\frac{m}{m+M}\right)^2 v^2 = gh \quad \text{so} \quad v = \frac{m+M}{m}\sqrt{2gh} \quad \text{or} \quad h = \frac{v^2}{2g}\left(\frac{m}{m+M}\right)^2$$

giving the relation between the velocity of the bullet and the height the block and bullet rise. For this problem

$$h = \frac{(350\text{m/s})^2}{2\cdot 9.8\text{m/s}^2}\left(\frac{0.012}{5.012}\right)^2 = 3.6\text{cm}$$

In this problem the 0.012 can be neglected in comparison to 5.0. This is not always the case so we write $m+M$ as 5.012 as a reminder to include both $m+M$ in the calculation.

10-2 A 6.0g bullet is fired horizontally into a 2.8kg block resting on a horizontal surface with coefficient of friction 0.30. The bullet comes to rest in the block, and the block slides 0.65m before coming to a stop. What is the velocity of the bullet?

Solution:

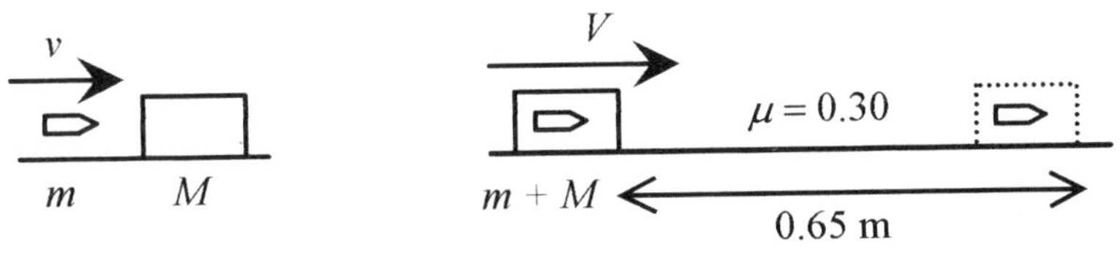

v m M V m + M μ = 0.30 0.65 m

Fig. 10-4

Assume the bullet comes to rest in the block before the block moves appreciably and that all the momentum in the bullet is transferred to the bullet-block combination.

$$mv = (m+M)V$$

Once the bullet-block combination is moving at V, the kinetic energy, $(1/2)(m+M)V^2$, **Goesinto** work to overcome friction $\mu(m+M)gx$

$$(1/2)(m+M)V^2 = \mu(m+M)gx \quad \text{or} \quad V^2/2 = \mu gx$$

Substituting

$$\frac{1}{2}\left(\frac{mv}{m+M}\right)^2 = \mu gx$$

or

$$v = \frac{m+M}{m}\sqrt{2\mu gx} = \frac{2.81}{0.060}\sqrt{2(0.30)(9.8\text{m/s}^2)0.65\text{m}} = 91\text{m/s}$$

Elastic Collisions

Now look at elastic collisions, where the particles bounce. In elastic collisions no energy is lost to permanent deformation of the particles. Write a conservation of momentum statement for the collision diagrammed in Fig. 10-5.

Fig. 10-5

$$m_A v_{A1} + m_B v_{B1} = m_A v_{A2} + m_B v_{B2}$$

Now write a conservation of energy statement.

$$\frac{1}{2} m_A v_{A1}^2 + \frac{1}{2} m_B v_{B1}^2 = \frac{1}{2} m_A v_{A2}^2 + \frac{1}{2} m_B v_{B2}^2$$

These two statements can be written as

$$m_A (v_{A1} - v_{A2}) = m_B (v_{B2} - v_{B1}) \quad \text{and} \quad m_A (v_{A1}^2 - v_{A2}^2) = m_B (v_{B2}^2 - v_{B1}^2)$$

Now divide the (rewritten) conservation of energy statement by the conservation of momentum statement to find

$$v_{A1} + v_{A2} = v_{B2} + v_{B1} \quad \text{or} \quad v_{A1} - v_{B1} = -(v_{A2} - v_{B2})$$

The term $v_{A1} - v_{B1}$ is the speed of approach, the speed of A relative to B measured by an observer on B. The term $v_{A2} - v_{B2}$ is the speed of departure. So in an elastic collision the speed of approach is equal to the speed of departure.

Now find expressions for the velocities after the collision in terms of the masses and the velocities before the collision. From the velocity statement (above)

$$v_{B2} = v_{A1} + v_{A2} - v_{B1}$$

From the conservation of momentum statement (above), substitute the equation for v_{B2}.

$$m_A v_{A1} + m_B v_{B1} = m_A v_{A2} + m_B (v_{A1} + v_{A2} - v_{B1})$$

and with a little algebra

$$v_{A2} = \frac{m_A - m_B}{m_A + m_B} v_{A1} + \frac{2m_B}{m_A + m_B} v_{B1}$$

With these two statements v_{A2} and then v_{B2} can be predicted from the initial masses and velocities.

Look at some special cases:

If $m_A = m_B$, the masses are equal, then $v_{A2} = v_{B1}$ and $v_{B2} = v_{A1}$

The particles exchange velocities. This is what happens in billiards!

If $v_{B1} = 0$, the struck mass is at rest, then

$$v_{A2} = \frac{m_A - m_B}{m_A + m_B} v_{A1} \quad \text{and} \quad v_{B2} = v_{A1} + \frac{m_A - m_B}{m_A + m_B} v_{A1} = \frac{2m_A}{m_A + m_B} v_{A1}$$

For some realistic cases where the masses are equal or one particle is at rest the resulting expressions for the velocities are easily calculated.

10-3 A $1.5\,\text{kg}$ block traveling at $6.0\,\text{m/s}$ strikes a $2.5\,\text{kg}$ block at rest. After an elastic collision, what are the velocities of the blocks?

Solution: This is the special case where the struck block is at rest so

$$v_{A2} = \frac{m_A - m_B}{m_A + m_B} v_{A1} = \frac{1.5 - 2.5}{4.0} 6.0 \frac{\text{m}}{\text{s}} = -1.5 \frac{\text{m}}{\text{s}}$$

$$v_{B2} = \frac{2m_A}{m_A + m_B} v_{A1} = \frac{2 \cdot 1.5}{4.0} 6.0 \frac{\text{m}}{\text{s}} = 4.5 \frac{\text{m}}{\text{s}}$$

The striking block rebounds at $(-)$ $1.5\,\text{m/s}$, and the struck block moves off at $4.5\,\text{m/s}$.

10-4 A $1.0\,\text{kg}$ steel ball is attached to a light-weight $1.0\,\text{m}$ long rod pivoted at the other end. The ball is released at the horizontal and strikes a $3.0\,\text{kg}$ steel block resting on a surface with coefficient of friction 0.25. How far does the block travel?

Solution: This problem is similar to the ballistic pendulum problem in that conservation of energy and conservation of momentum and then work-energy have to be applied correctly. From the description of the collision assume that it is elastic.

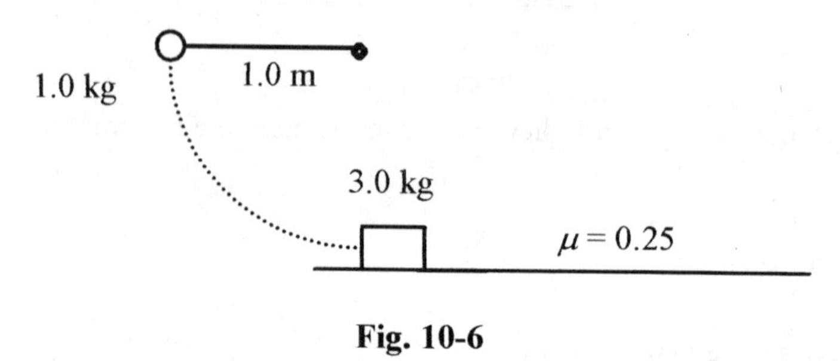

Fig. 10-6

First calculate the velocity of the ball as it hits the block. Potential energy **Goesinto** kinetic energy.

$$mgh = (1/2)mv^2 \quad \text{or} \quad v = \sqrt{2gh} = \sqrt{2(9.8\,\text{m/s}^2)1.0\,\text{m}} = 4.4\,\text{m/s}$$

The collision is elastic and $v_{B1} = 0$, so

$$v_{B2} = \frac{2m_A}{m_A + m_B} v_{A1} = \frac{2m_A}{m_A + m_B}\sqrt{2gh} = \frac{1}{2}\sqrt{2gh} = 2.2\,\text{m/s}$$

This gives the initial velocity of the struck block. The kinetic energy **Goesinto** work against friction.

$$\frac{1}{2}m_B v_{B2}^2 = \mu m_B g L \quad \text{or} \quad L = \frac{v_{B2}^2}{2\mu g} = \frac{1}{2\mu g}\frac{2gh}{4} = \frac{h}{4\mu} = \frac{1.0\,\text{m}}{4 \cdot 0.25} = 1.0\,\text{m}$$

Go back over this problem and see where energy analysis is applied, conservation of momentum is applied, and work-energy analysis applied. Knowing what laws to apply where in the problems is the hard part of collision problems.

Impulse

Impulse is the name given to a force that acts for a very short period of time. A struck baseball, golf ball, or tennis ball are examples of impulses or impulse forces. In most impulses, it is impossible to graph force versus time; though we can often estimate how the force varies with time.

Force is defined in terms of change in momentum as $F = \dfrac{\Delta(mv)}{\Delta t}$

If the force is time dependent, $F(t)$, then rewriting, $\Delta(mv) = F(t)\Delta t$

With calculus notation this statement would be written as $\int dp = \int F(t)dt$

The left side is the change in momentum, and the right side is the area under the $F(t)$ versus t curve. This integral is called the **impulse** or **impulse integral**. If the curve of $F(t)$ versus t were a parabola (reasonable for an impulse), then the right side of the equation would be the area under the curve in Fig. 10-7.

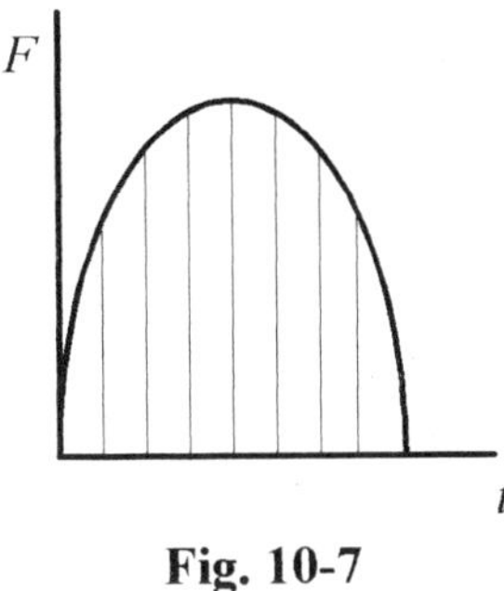

Fig. 10-7

10-5 A 1.2 kg croquet ball moving at 2.0 m/s is struck from behind by the impulse force shown in Fig. 10-8. What is the final velocity of the croquet ball?

Solution: The initial momentum is $p_i = 1.2\,\text{kg}(2.0\,\text{m/s}) = 2.4\,\text{kg}\cdot\text{m/s}$

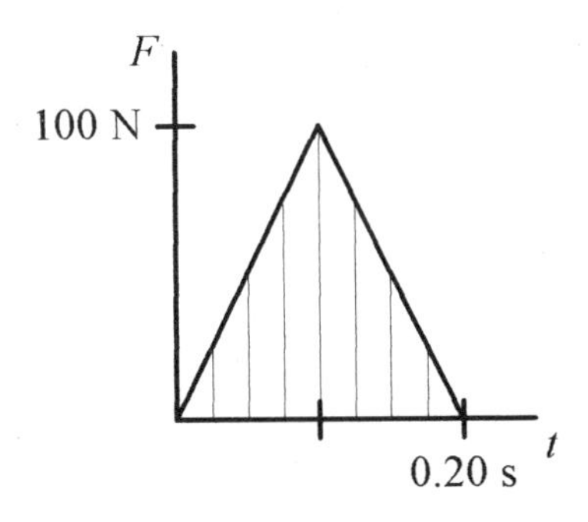

Fig. 10-8

The impulse integral, the difference in momentum, can be calculated by inspection. The area under the curve (one-half the base times the height) is

$$\Delta(mv) = \int F(t)dt = (1/2)(0.20\,\text{s})100\,\text{N} = 10\,\text{kg}\cdot\text{m/s}$$

The final momentum then is $p_f = 12.4\,\text{kg}\cdot\text{m/s}$

and the final velocity is $v_f = \dfrac{12.4\,\text{kg}\cdot\text{m/s}}{1.2\,\text{kg}} = 10.3\dfrac{\text{m}}{\text{s}}$

CHAPTER 11

ROTATIONAL MOTION

The study of rotational motion begins with the kinematics of the motion, that is, the relation between angle, θ, angular velocity, ω, and angular acceleration, α. Start with a point rotating about the origin of a coordinate system, usually an x - y system, with r - θ superimposed.

Begin with the definition of radian measure as angle equals arc length over radius. This gives the angle as a pure number, or radians. Radian is a phantom unit, sometimes it is used and sometimes not. For example, for an arc length of $4\,\text{m}$ on a circle of radius $2\,\text{m}$ the angle in radians is $4\,\text{m}/2\,\text{m} = 2$. This is usually written as $2\,\text{rad}$. In a problem involving canceling units this can add confusion because rad is length over length or unity and does not cancel with anything. So rad may appear in a problem, then disappear in the answer, or vice versa, thus fitting the description "phantom." When radian is used in a problem, the unit is for clarity and will not necessarily be in the final answer.

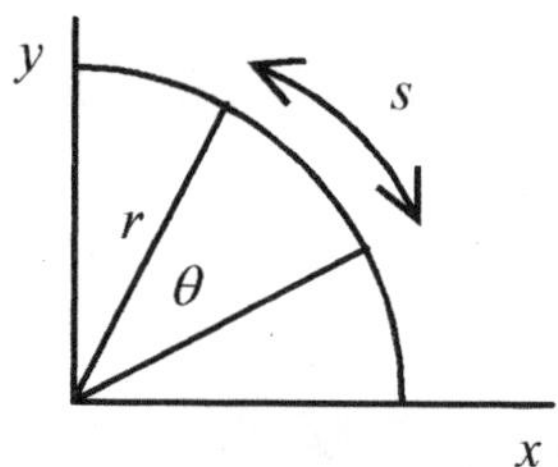

Fig. 11-1

The kinematic angular equations can be written as direct analogs to the linear ones.

Constant linear acceleration	Constant angular acceleration
$v = v_o + at$	$\omega = \omega_o + \alpha t$
$x = \dfrac{v_o + v}{2}t$	$\theta = \dfrac{\omega_o + \omega}{2}t$
$x = v_o t + (1/2)at^2$	$\theta = \omega_o t + (1/2)\alpha t^2$
$v^2 = v_o^2 + 2ax$	$\omega^2 = \omega_o^2 + 2\alpha\theta$

Another set of relationships in rotational motion is the one between linear motion and angular motion. These relate the motion of a particle, or point, along a circular path to the angular motion. Position along the circular path is called the linear position or rim position. These relationships come out of the definition of radian measure, $\theta = s/r$ or $s = r\theta$. For constant r, a small change in s is related to a small change in θ by $\Delta s = r\Delta\theta$; and if the change is over time, then $\Delta s/\Delta t = r(\Delta\theta/\Delta t)$ or $v = r\omega$, relating the linear velocity, the velocity of a point on the rim, to the angular velocity. For a small change in linear velocity $\Delta v = r\Delta\omega$; and if this occurs over time $\Delta v/\Delta t = r(\Delta\omega/\Delta t)$ or $a = r\alpha$, relating the linear acceleration, the acceleration of a point on the rim, to the angular acceleration. Formal definitions of angular velocity and angular acceleration are in the next chapter.

There is one more relationship in angular motion. If a point is rotating on a circle with constant angular velocity ω, the linear velocity is numerically constant. But velocity is a vector; and while the number associated with the linear velocity is not changing, the direction is changing. Refer to Fig. 11-2.

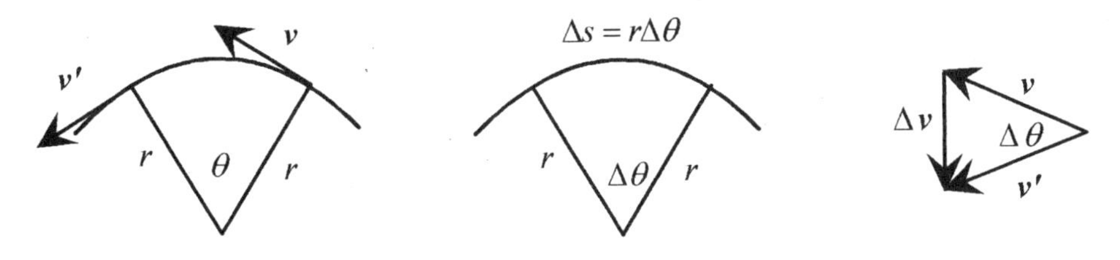

Fig. 11-2

The vectors $\boldsymbol{v}$ and $\boldsymbol{v}'$ are separated by an angle θ. The acceleration is the difference between the vectors. Remember, the vectors $\boldsymbol{v}$ and $\boldsymbol{v}'$ are at right angles to the radius, so the angle between $\boldsymbol{v}$ and $\boldsymbol{v}'$ is the same as the angle between the r's. Drawing the figure for the r's and another figure for the v's, with the v's placed tail to tail, we have similar triangles (isosceles triangles with the same angles). For small angles, Δs approaches a straight line so comparing the similar triangles write

$$\frac{\Delta v}{v} \approx \frac{\Delta s}{r} \quad \text{and from this relation write} \quad \frac{\Delta v}{\Delta t} = \frac{v}{r}\frac{\Delta s}{\Delta t}$$

In the limit $\Delta s / \Delta t$ is v and $\Delta v / \Delta t$ is a, so the acceleration is

$$a = \lim_{\Delta t \to 0} \frac{\Delta v}{\Delta t} = \lim_{\Delta t \to 0} \frac{v}{r}\frac{\Delta s}{\Delta t} = \frac{v^2}{r} \qquad \text{or} \qquad a = \frac{v^2}{r}$$

The acceleration vector points inward along r. Refer to Fig. 11-2 and note that in the limit as $\Delta t \to 0$, $\Delta s \to 0$, and $\Delta v \to 0$, $\Delta\boldsymbol{v}$ and $\boldsymbol{a}$ point toward the center of the circle. These are all the relations necessary to do problems in rotational motion.

Use this sample problem to become familiar with manipulating revolutions, revolutions per minute, and radians. A wheel is rotating on its axis at 100 revolutions per minute. This is usually written as $100\ \text{rev}/\text{min}$. The angle turned through in each minute is 100 times 2π (radians). This little calculation is often missed in rotation problems. Each revolution represents an angle of 2π measured in radians.

The angular speed is $\omega = 2\pi \cdot 100\,\text{rad/min}$,

Converting to seconds $\omega = (2\pi \cdot 100\,\text{rad/min})(1\,\text{min}/60\,\text{s}) = (10\pi/3)(\text{rad/s})$.

Assume the wheel slows, under constant acceleration, to zero in 3.0 minutes.

Calculate the acceleration as

$$\alpha = \frac{\omega - \omega_o}{\Delta t} = \frac{0 - (10\pi/3)\,(\text{rad / s})}{(3.0\,\text{min})(60\,\text{s/min})} = -\frac{\pi}{54}\frac{\text{rad}}{\text{s}^2}$$

Angular velocity is usually written with the units (rad/s) or (1/s) and angular acceleration as (rad / s^2) or $(1/\text{s}^2)$, which is why radian is called a phantom unit. Radians and revolutions are usually introduced as units for convenience in understanding a problem and not as units to be carried and manipulated, as is the case with kg, m, s, etc. Be very careful on this point. It can cause you trouble!

The angle turned through is the average angular velocity times the time.

$$\theta = \frac{\omega_o + \omega}{2} t = \frac{10\pi\,\text{rad} / 3\text{s}}{2} 180\text{s} = 300\pi\,\text{rad}$$

or

$$\theta = \omega_o t + \frac{1}{2}\alpha t^2 = \frac{10\pi\,\text{rad}}{3\text{s}} 180\text{s} + \frac{1}{2}\left(-\frac{\pi}{54}\frac{\text{rad}}{\text{s}^2}\right)(180\text{s})^2 = 300\pi\,\text{rad}$$

$$300\pi\,\text{rad}\frac{1\,\text{rev}}{2\pi\,\text{rad}} = 150\,\text{rev} \quad \text{and} \quad 150\,\text{rev}\frac{360^o}{1\,\text{rev}} = 54000^o$$

11-1 An automobile is traveling at $60\,\text{km/hr}$. Calculate the angular velocity of the 0.35m radius wheels.

Solution: The linear velocity of a point on the wheel is $\dfrac{60\text{km}}{\text{hr}}\dfrac{1\text{hr}}{3600\text{s}}\dfrac{10^3\,\text{m}}{\text{km}} = \dfrac{100}{6}\dfrac{\text{m}}{\text{s}}$

The linear velocity of a point on the wheel is the same as the velocity of the axle.

The angular velocity is $\omega = \frac{v}{r} = \left[\frac{100\text{m}}{6\text{s}}\right]\frac{1}{0.35\text{m}} = \frac{47.6\,\text{rad}}{\text{s}}$

Adding radians in the units makes the meaning clearer.

11-2 If the automobile of problem 11-1 accelerates uniformly from $60\,\text{km/hr}$ to $80\,\text{km/hr}$ in $3.0\,\text{s}$, what is the angular acceleration?

Solution: This acceleration is the acceleration of the point where the wheel meets the road. First calculate the change in velocity.

$$\Delta v = (80-60)\frac{\text{km}}{\text{hr}} = 20\frac{\text{km}}{\text{hr}}\frac{1\,\text{hr}}{60\,\text{min}}\frac{1\,\text{min}}{60\,\text{s}}\frac{10^3\,\text{m}}{\text{km}} = \frac{200}{36}\frac{\text{m}}{\text{s}}$$

This is the change in velocity of the point on the wheel. The linear acceleration, is

$$a = \frac{\Delta v}{\Delta t} = \left[\frac{200}{36}\frac{\text{m}}{\text{s}}\right]\frac{1}{3.0\,\text{s}} = 1.85\frac{\text{m}}{\text{s}^2}$$

The angular acceleration is

$$\alpha = \frac{a}{r} = \left[\frac{1.85\,\text{m}}{\text{s}^2}\right]\frac{1}{0.35\,\text{m}} = 5.29\frac{1}{\text{s}^2} = 5.29\frac{\text{rad}}{\text{s}^2}$$

11-3 A disk rotating at $30\,\text{rad/s}$ slows to $20\,\text{rad/s}$ while turning through 60 revolutions. How long does this take?

Solution: Since the acceleration is constant we can use $\theta = \frac{\omega_o + \omega}{2}t$ or

$$t = \frac{2\theta}{\omega_o + \omega} = \frac{2\cdot 2\pi\cdot 60\,\text{rad}}{30(\text{rad/s}) + 20(\text{rad/s})} = \frac{240\pi}{50}\text{s} = 15.1\,\text{s}$$

11-4 For the acceleration of the disk in problem 11-3, how many revolutions does it take for the disk to stop?

Solution: First, calculate $\alpha = \frac{\Delta\omega}{\Delta t} = \frac{-10(\text{rad/s})}{15.1\,\text{s}} = -0.662\frac{\text{rad}}{\text{s}^2}$

Now use $\omega^2 = \omega_o^2 + 2\alpha\theta$ with $\omega = 0$ and $\omega_o = 30\,\text{rad/s}$ to find θ:

$$\omega_o^2 = -2\alpha\theta \quad \text{or} \quad \theta = -\frac{\omega_o^2}{2\alpha} = \frac{(30\,\text{rad/s})^2}{2\cdot(0.662\,\text{rad/s}^2)} = 680\,\text{rad} \qquad 680\,\text{rad}\frac{1\,\text{rev}}{2\pi\,\text{rad}} = 108\,\text{rev}$$

Applications

From kinematics we can go to some more complicated rotational problems.

11-5 Consider a space station in the form of a donut with a rectangular cross section connected by spokes to a central axis. The "floor" is the inside of the outer wall. How fast would a 300 m radius station have to rotate to duplicate the "acceleration due to gravity" on the surface of the earth?

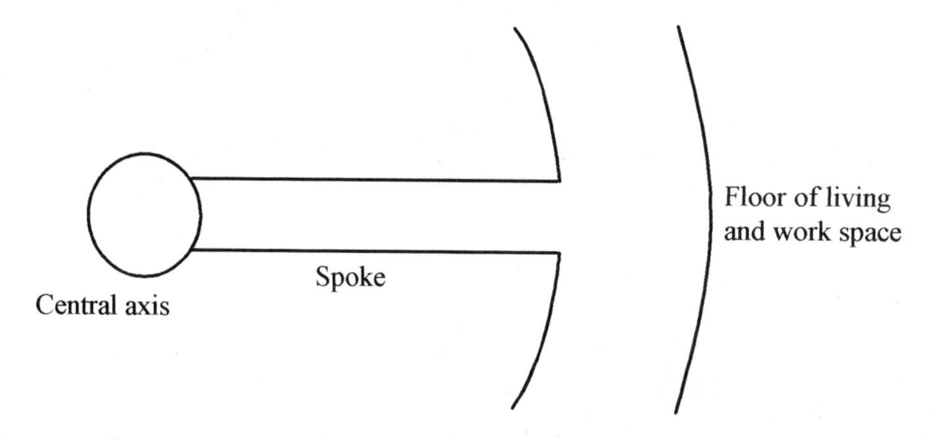

Fig. 11-3

Solution: The required acceleration is the v^2/r acceleration associated with rotational motion. Since v is related to ω by $v = r\omega$, then the required acceleration $a = v^2/r = r\omega^2$. Set this acceleration equal to $9.8\,\text{m/s}^2$ and solve for ω.

$$\omega = \sqrt{\frac{a}{r}} = \sqrt{\frac{9.8\,\text{m/s}^2}{300\,\text{m}}} = 0.181\frac{\text{rad}}{\text{s}}$$

The frequency in revolutions per minute is

$$f = \frac{\omega}{2\pi} = \frac{1\,\text{rev}}{2\pi\,\text{rad}}\frac{0.181\,\text{rad}}{\text{s}} = \frac{0.0288\,\text{rev}}{\text{s}}\frac{60\,s}{\text{min}} = 1.72\frac{\text{rev}}{\text{min}}$$

Notice that in this last calculation "rev" was introduced for clarity.

11-6 If someone of weight $800\,\mathrm{N}$ (on earth) living in the space station described in problem 11-5 were to move one-half the radius in toward the axis of rotation, what would be their weight at this radius?

Solution: Their weight, what a force meter placed between them and a "floor" would read, in this environment would be

$$W = \frac{mv^2}{r} = mr\omega^2 = \frac{800\,\mathrm{N}}{9.8(\mathrm{m/s^2})} 150\,\mathrm{m}\left(0.18\frac{1}{\mathrm{s}}\right)^2 = 400\,\mathrm{N}$$

If the person in the space station were placed in a "black box" where they were unaware of their surroundings, there is no experiment they could perform to determine whether their weight were produced by the mv^2/r of the space station or mg of the earth or any other planet.

Consider astronauts in earth orbit. From the reference frame of the astronauts the mv^2/r force is balanced by the gravitational force of the earth. This lack of force, or weightlessness, has a dramatic effect on the bodies of earth orbiters who typically "grow" 2 to $3\,\mathrm{cm}$ in earth orbit.

Many physics textbooks treat all problems involving centripetal force from an external reference point. This may sometimes be confusing. Keep this clear in your mind; from the reference point of the external observer there is no "centrifugal" force, or force that acts radially out. However, in many instances the most practical way of solving uniform circular motion problems is from the point of view of the object being rotated. Operationally this means to treat them as force balance problems in which mv^2/r acts radially out. Look back to problem 5-16 where the coin is rotating on the turntable. If you were riding on the coin at constant angular velocity you would say there is no unbalanced force in the radial direction; the mv^2/r force acting out is balanced by the μN force acting in.

11-7 Consider a mass of $2.0\,\mathrm{kg}$ being whirled in a sling in a horizontal circle. The period of rotation is $1.0\,\mathrm{s}$, and the radius $1.0\,\mathrm{m}$. What is the tension in the cord?

Solution: The force that makes the mass travel in a circle must be equal to mv^2/r. The velocity of the mass is the distance traveled (one circumference) divided by the time to travel one circumference (the period) $v = \dfrac{2\pi r}{T} = \dfrac{2\pi \cdot 1.0\,\mathrm{m}}{1.0\,\mathrm{s}} = 2.0\pi\dfrac{\mathrm{m}}{\mathrm{s}}$

The force (or tension) in the cord is $F = \frac{mv^2}{r} = \frac{2.0\,\text{kg}}{1.0\,\text{m}}\left(2.0\pi\frac{\text{m}}{\text{s}}\right)^2 = 79.0\,\text{N}$

Be aware of how tension in the cord is interpreted. If you are whirling the cord, the tension, from your point of view, is radially out. From the point of view of the mass in the sling, however, the tension is pointing in. This is similar to force problems with blocks where the tension in the connecting string acts one way on one block and the other way on another block. See for example problem 5-7 in Chapter 5, Force.

In linear motion, something moving at constant velocity is not acted on by an unbalanced force. In circular motion, something moving in uniform circular motion is subject to a force; the mv^2/r force to produce the acceleration necessary to make it move in a circle.

To help keep this straight in your mind take a "thought trip" in an airplane flying in a vertical bank (wings vertical) in a horizontal circle. A force meter between you and the seat of the airplane would read a number equal to your m times v^2/r. The force acting on you is toward the center of the circle, but you exert an equal (in magnitude) force on the seat that is radially out of the circle. In some airplanes the force can be so great as to cause the blood in the pilot to remain in the lower part of his body resulting in loss of oxygen (carried by the blood) to the brain and "blackout."

Make a slight variation in this "thought trip" by having the airplane move in a vertical circle. Now the force meter between you and the seat reads mg more or less than mv^2/r depending on whether you are on the top or the bottom of the loop.

11-8 Take the same mass, radius, and period as in problem 11-7 but with the rotation in a vertical circle. What is the force between the mass and the sling?

Solution: In this case gravity has to be taken into account. The mass exerts a force on the sling radially out and equal to mv^2/r. Gravity exerts a force mg down. Figure 11-4 shows the relative direction of these forces at four points on the circle.

The gravitational force is $mg = (2.0\,\text{kg})9.8\,\text{m/s}^2 = 19.6\,\text{N}$

The force between the mass and the sling as shown in Fig. 11-4 is $79.0 + 19.6 = 98.6\,\text{N}$ at the bottom of the circle and $79.0 - 19.6 = 59.4\,\text{N}$ at the top of the circle.

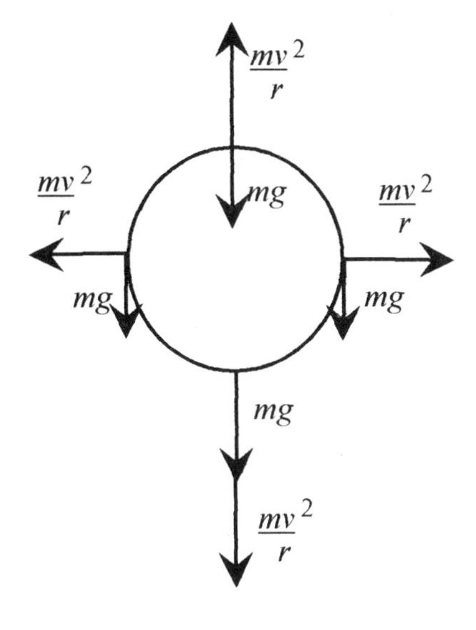

Fig. 11-4

11-9 For the situation in problem 11-8 calculate the minimum velocity and period of rotation to keep the mass from "falling out" at the top of the circle.

Solution: Zero tension in the cord occurs (at the top) when

$$\frac{mv_{min}^2}{r} = mg \qquad \text{or} \qquad v_{min} = \sqrt{rg}$$

Notice that the mass drops out of the equation. Using the 1.0 m radius

$$v_{min} = \sqrt{1.0\,\text{m}(9.8\,\text{m}/\text{s}^2)} = 3.13\,\text{m}/\text{s}$$

The period is calculated by replacing the velocity by $2\pi r / T$

$$\frac{2\pi r}{T} = \sqrt{rg} \qquad \text{or} \qquad T = 2\pi\sqrt{\frac{r}{g}} = 2\pi\sqrt{\frac{1.0\,\text{m}}{9.8(\text{m}/\text{s}^2)}} = 2.0\,\text{s}$$

This result can be checked by whirling a mass in a vertical circle and observing that the mass "falls out" of the circle at a period of over 2.0 s. (The length from shoulder to ground is a little over 1.0 m for most people so this is an easy experiment to perform. For a little adventure try it with a container of liquid!)

11-10 A glider pilot wishing to fly in a vertical loop dives to attain a speed sufficient to keep the glider from "falling out" of the top of the loop. What is the minimum entry speed for a 200 m radius loop?

Solution: This problem is solved with energy analysis. At the bottom of the loop the glider must have sufficient kinetic energy to climb the diameter of the loop and have enough left to satisfy the $mv^2/r = mg$ condition. Refer to Fig. 11-5 to write the energy balance statement and the criteria for the glider not "falling out" of the loop.

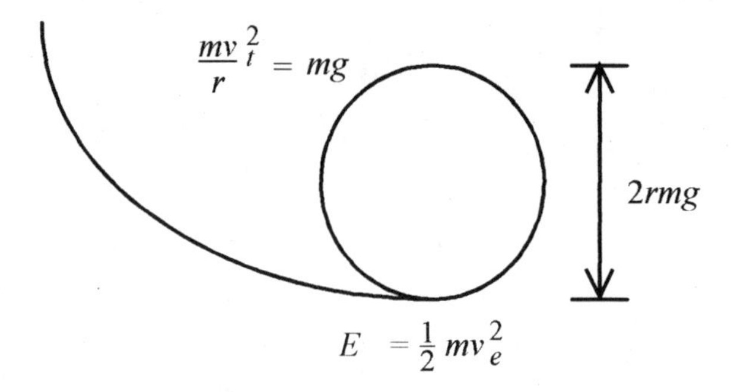

Fig. 11-5

The energy statement is in word form: Kinetic energy at entry equals the potential energy to reach the top of the loop plus the kinetic energy at the top of the loop

$$\frac{1}{2}mv_e^2 = 2rmg + \frac{1}{2}mv_t^2$$

The minimum velocity condition is $\frac{mv_t^2}{r} = mg$ or $v_t = \sqrt{rg}$.

Substituting for v_t the minimum energy statement is $\frac{1}{2}mv_e^2 = 2rmg + \frac{1}{2}mrg = \frac{5}{2}mrg$ yielding a minimum entry speed of $v_e = \sqrt{5rg}$

For a 200 m radius loop (very large) the minimum v_e for the glider is

$$v_e = \sqrt{5(200\,\text{m})9.8\,\text{m/s}^2} = 99\,\text{m/s} = 356\,\text{km/hr}$$

For a 100 m radius loop $v_e = 70\,\text{m/s} = 252\,\text{km/hr}$

11-11 A grinding wheel of 30 cm diameter is rotating with angular velocity of 3.0(rad / s) and slowing under constant acceleration of −3.0(rad / s). Calculate everything possible about the motion at several different times.

Solution: Circular motion problems even more than linear motion problems, often present a challenge as to how to proceed from the data to the specific question. When confused about the specific route to follow in a problem, the best thing to do is calculate something simple and let this first calculation lead you to others and ultimately the answer. After doing the problem once the "hard way" you will learn more direct routes through the problem. Given the data in this problem start calculating some simple things.

What is the angular velocity at 0.40 s? Use

$$\omega = \omega_o + \alpha t = 3.0(\mathrm{rad/s}) - 3.0(\mathrm{rad/s^2})0.40\,\mathrm{s} = 1.8(\mathrm{rad/s})$$

What angle has been turned through in this 0.40 s? Use

$$\theta = \omega_o t + (1/2)\,\alpha t^2 = 3.0(\mathrm{rad/s})0.40\,\mathrm{s} - (1/2)3.0(\mathrm{rad/s^2})(0.16\,\mathrm{s^2}) = 0.96\ \mathrm{rad}$$

What is the velocity of a point on the rim at 0.20 s? First find the angular velocity.

$$\omega = \omega_o + \alpha t = 3.0(\mathrm{rad/s}) - 3.0(\mathrm{rad/s^2})(0.20\,\mathrm{s}) = 2.4(\mathrm{rad/s})$$

And then find the linear velocity from $v = r\omega = 15\times10^{-2}\,\mathrm{m}\cdot2.4(\mathrm{rad/s}) = 0.36\,\mathrm{m/s}$.

With this acceleration, how long does it take for the wheel to come to rest? Use

$$\omega = \omega_o + \alpha t \quad \text{or} \quad 0 = 3.0(1/\mathrm{s}) - 3.0(1/\mathrm{s^2})\,t \quad \text{or} \quad t = 1.0\,\mathrm{s}$$

How much angle is turned through in this time? Use

$$\omega^2 = \omega_o^2 + 2\alpha\theta \quad \text{or} \quad 0 = 9.0(\mathrm{rad^2/s^2}) - 2\cdot3.0(\mathrm{rad/s^2})\,\theta \quad \text{or} \quad \theta = 1.5\ \mathrm{rad}$$

How far does a spot on the rim travel in this time? Use

$$s = r\theta = 15\times10^{-2}\,\mathrm{m}\cdot1.5(\mathrm{rad}) = 0.225\,\mathrm{m}$$

CHAPTER 12

ROTATIONAL DYNAMICS

Because we take another view of rotational motion in the beginning of this chapter, you may find it helpful to first review the discussion of rotational motion in the previous chapter.

Consider the rotation of a solid about some point in a plane. The rotation is counterclockwise in conformity with trigonometry. The angle is again defined as arc length over radius, the standard definition appropriate to radian measure $\theta = s/r$.

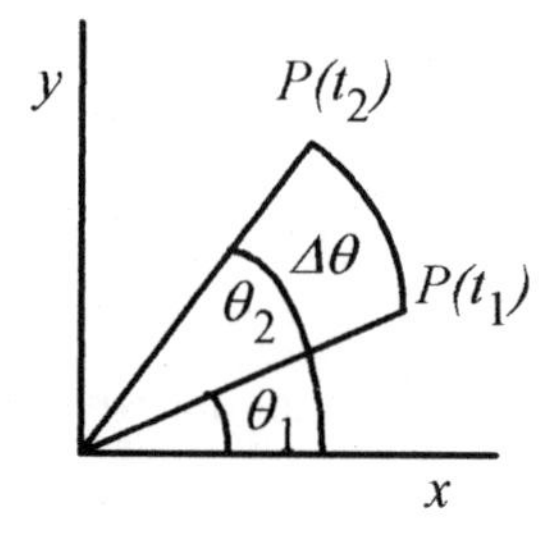

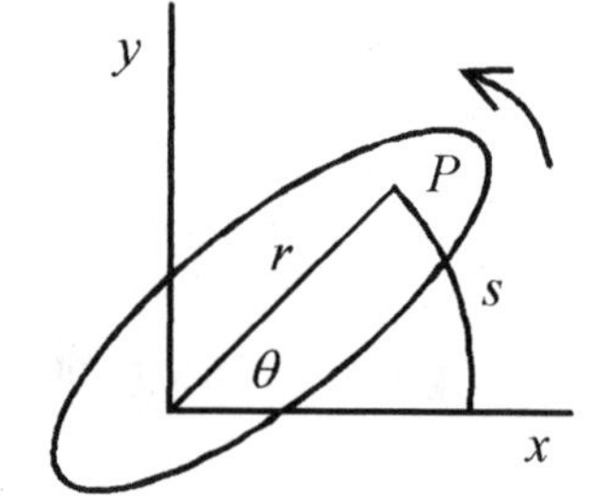

Fig. 12-1

The average angular speed is defined as $\bar{\omega} = \dfrac{\theta_2 - \theta_1}{t_2 - t_1} = \dfrac{\Delta\theta}{\Delta t}$

with the instantaneous angular speed defined by $\omega = \lim\limits_{\Delta t \to 0} \dfrac{\Delta\theta}{\Delta t} = \dfrac{d\theta}{dt}$

Similarly, the average angular acceleration is defined as $\bar{\alpha} = \dfrac{\omega_2 - \omega_1}{t_2 - t_1} = \dfrac{\Delta\omega}{\Delta t}$

with the instantaneous acceleration defined as $\alpha = \lim\limits_{\Delta t \to 0} \dfrac{\Delta\omega}{\Delta t} = \dfrac{d\omega}{dt}$

(C) The relationships between the point and the angle all start with $s = r\theta$; then with successive derivatives comes

$$\frac{ds}{dt} = r\frac{d\theta}{dt} \qquad \text{or} \qquad v = r\omega$$

and

$$\frac{dv}{dt} = r\frac{d\omega}{dt} \quad \text{or} \quad a = r\alpha$$

Remember that the radial, or center-directed, acceleration is $a_{rad} = v^2/r = r\omega^2$.

A force has to be associated with the angular acceleration. Force, however, does not lend itself well to motion of a particle constrained to move in a circle, because most forces are not tangent to a circle. Torque works much better. Torque is defined in a rather unusual manner. Unusual, that is, until its utility is seen through experience in rotational dynamics.

Torque is the vector product of $\boldsymbol{r}$, the vector from the axis of rotation to the rotating point, and $\boldsymbol{F}$, the force applied at that point.

$$\tau = \boldsymbol{r} \times \boldsymbol{F} \tag{12-1}$$

You may want to review the definitions of the cross product in the Introduction, Mathematical Background, before continuing. Most problems in torque can be done with the geometric interpretation of the cross product. Torque, $\boldsymbol{r} \times \boldsymbol{F}$, is the product of r and the component of F perpendicular to r or $|r| \cdot |F| \sin\theta$ as shown on the left in Fig. 12-2.

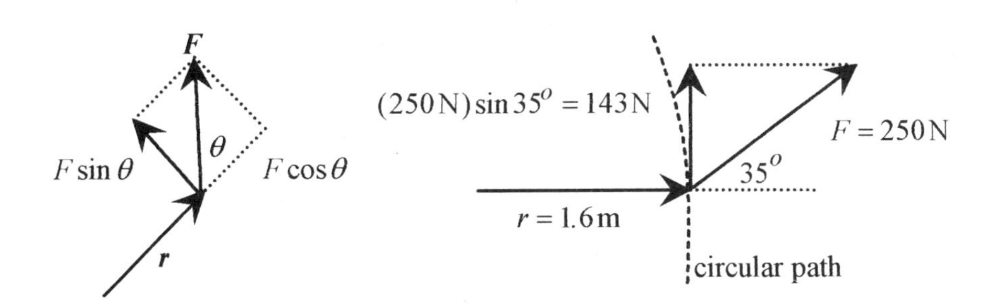

Fig. 12-2

The torque vector is perpendicular to the plane of $\boldsymbol{r}$ and $\boldsymbol{F}$ and follows the right-hand rule of rotating $\boldsymbol{r}$ into $\boldsymbol{F}$. In this case the torque is out of the paper.

12-1 Calculate the torque on a 2.5 kg mass constrained to rotate in a 1.6 m radius circle with a 250 N force applied at a (constant) angle of 35° between $\boldsymbol{r}$ and $\boldsymbol{F}$.

Solution: The problem is depicted in Fig. 12-2. The definition of $\boldsymbol{r} \times \boldsymbol{F}$ gives the direction of the torque as out of the paper. Application of the crossing of $\boldsymbol{r}$ into $\boldsymbol{F}$ is similar to generating a right-handed coordinate system by rotating x into y with the right hand with the thumb giving the direction of z. The magnitude of the torque is

$$T = |r||F|\sin\theta = 1.5\,\text{m}(250\,\text{N})\sin 35^o = 215\,\text{N}\cdot\text{m}$$

Torque is $\boldsymbol{r}$ cross force, and similarly, angular momentum is $\boldsymbol{r}$ cross momentum or

$$\boldsymbol{L} = \boldsymbol{r} \times \boldsymbol{p} \qquad \textbf{(12-2)}$$

Angular momentum is a vector in a direction perpendicular to the plane of $\boldsymbol{r}$ and $\boldsymbol{p}$ and magnitude equal to the product of r and the component of p perpendicular to r.

12-2 For the situation of problem 12-1 add that the linear velocity is $2.8\,\text{m/s}$ and find the angular momentum.

Solution: The linear momentum is $mv = 2.5\,\text{kg}(2.8\,\text{m/s}) = 7.0\,\text{kg}\cdot\text{m/s}$. The linear velocity is the tangential velocity, which is always at right angles to the radius; so the direction of the angular momentum vector is out of the paper, and the magnitude of the cross product is simplified because the sine of 90° is one.

$$L = rp = 1.6\,\text{m}(7.0\,\text{kg}\cdot\text{m/s}) = 11.2\,\text{kg}\cdot\text{m}^2/\text{s}^2$$

(C) With these definitions, several relationships can be derived. Start with

$$\tau = \boldsymbol{r} \times \boldsymbol{F} = \boldsymbol{r} \times \frac{d\boldsymbol{p}}{dt} \quad \text{and} \quad \frac{d\boldsymbol{L}}{dt} = \boldsymbol{r} \times \frac{d\boldsymbol{p}}{dt} + \frac{d\boldsymbol{r}}{dt} \times \boldsymbol{p}$$

but $\frac{d\boldsymbol{r}}{dt} = \boldsymbol{v}$ and $\boldsymbol{p} = m\boldsymbol{v}$ so $\frac{d\boldsymbol{r}}{dt} \times \boldsymbol{p} = \boldsymbol{v} \times m\boldsymbol{v}$ and $\boldsymbol{v} \times \boldsymbol{v} = 0$ thus

$$\frac{d\boldsymbol{L}}{dt} = \boldsymbol{r} \times \frac{d\boldsymbol{p}}{dt} \quad \text{and} \quad \tau = \frac{d\boldsymbol{L}}{dt}$$

Torque is the time derivative of angular momentum as force is the time derivative of linear momentum. Now, by analog, several things follow.

The total angular momentum of a system is the sum of the angular momentum of the individual pieces. Internal torques are equal and opposite, so only external torques change the angular momentum of the system. The statement

$$\frac{d\boldsymbol{L}}{dt} = \tau_{ext} \quad \text{is parallel to} \quad \frac{d\boldsymbol{P}}{dt} = \boldsymbol{F}_{ext}$$

If the derivative of the total angular momentum is zero, the condition for no external torques, then the angular momentum is a constant.

Now look to the kinetic energy associated with rotating particles. Each particle, or piece, of a rotating mass has a linear speed $v = r\omega$, and the KE for each piece is a sum

$$KE = \frac{1}{2}(m_1 v_1^2 + m_2 v_2^2 + \ ...) = \frac{1}{2}(m_1 r_1^2 + m_2 r_2^2 + \ ...)\omega^2$$

This sum of the mr^2's for the collection of pieces is called the rotational inertia, I, so that the KE can be written compactly as

$$KE = \frac{1}{2} I\omega^2$$

The rotational inertia, I, is easy to calculate for individual particles and has been calculated for many shapes. Tabulations are found in most physics books and some mathematical tables.

12-3 Calculate the rotational inertia for one $2.0\,\text{kg}$ mass at the end of a $3.0\,\text{m}$ rod of negligible mass and then the rotational inertia of a dumbbell consisting of two $2.0\,\text{kg}$ masses on the ends of a $6.0\,\text{m}$ long rod of negligible mass pivoted about the center of the rod.

Solution: For the one mass, the sum of the mr^2's is just $2.0\,\text{kg}(9.0\,\text{m}^2) = 18\,\text{kg}\cdot\text{m}^2$.

For the dumbbell, there are two masses, and the rotational inertia would be $36\,\text{kg}\cdot\text{m}^2$.
Notice that if the dumbbell were rotated about one mass then the rotational inertia would be $2.0\,\text{kg}(36\,\text{m}^2) = 72\,\text{kg}\cdot\text{m}^2$. This is of course neglecting the extent of the mass close around the rotating axis.

The total momentum of a rotating mass (point) is mvr. If the mass is rotating, then v is perpendicular to r leading to a simple statement of angular momentum.

Angular momentum is $L = mvr = mr^2\omega = I\omega$.

By definition $\tau = \dfrac{dL}{dt} = I\dfrac{d\omega}{dt} = I\alpha$.

The power is $\dfrac{d(KE)}{dt} = \dfrac{d}{dt}\left(\dfrac{1}{2}I\omega^2\right) = I\omega\dfrac{d\omega}{dt} = I\omega\alpha = \tau\omega$.

Conservation of Angular Momentum

Angular momentum, as a property of the motion, is conserved and is a powerful tool in solving certain problems in rotational motion. First, calculate a torque using the vector form for position and force and the determinant for the cross product. See the Introduction, Mathematical Background for a review of this definition of cross products.

$$\boldsymbol{r} = x\boldsymbol{i} + y\boldsymbol{j} + z\boldsymbol{k} \quad \text{and} \quad \boldsymbol{F} = F_x\boldsymbol{i} + F_y\boldsymbol{j} + F_z\boldsymbol{k} \quad \text{so}$$

$$\tau = \boldsymbol{r} \times \boldsymbol{F} = \begin{vmatrix} \boldsymbol{i} & \boldsymbol{j} & \boldsymbol{k} \\ x & y & z \\ F_x & F_y & F_z \end{vmatrix} = (F_z y - F_y z)\boldsymbol{i} + (F_x z - F_z x)\boldsymbol{j} + (F_y x - F_x y)\boldsymbol{k} \qquad \textbf{(12-3)}$$

Most problems do not require as extensive a calculation as this equation would indicate. For a position and force vector in the x-y plane the torque reduces to a single z component.

12-4 Calculate the angular momentum of a $3.0\,\text{kg}$ mass at $3.0\,\text{m}$ in the x-direction and $-2.0\,\text{m}$ in the y-direction and with velocity components of $20\,\text{m/s}$ in the x-direction and $-30\,\text{m/s}$ in the y-direction?

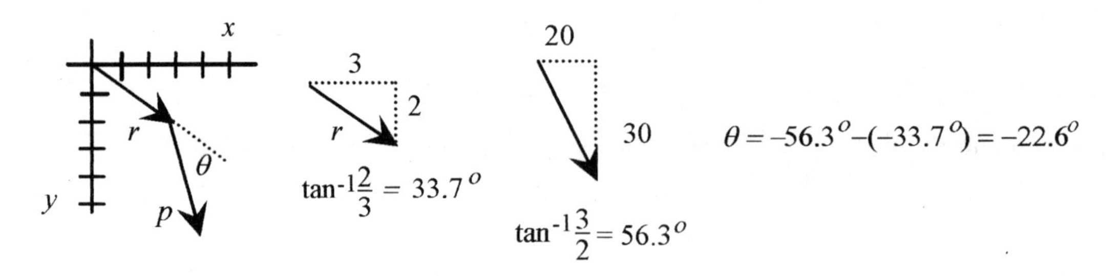

Fig. 12-3

Solution: Write $\boldsymbol{r}$ and $\boldsymbol{p}$ in vector form: $\boldsymbol{r} = (3.0\boldsymbol{i} - 2.0\boldsymbol{j})\,\text{m}$ $\quad \boldsymbol{p} = (60\boldsymbol{i} - 90\boldsymbol{j})\,\text{kg}\cdot\text{m/s}$
The angular momentum is

$$\boldsymbol{L} = \boldsymbol{r} \times \boldsymbol{p} = \begin{vmatrix} \boldsymbol{i} & \boldsymbol{j} & \boldsymbol{k} \\ 3.0 & -2.0 & 0 \\ 60 & -90 & 0 \end{vmatrix} = (-270 + 120)\boldsymbol{k} = -150\boldsymbol{k}\,\frac{\text{kg}\cdot\text{m}^2}{\text{s}}$$

The angular momentum vector is $150\,\text{kg}\cdot\text{m}^2/\text{s}$ pointed into the paper.

Second Solution: Now redo the problem with angle and magnitude. The angle between $\boldsymbol{r}$ and $\boldsymbol{p}$ is -22.6^o. Figure 12-3 shows the trigonometric relations. Note that the angle between $\boldsymbol{r}$ and $\boldsymbol{p}$ is negative. The fact that the angle is negative can be easily missed. Rather than rely on the sign of the angle, the better way to determine the direction of the angular momentum vector is to use the right-hand rule and cross $\boldsymbol{r}$ into $\boldsymbol{p}$ with the thumb giving the direction of $\boldsymbol{L}$. The magnitude of L *is*

$$L = |r|\cdot|p|\sin\theta = \sqrt{13}\sqrt{11700}\,\sin(-22.6^o) = -150\,\text{kg}\cdot\text{m}^2/\text{s}$$

Applications

Now to a few problems with rotating hoops and cylinders using rotational dynamics, conservation of energy, and conservation of angular momentum.

12-5 A hoop of mass $1.0\,\text{kg}$ and radius $0.25\,\text{m}$ is rotating in a horizontal plane with angular momentum $4.0\,\text{kg}\cdot\text{m}^2/\text{s}$. A lump of clay of mass $0.20\,\text{kg}$ is placed (gently) on the hoop. What happens to the angular velocity of the hoop?

Solution: The moment of inertia for a hoop (from the table in your text) is $I = mr^2$.

$$I = 1.0\,\text{kg}(0.25\,\text{m})^2 = 0.0625\,\text{kg}\cdot\text{m}^2$$

The angular momentum, L, is $I\omega$, so the initial angular velocity can be calculated from $L = I\omega_i$

$$4.0\,\text{kg}\cdot\text{m}^2/\text{s} = (0.0625\,\text{kg}\cdot\text{m}^2)\,\omega_i \qquad \text{or} \qquad \omega_i = 64\,\text{rad/s}$$

Since the lump of clay is placed (gently) on the hoop (The hoop is rotating in the horizontal plane, and the clay is placed on the hoop vertically.), placing the clay only adds mass to the hoop. This additional mass adds "another" moment of inertia mr^2, so with the angular momentum the same (no external torque has acted to change it)

$$4.0\,\text{kg}\cdot\text{m}^2/\text{s} = \left[1.0\,\text{kg}(0.25\,\text{m})^2 + 0.2\,\text{kg}(0.25\,\text{m})^2\right]\omega_f \qquad \text{or} \qquad \omega_f = 53.3\,\text{rad/s}$$

12-6 An amusement park game consists of a paddle wheel arrangement where you shoot at the paddles with a pellet gun, thereby turning the wheel. The paddle wheel is set in motion with an initial angular momentum of $200\,\text{kg}\cdot\text{m}^2/\text{s}$ and angular frequency of $4.0\,\text{rad}/\text{s}$. You shoot eight $40\,\text{g}$ pellets at a speed of $200\,\text{m}/\text{s}$ at the paddles. The pellets hit the paddles at $0.80\,\text{kg}$ radius, stick, and impart all their momentum to the wheel. Find the new angular momentum and the new angular frequency.

Solution: First calculate the initial moment of inertia of the paddle wheel from $L = I\omega$.

$$200\,\text{kg}\cdot\text{m}^2/\text{s} = I(4.0\,\text{rad/s}) \qquad \text{or} \qquad I = 50\,\text{kg}\cdot\text{m}^2$$

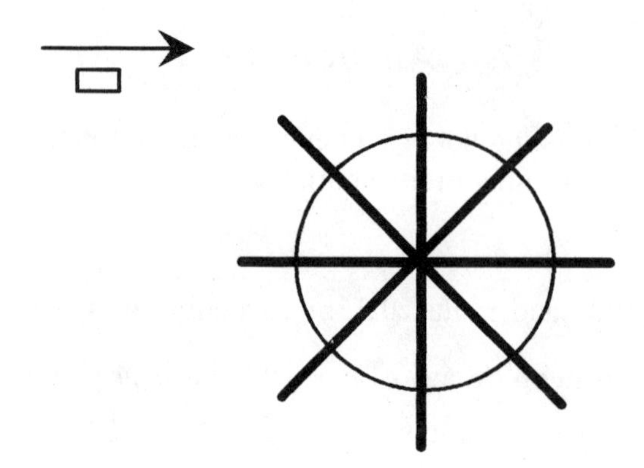

Fig. 12-4

Now calculate the ω after the wheel has absorbed the (momentum in the) eight pellets. The angular momentum of each pellet is the linear momentum times the 0.80 m radius.

$$L_{\text{each pellet}} = mvr = 0.040\,\text{kg}(200\,\text{m/s})0.80\,\text{m} = 6.4\,\text{kg}\cdot\text{m}^2/\text{s}$$

For eight pellets the "absorbed" angular momentum is $L_{\text{all pellets}} = 51.2\,\text{kg}\cdot\text{m}^2/\text{s}$

The new angular momentum is the original plus this, or $251.2\,\text{kg}\cdot\text{m}^2/\text{s}$.

The I also has increased because of the additional mass at the 0.80 m radius. This "additional" I is

$$I_{\text{pellets}} = 8\cdot 0.040\,\text{kg}(0.80\,\text{m})^2 = 0.205\,\text{kg}\cdot\text{m}^2$$

The equation for calculating the new angular velocity ($L = I\omega$) is

$$251.2\,\text{kg}\cdot\text{m}^2/\text{s} = 50.2\,\text{kg}\cdot\text{m}^2\omega_f \quad \text{or} \quad \omega_f = 5.00\,\text{rad/s}$$

12-7 Now consider a 10 kg solid cylindrical drum with radius 1.0 m rotating about its cylindrical axis under the influence of a force produced by a 30 kg mass attached to a cord wound around the drum. What is the torque, moment of inertia, and angular acceleration?

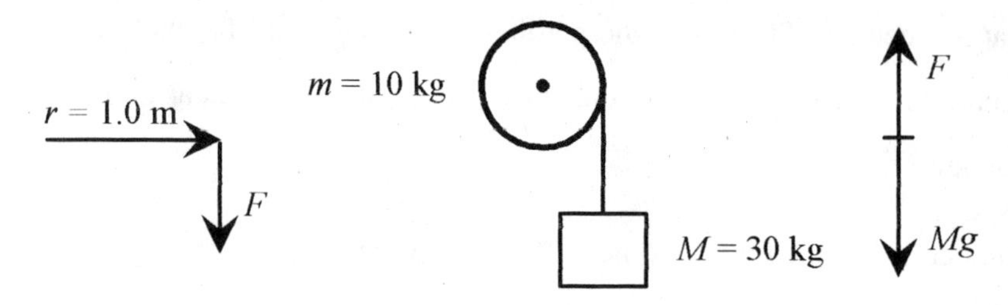

Fig 12-5

Solution: The moment of inertia for a solid drum rotating about its axis is $mr^2/2$.

$$I = mr^2/2 = 10\,\text{kg}(1.0\,\text{m})^2/2 = 5.0\,\text{kg}\cdot\text{m}^2$$

The torque on the drum is the tension in the cable times the radius of the drum. This tension is much like the tension encountered in the problems of multiple blocks sliding on tables as found in Chapter 5, Force. The system accelerates clockwise, so the unbalanced force on M makes it accelerate according to

$$Mg - F = Ma \quad \text{or} \quad F = M(g-a)$$

The torque on the drum makes it angularly accelerate according to $rF = I\alpha$.

The acceleration of M is the same as the tangential acceleration, $a = r\alpha$ so

$$rF = \left[\frac{mr^2}{2}\right]\frac{a}{r} \quad \text{or} \quad F = \frac{ma}{2} \quad \text{leading to} \quad \frac{ma}{2} = Mg - Ma \quad \text{or} \quad a = \frac{g}{1 + m/2M}$$

Evaluating $a = \dfrac{9.8\,\text{m/s}^2}{1 + 10/60} = 8.4\,\text{m/s}^2$.

The angular acceleration is $\alpha = \dfrac{a}{r} = \dfrac{8.4\,\text{m/s}^2}{1.0\,\text{m}} = 8.4\,\text{rad/s}^2$.

The torque is $\tau = I\alpha = (5.0\,\text{kg}\cdot\text{m}^2)8.4\,\text{rad/s}^2 = 42\,\text{N}\cdot\text{m}$.

12-8 When the $30\,\text{kg}$ mass in problem 12-7 has fallen (starting from rest) through $4.0\,\text{m}$, what is the linear velocity, angular velocity of the drum, and the time for this to occur?

Solution: This part of the problem is approached from an energy point of view. The $30\,\text{kg}$ mass falls through a distance of $4.0\,\text{m}$. The work performed by gravity **Goesinto** translational KE of the $30\,\text{kg}$ mass and rotational KE of the cylinder. The main difficulty in problems like this is remembering that some of the work performed results in rotational KE and some in translational KE.

$$Mgh = \frac{1}{2}Mv^2 + \frac{1}{2}I\omega^2 = \frac{1}{2}Mv^2 + \frac{1}{2}\left(\frac{mr^2}{2}\right)\frac{v^2}{r^2} \quad \text{or} \quad Mgh = v^2\left(\frac{M}{2} + \frac{m}{4}\right)$$

Putting in the numbers

$$30\,\text{kg}\frac{9.8\text{m}}{\text{s}^2}4.0\,\text{m} = v^2\left(\frac{30\,\text{kg}}{2} + \frac{10\,\text{kg}}{2}\right) \quad \text{or} \quad v^2 = 59\frac{\text{m}^2}{\text{s}^2} \quad \text{or} \quad v = 7.7\frac{\text{m}}{\text{s}}$$

From $v = r\omega$ the angular velocity is $\omega = \frac{v}{r} = \frac{7.7\,\text{m/s}}{1.0\,\text{m}} = 7.7\frac{\text{rad}}{\text{s}}$.

The time for this system to reach this velocity is from $\omega = \omega_o + \alpha t$.

$$t = \frac{\omega}{\alpha} = \frac{7.7(\text{rad / s})}{8.4(\text{rad / s}^2)} = 0.92\,\text{s}$$

12-9 A solid sphere is constrained to rotate about a vertical axis passing through the center of the sphere. A cord is wrapped around what would be the equator, passes over a pulley of negligible mass, and is attached to a mass that is allowed to fall under the influence of gravity. Write a conservation of energy statement for the system.

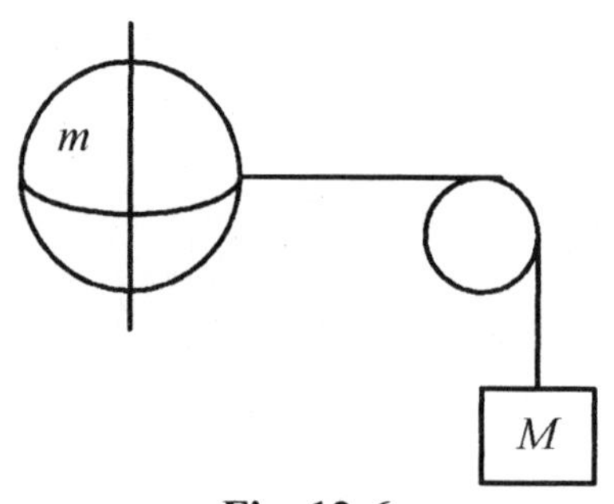

Fig. 12-6

Solution: If the mass falls a distance h, then the energy gained by the system is Mgh. This energy **Goesinto** translational kinetic energy of the mass and rotational kinetic energy of the sphere. A conservation of energy statement is

$$Mgh = \frac{1}{2}I\omega^2 + \frac{1}{2}Mv^2$$

The velocity of a point on the string and the velocity of the mass are the same and are related to the angular velocity through $v = r\omega$, so the energy statement can be written in terms of v. Additionally, the formula for $I = 2mr^2/5$ can be substituted so

$$Mgh = \frac{1}{2}I\frac{v^2}{r^2} + \frac{1}{2}Mv^2 = \left(\frac{m}{5} + \frac{M}{2}\right)v^2$$

Note that the energy statement could have been written in terms of ω rather than v.

12-10 For problem 12-9 take the mass of the sphere as $m = 2.0\,\text{kg}$, $r = 0.30\,\text{m}$, the hanging mass as $M = 0.80\,\text{kg}$, and the height the mass falls through as $h = 1.5\,\text{m}$. Find the velocity of the hanging mass, and the angular momentum of the sphere.

Solution: The velocity of the mass, the rope, and a point on the equator of the sphere (all the same velocity) is from the energy statement

$$0.80\,\text{kg}\frac{9.8\,\text{m}}{\text{s}^2}1.5\,\text{m} = \left(\frac{2.0\,\text{kg}}{5} + \frac{0.80\,\text{kg}}{2}\right)v^2 \qquad \text{or} \qquad v = 3.8\frac{\text{m}}{\text{s}}$$

The angular velocity of the sphere is $\omega = \dfrac{v}{r} = \dfrac{3.8\,\text{m/s}}{0.30\,\text{m}} = 13\dfrac{\text{rad}}{\text{s}}$

The angular momentum of the sphere is

$$L = I\omega = \left[\frac{2 \cdot 2.0\,\text{kg}(0.30\,\text{m})^2}{5}\right]\left[\frac{13\,\text{rad}}{\text{s}}\right] = 0.94\frac{\text{kg}\cdot\text{m}^2}{\text{s}}$$

As an added exercise give the pulley in this problem a mass and a radius and go through the problem again. Remember that I and ω will be different for the pulley.

CHAPTER 13

EQUILIBRIUM

The only difficulty you will encounter in doing equilibrium problems is lack of familiarity with the mechanics of doing the problems. This is overcome by doing problems. The theory is simple. If something is not moving, that is, it is in equilibrium, then the sum of the forces on it must be zero. Likewise if it is not rotating, then the sum of the torques must be zero.

For problems in two dimensions these laws are stated very simply:

$$\sum F_x = 0 \quad \sum F_y = 0 \quad \sum \tau = 0$$

We start off with some simple problems requiring force analysis only and progress to problems involving torques.

13-1 Hang a 50 kg mass with ropes making angles of 30° and 45° as shown in Fig. 13-1. Calculate the tension in the ropes.

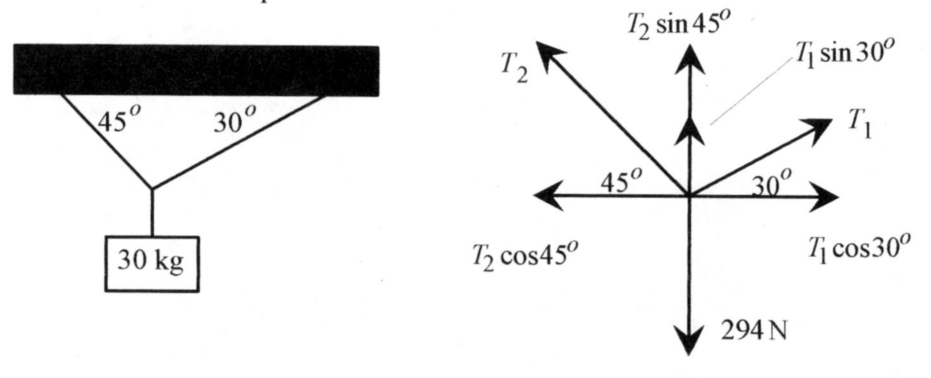

Fig. 13-1

Solution: Note that all the forces come together at the knot in the rope so draw a force diagram about this point. The only laws to apply are for equilibrium in the x and y directions.

$$\sum F_x = 0: \quad T_1 \cos 30^o = T_2 \cos 45^o \qquad \sum F_y = 0: \quad T_1 \sin 30^o + T_2 \sin 45^o = 294\,\text{N}$$

This provides two equations in two unknowns. Because $\sin 45^o = \cos 45^o$ rewrite

$$T_1(\sin 30^o + \cos 30^o) = 294\,\text{N} \quad \text{or} \quad T_1 = 215\,\text{N}$$

and

$$T_2 \cos 45^o = (215\,\text{N})\cos 30^o \quad \text{or} \quad T_2 = 263\,\text{N}$$

As an exercise work through this problem with different angles.

13-2 In the arrangement shown in Fig. 13-2, what is the minimum coefficient of friction to prevent the $8.0\,\text{kg}$ mass from sliding?

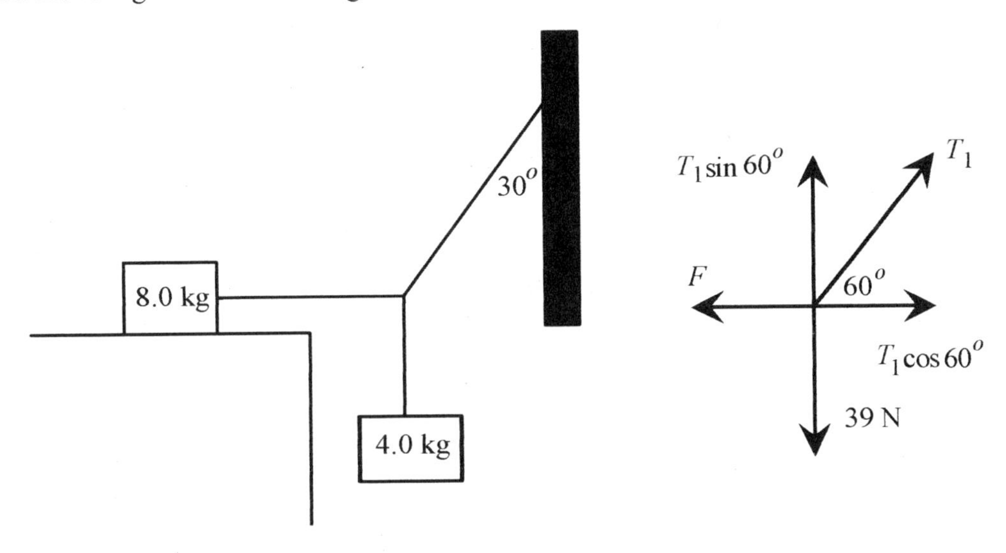

Fig. 13-2

Solution: The forces come together where the cords come together. This point is not moving so apply the equilibrium conditions here. The vector diagram at this point is also in Fig. 13-2.

The equilibrium conditions are:

$$\sum F_x = 0: \quad T_1 \cos 60^o = F \qquad \sum F_y = 0: \quad T_1 \sin 60^o = 39\,\text{N}$$

Solving for T_1: $\quad T_1 = \dfrac{39\,\text{N}}{\sin 60^o} = 45\,\text{N}$

The force is: $\quad F = 45\,\text{N} \cos 60^o = 22.5\,\text{N}$

This force is supplied by the frictional force μmg, so $\quad \mu = \dfrac{22.5\,\text{N}}{8.0\,\text{kg} \cdot 9.8(\text{m/s}^2)} = 0.29$

This is the minimum coefficient of friction that will prevent the system from moving.

13-3 A 500N diver is on the end of a 4.0m diving board of negligible mass. The board is on pedestals as shown in Fig. 13-3. What are the forces that each pedestal exerts on the diving board?

Solution: There are no forces in the *x*-direction, but in the *y*-direction the pedestal at A and the diver are acting down while the force at pedestal B is acting up. Pedestal B is in compression while pedestal A is in tension. Draw the vector diagram and write the equations.

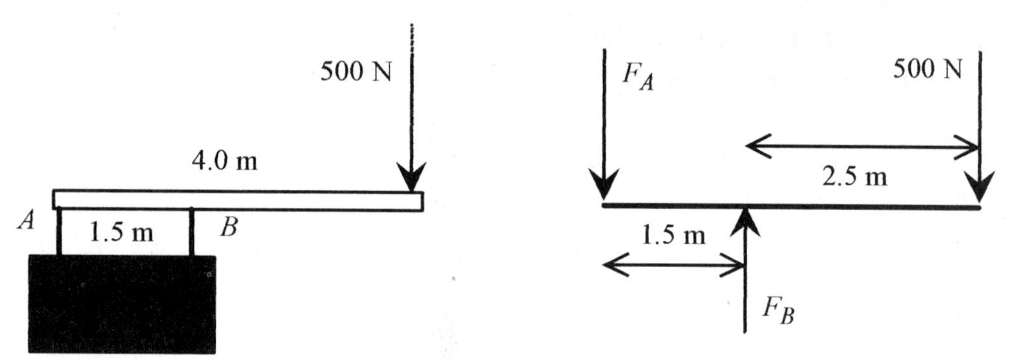

Fig. 13-3

$$\sum F_y = 0: \quad F_A + 500\text{N} = F_B$$

In this problem apply $\sum \tau = 0$ about pedestal A. Taking the rotation point at pedestal A, the diver produces a clockwise torque, and pedestal B an anti-clockwise torque that are in equilibrium. Note that this is not the only possible rotation point.

$$\sum \tau = 0 \quad 500\text{N} \cdot 4.0\text{m} = F_B \cdot 1.5\text{m}$$

Solve for F_B: $F_B = \dfrac{500\text{N} \cdot 4.0\text{m}}{1.5\text{m}} = 1330\text{N}$

Solve for F_A: $F_A = F_B - 500\text{N} = 1330\text{N} - 500\text{N} = 830\text{N}$

As an exercise write the $\sum \tau = 0$ about pedestal B and the diver.

13-4 Place a 7.0m uniform, 150N ladder against a frictionless wall at an angle of 75°. What are the reaction forces at the ground and wall and the minimum coefficient of friction of the ground?

Solution: Figure 13-4 shows the ladder with the 150N acting down at the center of the ladder and the sides of the triangle formed by the ladder, wall, and ground. The vector portion of the figure shows the two components of the reaction force at the ground. The reaction force at the ground is not necessarily in the direction of the ladder. This (vector) force can be written in terms of a force (magnitude) at an angle (with horizontal and vertical components) or directly in component form as done here. Writing the force in component force helps to avoid the temptation to place the force at the same angle as the ladder. Either way requires two variables. Since the wall is frictionless, the reaction at the wall is entirely horizontal. Refer to Fig. 13-4 and write the equilibrium conditions.

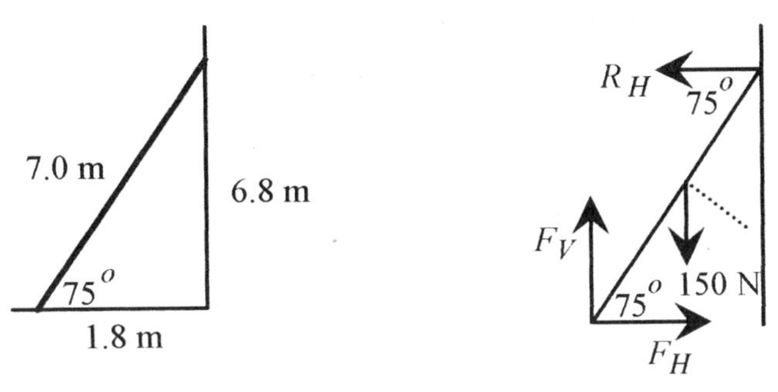

Fig. 13-4

$$\sum F_x = 0: \quad F_H = R_H \qquad \sum F_y = 0: \quad F_V = 150\,\text{N}$$

$\sum \tau = 0$: The torque on the ladder is taken about the point where the ladder contacts the ground. Note that this choice eliminates two variables from the torque statement. Torque is the (component of the) force at right angles to the lever arm times that lever arm. The $150\,\text{N}$ weight has a component $150\,\text{N}\cos 75^o$ times $3.5\,\text{m}$. The reaction at the wall has a component $R_H \cos 15^o$ times $7.0\,\text{m}$ so the sum of the torques equal zero statement is

$$(150\,\text{N}\cos 75^o)3.5\,\text{m} = (R_H \cos 15^o)7.0\,\text{m} \qquad \text{or} \qquad 150(0.90)\,\text{N}\cdot\text{m} = R_H(6.8)\,\text{N}\cdot\text{m}$$

so

$$R_H = (75\,\text{N})(\cos 75^o / \cos 15^o) = 20\,\text{N} \qquad \text{and} \qquad F_H = 20\,\text{N}$$

F_V and F_H are related via the coefficient of friction, $F_H = \mu F_V$ so $\mu = 20/150 = 0.13$

Second Solution: There is another way to calculate the torques. Notice that the $150\,\text{N}$ force is horizontally $0.90\,\text{m}$ away from the pivot point. If the $150\,\text{N}$ force vector is moved vertically so its tail is on a horizontal line from the pivot point, then the torque is easily written as $(150\,\text{N})(0.90\,\text{m})$. A similar operation can be performed on R_H yielding a torque $R_H(6.8\,\text{m})$. Vectors maintain their length (magnitude) and orientation (angle) but can be moved about a vector diagram for (our) convenience. If the torque statement is written in this way it is identical to the torque statement in the first solution.

$$150\,\text{N}\cdot 0.90\,\text{m} = R_H 6.8\,\text{m}$$

As an exercise redo the problem with a person of known weight positioned a specific distance up the ladder.

13-5 Place an 80 N ladder 8.0 m long on a floor and against a frictionless roller on a 5.0 m high wall. The minimum angle for equilibrium is 70°. Find the coefficient of friction between the ladder and the floor.

Solution: The roller is free to rotate so the reaction force is normal to the ladder. This reaction force can be viewed as having vertical and horizontal components. The reaction force at the floor is written as vertical and horizontal components. The geometry and vector diagram are shown in Fig. 13-5. The length to the roller is obtained from the geometry: $\sin 70^o = 5.0\,\text{m}/L$ or $L = 5.3\,\text{m}$

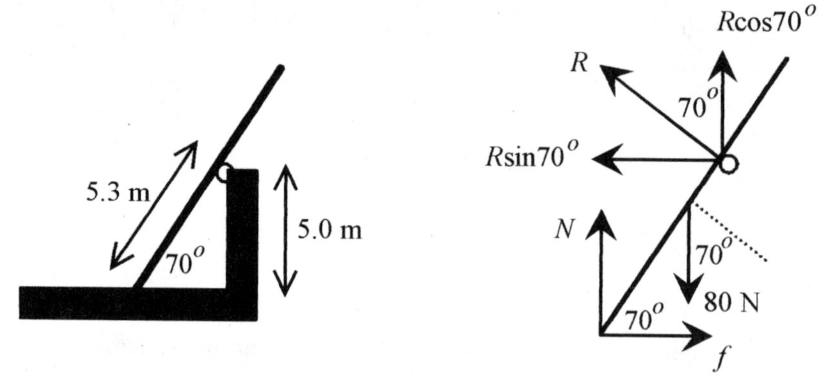

Fig. 13-5

Write the force equilibrium equations.

$$\sum F_x = 0: \quad f = R\sin 70^o \qquad \sum F_y = 0: \quad N + R\cos 70^o = 80\,\text{N}$$

The torque about the point on the floor is produced by the reaction force at the roller and the weight of the ladder taken at its midpoint.

$$\sum \tau = 0: \quad (80\,\text{N}\cos 70^o)4.0\,\text{m} = R \cdot 5.3\,\text{m}$$

Rewriting with $f = \mu N$

$$\mu N = R\sin 70^o \qquad N + R\cos 70^o = 80\,\text{N} \qquad (320\,\text{N}\cdot\text{m})\cos 70^o = (5.3\,\text{m})R$$

Solve these equations from last to first to obtain

$$R = 20.6\,\text{N} \qquad N = 73.0\,\text{N} \qquad \mu = 0.26$$

13-6 A 20 kg beam 4.0 m long is pivoted at a wall and supported by a cable as shown in Fig. 13-6. What is the force at the wall and the tension in the cable?

Solution: There are two points of possible confusion in this problem. First, the mass of the beam is taken as down at the center of the beam. Second, the reaction of the wall, at the pivot, has two components, one up the wall (holding the end of the beam up) and one out of the wall (keeping the beam from sinking into the wall).

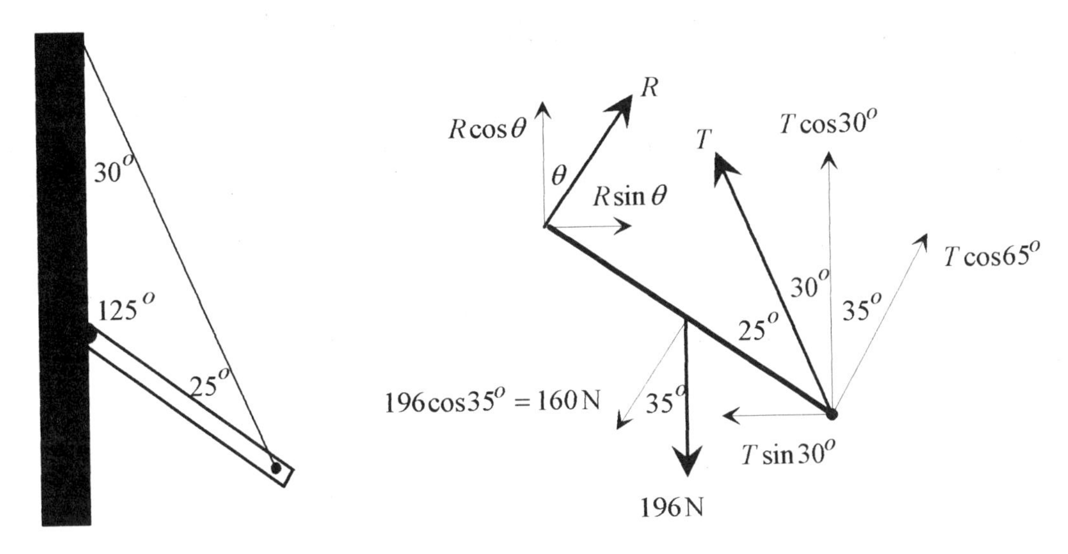

Fig. 13-6

Set up the vector diagram at the pivot and the point where the cable is attached. The reaction at the wall is shown with components up the wall and normal to the wall. The weight of the beam is shown as down and from the center of the beam with a component at a right angle to the beam. The tension in the cable is broken up into vertical and horizontal components (for the sum of the forces laws) and a component at right angles to the beam (for the sum of the torques laws). Applying the sum of the forces laws means looking at the vector diagram and adding up all the vectors in the horizontal or vertical directions.

$$\sum F_y = 0: \quad R\cos\theta + T\cos 30^o = 196\,\mathrm{N} \qquad \sum F_x = 0: \quad R\sin\theta = T\sin 30^o$$

In applying the sum of the torques, it is convenient to take the pivot point at the wall. Since the angle θ is unknown, the least complicated sum of the torques statement will come from the tension in the cable and the weight of the beam.

Using the components at right angles $\sum \tau = 0: \quad (T\cos 65^o)4.0\,\mathrm{m} = 160\,\mathrm{N}\cdot 2.0\,\mathrm{m}$

Solve the last equation for the tension in the cable $T = \dfrac{160\,\mathrm{N}}{2\cos 65^o} = 189\,\mathrm{N}$

Use this value to solve for

$$R\sin\theta = 189\,\mathrm{N}\sin 30^o = 95\,\mathrm{N} \qquad \text{and} \qquad R\cos\theta = 196\,\mathrm{N} - 189\,\mathrm{N}\cos 30^o = 32\,\mathrm{N}$$

Now divide these two equations $\dfrac{R\sin\theta}{R\cos\theta} = \tan\theta = \dfrac{95}{32} = 3.0 \quad \text{or} \quad \theta = 71^o$

and finally, R is obtained from one of the original equations $R = \dfrac{189\,\text{N}\sin 30^o}{\sin 71^o} = 100\,\text{N}$

The main points in this problem are the detailed vector diagram and the choice of the wall as the pivot point in calculating the torques.

In the early problems, construction of the vector diagram was confined to one point where the forces all "came together." In problems where there is a strut or ladder or some other extended object the common mistake is to miss a force or torque and have an incomplete vector diagram.

13-7 Set up a 300 N strut 14 m long with a cable as shown in Fig. 13-7. Find the tension in the cable and the reaction force at the pivot.

Solution: The reaction force at the pivot is set up with vertical and horizontal components. The weight of the strut is taken at the center of the strut with the 2000 N weight at the end. The tension is written along with all the components needed to write the force and torque equations.

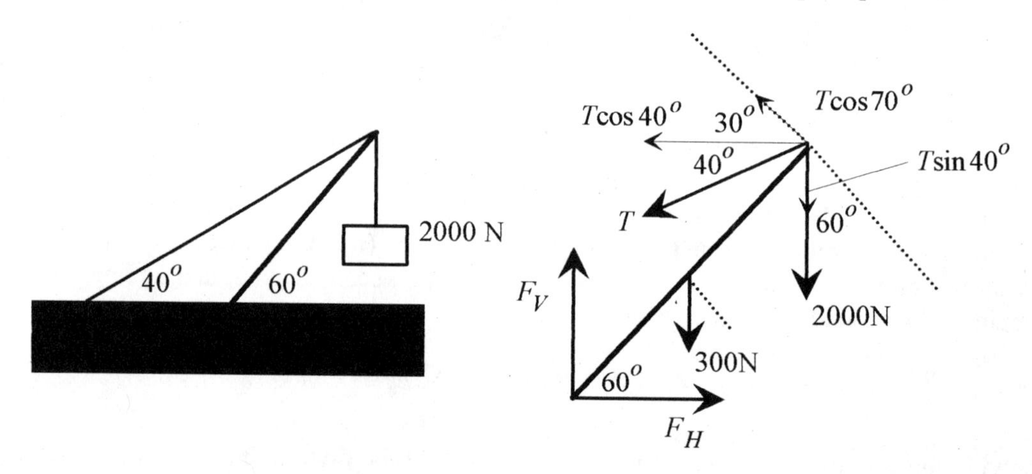

Fig. 13-7

Apply the sum of the forces laws.

$$\sum F_x = 0: \quad F_H = T\cos 40^o \qquad \sum F_y = 0 \quad F_V = 2000\,\text{N} + 300\,\text{N} + T\sin 40^o$$

Apply the sum of the torques law about the pivot point.

$$\sum \tau = 0: \quad (300\,\text{N}\cos 60^o)7.0\,\text{m} + (2000\,\text{N}\cos 60^o)14\,\text{m} = (T\cos 70^o)14\,\text{m}$$

Note that the length of the strut does not influence the calculation.

Solve the torque equation for T.

$$2150\,\mathrm{N}\cos 60^{o} = T\cos 70^{o} \qquad \text{or} \qquad T = 3140\,\mathrm{N}$$

Use this value to find F_H and F_V.

$$F_H = 3140\,\mathrm{N}\cos 40^{o} = 2410\,\mathrm{N} \qquad \text{and} \qquad F_V = 2300\,\mathrm{N} + 3140\,\mathrm{N}\sin 40^{o} = 4320\,\mathrm{N}$$

The reaction at the floor can be written in terms of magnitude and angle.

$$F = \sqrt{2410^2 + 4320^2} = 4950\,\mathrm{N} \qquad \theta = \tan^{-1}(4320/2410) = 61^{o}$$

Another feature of equilibrium problems is that an object hung by two cords has its center of mass along a vertical line determined by the (extension of) intersection of those cords. The argument is based on the fact that a body hung by a cord has its center of mass along the extension of that cord. If the center of mass were not on this line, a torque would exist and the body would move. Likewise if the center of mass were not on the intersection of the two cords, a torque would exist and the body would move.

13-8 Consider a non-uniform bar of $200\,\mathrm{N}$ weight and $4.0\,\mathrm{m}$ length suspended horizontally between two walls where the angles between the walls and cables are 37° and 50°. What are the tensions in the cables?

Solution: The extensions of the cables specify the center of mass. Looking just to the center of mass and using the common side of the triangle write

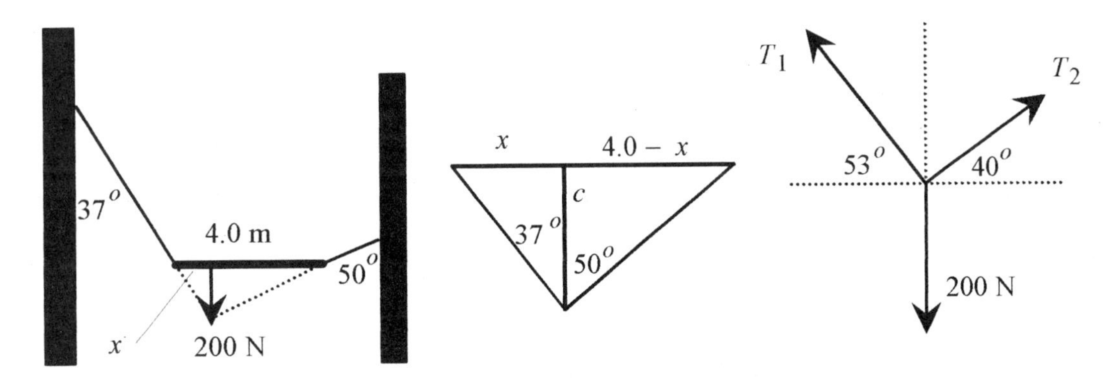

Fig. 13-8

$$\tan 37^{o} = \frac{x}{c} \qquad \text{and} \qquad \tan 50^{o} = \frac{4.0 - x}{c}. \qquad \text{Solve for } c \text{ and set equal} \qquad \frac{4.0 - x}{\tan 50^{o}} = \frac{x}{\tan 37^{o}}$$

and $4.0\tan 37^o - x\tan 37^o = x\tan 50^o$ or $x = \dfrac{4.0\tan 37^o}{\tan 37^o + \tan 50^o} = 1.55\,\text{m}$.

Draw a vector diagram at the center of mass line. Write the equilibrium conditions as

$$T_1 \cos 53^o = T_2 \cos 40^o \quad \text{and} \quad T_1 \sin 53^o + T_2 \sin 40^o = 200\,\text{N}$$

Rewriting

$$T_1 \cos 53^o - T_2 \cos 40^o = 0 \quad \text{and} \quad T_1 \sin 53^o + T_2 \sin 40^o = 200\,\text{N}$$

Multiplying and adding

$$T_1 \cos 53^o \sin 40^o - T_2 \cos 40^o \sin 40^o = 0$$

$$T_1 \sin 53^o \cos 40^o + T_2 \sin 40^o \cos 40^o = 200\,\text{N} \cos 40^o$$

and

$$T_1 = \frac{200\,\text{N} \cos 40^o}{\cos 53^o \sin 40^o + \sin 53^o \cos 40^o} = \frac{200\,\text{N} \cos 40^o}{\sin 93^o} = 153\,\text{N}$$

A sum of two angles trigonometric identity (see the Introduction, Mathematical Background) was used in the denominator.

$$T_2 = \frac{153\,\text{N} \cos 53^o}{\cos 40^o} = 120\,\text{N}$$

CHAPTER 14

GRAVITY

The basic law of gravity describes the force of one mass on another

$$F = G\frac{m_1 m_2}{r^2} \quad \textbf{(14-1)}$$

where G is a constant that depends on the units of the m's, the masses, and r, the distance between centers. In SI units the gravitational constant is $G = 6.7 \times 10^{-11}\,\mathrm{N \cdot m^2 / kg^2}$. The gravitational force acts along the line of centers and is an action-reaction pair. (The force on both masses is the same.) Gravitational forces are vectors and superpose (add) as vectors.

The gravitational force of attraction for a mass of small (compared to the earth) radius on the surface of the earth is known as the weight of that mass on the earth and is

$$F = W = G\frac{m m_E}{r_E^2} = mg \quad \textbf{(14-2)}$$

with $g = Gm_E / r_E^2$ being called the acceleration due to gravity. Weight then is the familiar $W = mg$. Acceleration units (the units of g) times mass produce force or weight units. Also, "g" is a vector that points between the center line of the masses involved.

14-1 What is the gravitational force of attraction between two $7.5\,\mathrm{kg}$ bowling balls with $0.50\,\mathrm{m}$ between centers?

Solution: Use the gravitational force equation

$$F = G\frac{m_1 m_2}{r^2} = (6.7 \times 10^{-11}\,\mathrm{N \cdot m^2/kg^2})\frac{7.5\,\mathrm{kg} \cdot 7.5\,\mathrm{kg}}{(0.50\,\mathrm{m})^2} = 1.5 \times 10^{-8}\,\mathrm{N}$$

Each ball is attracted to the other on a line between their centers.

Gravitational Potential

(C) Following the general form for potential energy, the gravitational potential is defined as the work performed by the gravitational force when r increases from r_1 to r_2.

$$U = \int F dr = -\int_{r_1}^{r_2} \frac{Gmm_E}{r^2} dr = -Gmm_E \int_{r_1}^{r_2} \frac{dr}{r^2}$$

The minus sign is in the integral because the force points opposite to the direction of increasing r. The force decreases as r increases. The integral definition is necessary because the force depends on r. Performing the integration

$$U = Gmm_E \frac{1}{r}\Big|_{r_1}^{r_2} = Gmm_E \left[\frac{1}{r_2} - \frac{1}{r_1}\right]$$

Look at the signs in this expression. If r_1 is taken at the surface of the earth and r_2 above the earth then the gravitational potential at r_2 is positive (less negative) with respect to the earth.

Notice that the general expression for gravitational potential is

$$U = -G\frac{mm_E}{r} \tag{14-3}$$

The zero of gravitational potential is at $r = \infty$. The gravitational potential is negative at the surface of the earth and becomes more negative as we go in toward the center of the earth.

With the gravitational potential energy and the conservation of energy we can calculate the escape velocity. This is the velocity with which we would have to shoot something vertically up to completely remove it from the earth's gravitational pull.

14-2 Calculate the escape velocity on the surface of the earth.

Solution: The escape velocity is the minimum velocity for a mass to escape the gravitational attraction of the earth. Envision a mass being shot vertically up from the surface of the earth and write the energy statement for it at two levels, the surface of the earth and some height r further out from the center of the earth. The conservation of energy statement will read: The kinetic plus potential energy on the surface of the earth equals the kinetic plus potential energy at some point r.

$$\frac{1}{2}mv_1^2 - G\frac{mm_E}{r_E} = \frac{1}{2}m_2 v_2^2 - G\frac{mm_E}{r} \tag{14-4}$$

The escape velocity is that velocity that will produce zero velocity ($v_2 = 0$) when the mass is infinitely far away (when $r = \infty$, $Gmm_E / r = 0$). This reduces the equation to

$$\frac{v_1^2}{2} = \frac{Gm_E}{r_E} \quad \text{or} \quad v_1 = \sqrt{\frac{2Gm_E}{r_E}} \qquad \textbf{(14-5)}$$

The mass and radius of the earth are on the constants page of your text so

$$v_1 = \sqrt{\frac{2(6.7 \times 10^{-11}\,\mathrm{N \cdot m^2/kg^2})6.0 \times 10^{24}\,\mathrm{kg}}{6.4 \times 10^6\,\mathrm{m}}} = 1.1 \times 10^4\,\frac{\mathrm{m}}{\mathrm{s}}$$

As an exercise calculate the escape velocity for the moon of radius $1.7 \times 10^6\,\mathrm{m}$ and mass $7.4 \times 10^{22}\,\mathrm{kg}$.

14-3 What initial vertical speed is necessary to shoot a satellite to $300\,\mathrm{km}$ above the earth?

Solution: Use energy analysis and the same energy-balance equation as in the previous problem with the velocity at $300\,\mathrm{km}$ equal to zero.

$$\frac{1}{2}mv^2 - \frac{Gmm_E}{r_E} = -\frac{Gmm_E}{r_E + 3.0 \times 10^5\,\mathrm{m}} \quad \text{or} \quad v^2 = 2Gm_E\left[\frac{1}{r_E} - \frac{1}{r_E + 3.0 \times 10^5\,\mathrm{m}}\right]$$

Putting in the numbers

$$v^2 = 2 \cdot 6.7 \times 10^{-11}\,\frac{\mathrm{N \cdot m^2}}{\mathrm{kg^2}}(6.0 \times 10^{24}\,\mathrm{kg})\left[\frac{1}{6.4 \times 10^6\,\mathrm{m}} - \frac{1}{6.7 \times 10^6\,\mathrm{m}}\right]$$

or

$$v = 2.4 \times 10^3\,\mathrm{m/s}$$

Satellites

A satellite, whether artificial or moon, moves in a circular orbit about a larger planet. The gravitational force supplies the center-directed acceleration necessary to make the satellite move in a circular orbit. Mathematically stated the force balance is

$G\frac{m_1 m_2}{r^2} = \frac{m_2 v^2}{r}$ which reduces to $v^2 r = Gm_1$ with the right hand side a constant.

The orbit radius, period, and velocity are related through

$$\frac{2\pi r}{T} = v = \sqrt{\frac{Gm_1}{r}} \quad \text{or} \quad T = 2\pi r\sqrt{\frac{r}{Gm_1}} \qquad \textbf{(14-6)}$$

14-4 Calculate the speed, period, and radial acceleration of a satellite placed in orbit 400 km above the earth.

Solution: The orbit radius is $6.4\times10^6\,\text{m} + 0.4\times10^6\,\text{m} = 6.8\times10^6\,\text{m}$

The velocity is $v = \sqrt{\frac{Gm_E}{r}} = \left[\frac{6.7\times10^{-11}\,\text{N}\cdot\text{m}^2/\text{kg}^2\,(6.0\times10^{24}\,\text{kg})}{6.8\times10^6\,\text{m}}\right]^{1/2} = 7690\frac{\text{m}}{\text{s}}$

The period is $T = 2\pi\sqrt{\frac{r^3}{Gm_E}} = 2\pi\left[\frac{(6.8\times10^6\,\text{m})^3}{6.7\times10^{-11}\,\text{N}\cdot\text{m}^2/\text{kg}^2\,(6.0\times10^{24}\,\text{kg})}\right]^{1/2} = 93\,\text{min}$

The radial acceleration is

$$a_{rad} = \frac{v^2}{r} = \frac{Gm_E}{r^2} = \frac{6.7\times10^{-11}\,\text{N}\cdot\text{m}^2/\text{kg}^2\,(6.0\times10^{24}\,\text{kg})}{(6.8\times10^6\,\text{m})^2} = 8.7\frac{\text{m}}{\text{s}^2}$$

A geosynchronous satellite is one that is orbiting at an equator and always over the same spot on earth. Using the expression for the period of a satellite, find the height for a satellite with a period of 1.0 day.

Kepler's Laws

There are three Kepler's laws that govern planetary motion. Kepler based these laws on observations made before the invention of the telescope! They are:

I. All planets move in elliptical orbits with the sun at one focus.

II. A line from the sun to the planct sweeps out equal areas in equal lengths of time (see Fig. 14-1).

III. The square of the period is proportional to the cube of the semi-major axis of the ellipse.

(C) Kepler's second law is consistent with the angular momentum of the motion being a constant. Refer to Fig. 14-1. For small angles the area swept out by the arc is equal to the area of a triangle with the side $r\Delta\theta$ nearly perpendicular to the sides r. The area of the triangle is

$(1/2)(r\Delta\theta)r$. In calculus notation the differential area is $dA = (1/2)r^2 d\theta$, and the change in area with time is $\frac{dA}{dt} = \frac{1}{2}r^2\frac{d\theta}{dt} = \frac{1}{2}r^2\omega = \frac{1}{2}rv$

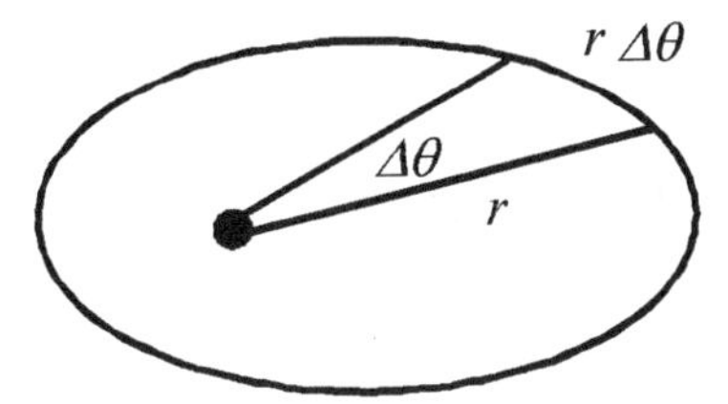

Fig. 14-1

The magnitude of the cross product of two vectors is proportional to the area of the trapezoid defined by the vectors (see the Introduction, Mathematical Background). Here the two vectors defining the area swept out are $\boldsymbol{r}$ and $\boldsymbol{v}$. Rewriting rv as the magnitude of the cross product of $\boldsymbol{r}$ and $\boldsymbol{v}$, or more conveniently $m\boldsymbol{v}$,

$$\frac{dA}{dt} = \frac{1}{2m}|\boldsymbol{r} \times m\boldsymbol{v}|$$

but $\boldsymbol{r} \times m\boldsymbol{v}$ is $\boldsymbol{L}$, the angular momentum, which is a constant of the motion. Kepler's second law is equivalent to conservation of angular momentum.

$$\frac{dA}{dt} = \frac{L}{2m}$$

Going back to rotational dynamics, $d\boldsymbol{L}/dt = \boldsymbol{r} \times \boldsymbol{F}$, and in planetary motion the force is gravitational and acts along r, so $\boldsymbol{r} \times \boldsymbol{F}$ must equal zero, and if the derivative of the angular momentum is zero then the angular momentum is a constant.

Equation 14-6 is based on conservation of energy and shows T as proportional to the 3/2 power of r. This is verification of Kepler's third law for the case of circular orbits.

14-5 The orbit radius of the satellite of problem 14-4 is 6.8×10^6 m, and the period is 93 min. The orbit radius of the moon is 3.8×10^8 m, and the period is 27.3 days. Is this consistent with Kepler's third law?

Solution: Kepler's third law states that T is a constant times the 3/2 power of the radius, so $T_m = kr_m^{3/2}$ and $T_s = kr_s^{3/2}$. If the constant is the same then $\dfrac{T_m}{T_s} = \left(\dfrac{r_m}{r_s}\right)^{3/2}$.

Putting in the numbers

$$\frac{T_m}{T_s} = \frac{27.3\,\text{day}}{93\,\text{min}}\frac{24\,\text{hr}}{\text{day}}\frac{60\,\text{min}}{\text{hr}} = 420 \quad \text{and} \quad \left(\frac{r_m}{r_s}\right)^{3/2} = \left(\frac{3.8\times10^{8}\,\text{m}}{6.8\times10^{6}\,\text{m}}\right)^{3/2} = 420$$

The satellite and moon orbits verify Kepler's third law for circular orbits.

CHAPTER 15

SIMPLE HARMONIC MOTION

Simple harmonic motion is the phrase used to describe a repetitive harmonic motion such as the motion of a mass oscillating up and down on the end of a spring, the rotation of a point moving at constant speed along a circle, or a pendulum in small (amplitude) oscillation. The most convenient visualization of an oscillating system is the mass-spring system. This is also the one most analyzed. Simple harmonic motion is often characterized by the period, the time for one oscillation, and the amplitude, the maximum excursion from equilibrium. (Period is the time for one repetition and is measured in seconds, while the frequency is the reciprocal of the period and is measured in Hertz or 1/s.) More complete analysis usually involves writing statements about the total energy of the system.

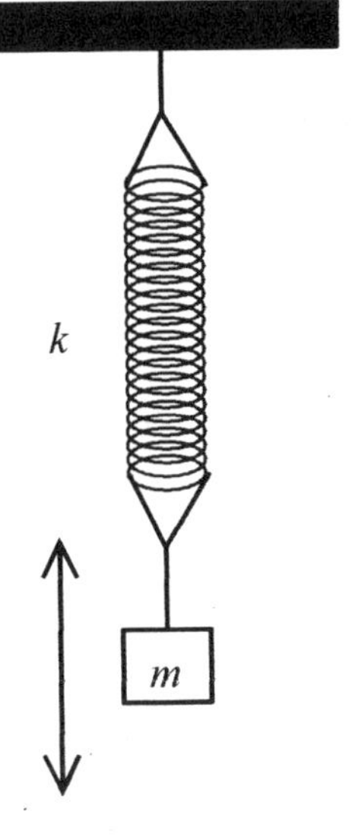

Fig. 15-1

For a first look at simple harmonic motion consider the analysis of a mass-spring system. The kinetic energy is $(1/2)mv^2$, the energy associated with the moving mass, and the potential energy of the system is $(1/2)kx^2$, the energy associated with the elongation and compression of the spring. Here we assume the spring is a Hook's law spring obeying $F = -kx$. You may want to review the discussion of the energy stored in a Hooke's law spring in Chapters 7 and 8.

(C) To describe the motion of the mass from a force point of view, take the force as $-kx$ and equate that to ma with a written as the second derivative of position.

$$-kx = m\frac{d^2x}{dt^2} \tag{15-1}$$

If you don't have much experience in calculus, this may look like a formidable equation. It is not. It has a very simple and obvious (once you've seen it) solution. Acceleration is the time derivative of velocity ($a = dv/dt$), and velocity is the time derivative of displacement ($v = dx/dt$) as defined in Chapter 1. Acceleration, then, can be written as the second derivative of the displacement.

The solution to this equation requires that the second derivative of a function of x with respect to time be proportional to the negative of the function. This reduces to the question: What function differentiated twice yields the negative of itself? The answer is a sine or cosine function or a linear combination of both. If you have not encountered this yet, see the table of derivatives in the Introduction, Mathematical Background.

The derivative of the sine function is the cosine function and the derivative of the cosine function is the negative of the sine function. Two derivatives of the sine function yield the negative of itself. This is also true of the cosine function. The solution can also be written as the sine or cosine of a constant times t plus a phase angle. $A\cos\omega t + B\sin\omega t$ and $C(\cos\ or\ \sin)(\omega t + \vartheta)$ are equivalent solutions.

Take as a general solution $x = A\cos\omega t + B\sin\omega t$

If the starting position ($t = 0$) is taken as an extreme of the motion, corresponding to maximum compression of the spring, then $B = 0$, and A is the amplitude of the motion. If the starting position were taken at the midpoint of the motion, then at $t = 0$ the amplitude would have to be zero; therefore, $A = 0$ and B would be the amplitude. (Remember $\sin 0 = 0$ and $\cos 0 = 1$.) The choice of initial condition determines whether the motion is described by a sine or cosine function.

Use as a solution $x = B\sin\omega t$. Take two time derivatives to obtain

$$\frac{d^2x}{dt^2} = -B\omega^2 \sin\omega t$$

and substitute in the force law statement, $-kx = -m(d^2x/dt^2)$

$$-k = -m\omega^2 \quad \text{or} \quad \omega = \sqrt{k/m} \tag{15-2}$$

In the statement $x = B\sin\omega t$, ω is $2\pi/T$. The 2π effectively scales the sine function, and the ratio t/T determines the fraction of 2π required for determining the value of the function. For example, at $t = T$ (one period) the sine function has gone through one cycle. At $t = T/2$ (one-half period) the sine function has gone through one-half cycle. Because of the 2π factor, ω is called the angular frequency. The term frequency applies to f in the relation

$$\omega = 2\pi(1/T) = 2\pi f \qquad \textbf{(15-3)}$$

In solving problems be careful to differentiate between frequency, f, and angular frequency, ω.

15-1 A mass-spring system has mass $2.0\,\text{kg}$, spring with constant $3.0\,\text{N}/\text{m}$, and amplitude of vibration $0.10\,\text{m}$. Calculate the angular frequency and frequency. Write expressions for the position, velocity, and acceleration as a function of time. Take $t = 0$ at the equilibrium position. Graph position, velocity, and acceleration as a function of time.

Solution: The angular frequency is $\omega = \sqrt{\dfrac{k}{m}} = \sqrt{\dfrac{3.0\,\text{N}}{2.0\,\text{kg}\cdot\text{m}}} = \sqrt{\dfrac{3}{2}}\dfrac{1}{\text{s}} = 1.2\dfrac{1}{\text{s}}$

The frequency is $f = \dfrac{1}{2\pi}\sqrt{\dfrac{k}{m}} = \dfrac{1}{2\pi}\sqrt{\dfrac{3.0\,\text{N}}{2.0\,\text{kg}\cdot\text{m}}} = \dfrac{1}{2\pi}\sqrt{\dfrac{3}{2}}\dfrac{1}{\text{s}} = 0.19\dfrac{1}{\text{s}}$

If $t = 0$ at the equilibrium position, then the displacement must be given by $x = B\sin\omega t$.

At $t = 0$ the sine function is zero, so this is the appropriate function to describe the motion. For an amplitude of $0.10\,\text{m}$ this expression should read

$$x = (0.10\,\text{m})\sin\sqrt{\frac{3}{2}}\frac{1}{\text{s}}t \qquad \text{or} \qquad x = (0.10\,\text{m})\sin 2\pi 0.19(1/\text{s})t$$

The velocity is the time derivative of x or $v = 0.10\sqrt{\dfrac{3}{2}}\dfrac{\text{m}}{\text{s}}\cos\sqrt{\dfrac{3}{2}}t$

At $t = 0$ the velocity, as the mass passes through the equilibrium position, is a maximum. The acceleration is another time derivative of the velocity or

$$a = -0.10\frac{3}{2}\frac{\text{m}}{\text{s}^2}\sin\sqrt{\frac{3}{2}}t$$

At $t = 0$ the acceleration is zero; the spring is neither elongated nor compressed (at the equilibrium position), so the force (and acceleration) of the mass is zero.

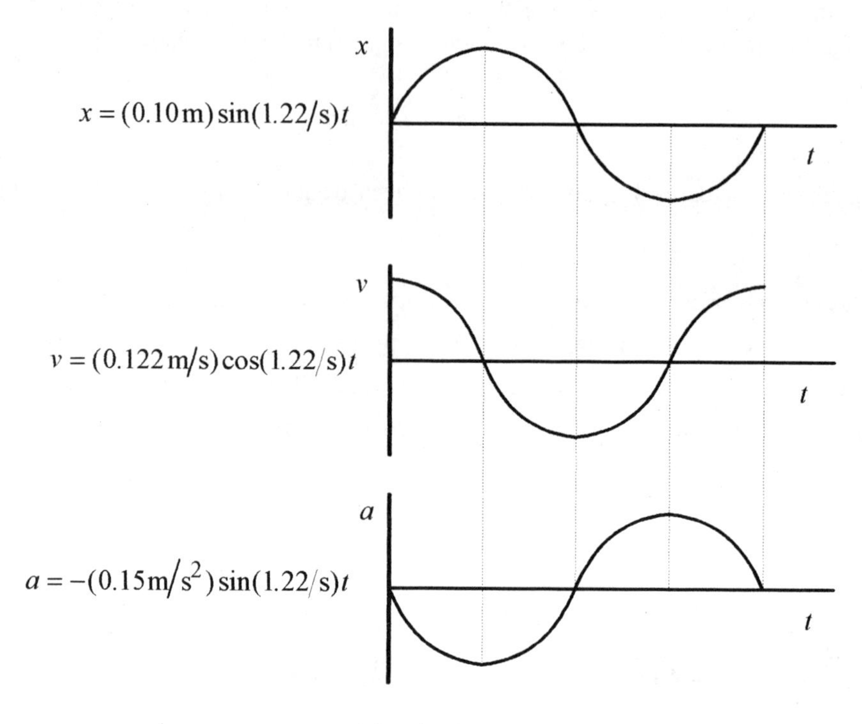

Fig. 15-2

These three curves graphed one under the other in Fig. 15-2 give a good picture of the motion from the position, velocity, and acceleration point of view.

Look at these graphs and visualize the motion. At the equilibrium position, displacement is zero, the velocity is a maximum, and acceleration is zero. As the mass moves to the positive maximum displacement, the velocity decreases to zero and the acceleration (down) increases to a maximum. At the maximum displacement, the velocity is zero and the acceleration is a maximum. As the mass starts down, the velocity is negative reaching maximum again as the mass passes through equilibrium. Go over this picture of the motion until the relationship between position, velocity, and acceleration is clear in your mind.

15-2 A piece of metal rests on top of a piston executing simple harmonic motion in the vertical plane with an amplitude of $0.20\,\text{m}$. At what frequency will the piece of metal "float" at the top of the cycle?

Solution: Take the motion to be described by $y = (0.20\,\text{m})\cos\omega t$.

Two derivatives produces the acceleration $a = -(0.20\,\text{m})\omega^2 \cos\omega t$.

The maximum acceleration, which is the constant $(0.20\,\text{m})\omega^2$, is at the extreme of the motion where the cosine term is one. At the bottom of the cycle, the piece of metal will experience an acceleration greater than g and at the top an acceleration less than g. The piece will "float when that acceleration is greater than g.

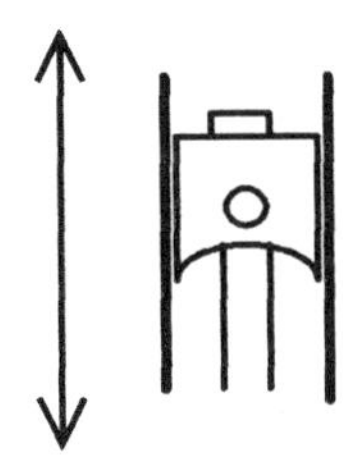

Fig. 15-3

$$(0.20\,\text{m})\omega^2 = g = 9.8\frac{\text{m}}{\text{s}^2} \quad \text{or} \quad \omega = \sqrt{\frac{9.8\,\text{m/s}^2}{0.20\,\text{m}}} = 7.0\frac{1}{\text{s}}$$

$$\text{or} \quad f = \frac{1}{2\pi}\sqrt{\frac{9.8\,\text{m/s}^2}{0.20\,\text{m}}} = 1.1\frac{1}{\text{s}}$$

15-3 A mass-spring system has a mass of $0.50\,\text{kg}$ and period of $1.0\,\text{s}$. The maximum velocity is $0.20\,\text{m/s}$. Find the frequency, angular frequency, constant of the spring, and displacement for this velocity and period.

Solution: The frequency is the reciprocal of the period or $f = 1.0(1/\text{s})$, and the angular frequency is $2\pi f$ or $2\pi(1/\text{s})$. The spring constant comes from the relation

$$\omega = \sqrt{\frac{k}{m}} \quad \text{or} \quad k = \omega^2 m = [2\pi(1/\text{s})]^2\,0.50\,\text{kg} = 20\frac{\text{kg}}{\text{s}^2}\left(\frac{\text{m}}{\text{m}}\right) = 20\frac{\text{N}}{\text{m}}$$

The general expression for velocity is $v = A\omega\cos\omega t$ where $A\omega$ is the maximum velocity. It makes no difference whether the function is a sine or cosine. The maximum velocity is still $A\omega$. The amplitude for this particular motion then is

$$A = \frac{v}{\omega} = \frac{0.20\,\text{m/s}}{2\pi(1/\text{s})} = 0.032\,\text{m}$$

15-4 An $0.80\,\text{kg}$ mass hangs from a spring. When an additional $0.20\,\text{kg}$ mass is added the spring elongates another $3.0\,\text{cm}$ What is the period of oscillation of the spring?

Solution: First determine the force constant of the spring from $F = -kx$

$$0.20\,\mathrm{kg}\frac{9.8\,\mathrm{m}}{\mathrm{s}^2} = k \cdot 3.0\times10^{-2}\,\mathrm{m} \quad \text{or} \quad k = \frac{0.20\,\mathrm{kg}\cdot 9.8\,\mathrm{m/s^2}}{0.03\,\mathrm{m}} = 65\frac{\mathrm{N}}{\mathrm{m}}$$

The period is the reciprocal of the frequency $\omega = \sqrt{\frac{k}{m}} = 2\pi f = \frac{2\pi}{T}$

so

$$T = 2\pi\sqrt{\frac{m}{k}} = 2\pi\sqrt{\frac{0.80\,\mathrm{kg}}{65\,\mathrm{N/m}}} = 0.70\,\mathrm{s}$$

15-5 An oscillating mass-spring system has displacement $10\,\mathrm{cm}$, velocity $-12\,\mathrm{m/s}$, and acceleration $-20\,\mathrm{m/s^2}$. What is the period of the system?

Solution: Since the displacement is not zero at $t = 0$, the motion cannot be described with a $\sin\omega t$ function. Since the velocity is not zero at $t = 0$, the motion in not at an extreme, and cannot be described with a $\cos\omega t$ function. Because of the initial conditions, a sine or cosine function of ωt plus a phase angle must be used. It makes no difference whether a sine or cosine function is used since the phase angle can absorb the 90° difference between a sine and a cosine function. Look at Fig. 15-4 and go over the logic in this paragraph until it is clear in your mind.

Take $x = A\cos(\omega t + \vartheta)$ then $v = -A\,\omega\sin(\omega t + \vartheta)$ and $a = -A\,\omega^2\cos(\omega t + \vartheta)$

Using the equations for x and a, at $t = 0$, $\omega^2 = -\frac{a(0)}{x(0)}$ and $\omega = \sqrt{\frac{20\,\mathrm{m/s^2}}{0.10\,\mathrm{m}}} = 14\frac{1}{\mathrm{s}}$

The period is from $\omega = \frac{2\pi}{T}$ or $T = \frac{2\pi}{\omega} = \frac{2\pi}{\sqrt{200}}\,\mathrm{s} = 0.44\,\mathrm{s}$

The phase angle comes from $v(0)$ and $a(0)$.

$$\frac{v(0)}{a(0)} = \frac{\sin\vartheta}{\omega\cos\vartheta} = \frac{1}{\omega}\tan\vartheta \quad \text{or} \quad \tan\vartheta = \omega\frac{v(0)}{a(0)} = \sqrt{2}\frac{1}{\mathrm{s}}\frac{-12\,\mathrm{m/s}}{-20\,\mathrm{m/s^2}} = 0.85 \quad \text{so} \quad \vartheta = 40^o$$

The specific functional relations are

$$x = A\cos(\omega t + 40^o),\quad v = -A\,\omega\sin(\omega t + 40^o) \quad \text{and} \quad a = -A\,\omega^2\cos(\omega t + 40^o)$$

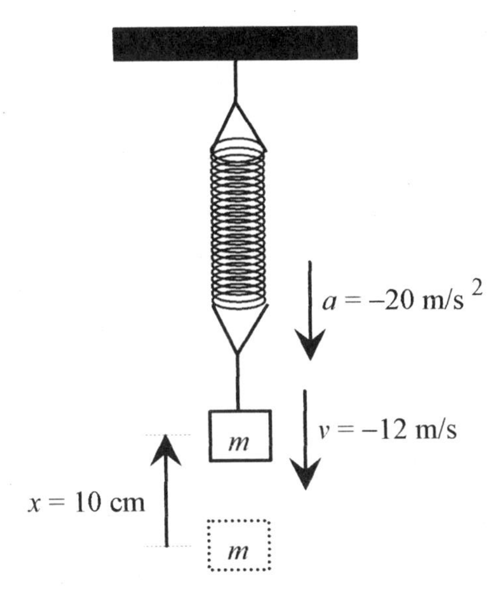

Fig. 15-4

Look at these relations and relate them to what the system is doing at $t = 0$. The amplitude is past its maximum positive excursion and on the way down corresponding to the cosine function at 40°. The velocity is negative and heading toward its most negative value as is the negative sine function at 40°. The acceleration is negative and tending toward zero just as is the negative cosine function at 40°.

Energy Analysis

Simple harmonic motion can be analyzed from an energy point of view. The energy stored in a spring is $kx^2/2$, and the energy stored in the mass by virtue of its velocity is $mv^2/2$. A total energy statement for the system then is

$$\frac{kx^2}{2} + \frac{mv^2}{2} = E \tag{15-4}$$

which is most easily evaluated by looking at the extrema of the motion,

$$\text{at } v = 0 \quad \text{where} \quad E = \frac{kx_{max}^2}{2} \qquad \text{or} \qquad \text{at } x = 0 \quad \text{where} \quad E = \frac{mv_{max}^2}{2}$$

15-6 Find the maximum velocity of a mass-spring system with mass $2.0\,\text{kg}$, spring constant $0.80\,\text{N/m}$, and amplitude of oscillation $0.36\,\text{m}$.

Solution: All the energy of the system is stored in the spring at maximum displacement so the total energy is

$$E = \frac{kx_{max}^2}{2} = \frac{0.80\,\text{N}}{\text{m}}\frac{(0.36\,\text{m})^2}{2} = 0.052\,\text{J}$$

This is the total energy in the system, and at zero displacement all this energy is in kinetic energy so

$$\frac{mv_{max}^2}{2} = 0.052\,\text{J} \qquad \text{or} \qquad v_{max} = \left[\frac{2 \cdot 0.052\,\text{J}}{2.0\,\text{kg}}\right]^{1/2} = 0.23\frac{\text{m}}{\text{s}}$$

It is relatively easy, and quite instructive, to show that the total energy of an oscillating system is a constant. Take an oscillating system with $x = A\sin\omega t$ and velocity $v = A\omega\cos\omega t$. Now place these expressions for x and y into the total energy statement

$$\frac{kx^2}{2} + \frac{mv^2}{2} = E \qquad \text{so} \qquad \frac{kA^2\sin^2\omega t}{2} + \frac{mA^2\omega^2\cos^2\omega t}{2} = E$$

But from the previous analysis $\omega^2 = k/m$ so the statement reads

$$\frac{kA^2}{2}\sin^2\omega t + \frac{kA^2}{2}\cos^2\omega t = E \qquad \text{and since} \qquad \sin^2\omega t + \cos^2\omega t = 1, \qquad \frac{kA^2}{2} = E$$

which shows that the total energy does not vary over time. The energy does not vary over one cycle, it goes back and forth from kinetic to potential in a manner such that the total (energy) remains a constant.

15-7 Find the amplitude, angular frequency, and frequency for a mass-spring system with mass $2.0\,\text{kg}$ that is oscillating according to $x = (2.0\,\text{m})\cos(6\pi t/\text{s})$. Write and graph the expressions for position, velocity, and acceleration. Finally, calculate the total energy of the system.

Solution: The amplitude, angular frequency, and frequency are determined by comparing $x = (2.0\,\text{m})\cos(6\pi t/\text{s})$ to the standard form $x = A\cos\omega t = A\cos 2\pi f t$.

The amplitude is $2.0\,\text{m}$. The angular frequency is $6\pi(1/\text{s})$. The frequency is $3(1/\text{s})$.
The expressions for position, velocity (first derivative), and acceleration (second derivative) are shown in Fig. 15-5 associated with the graphs.

In Fig. 15-5, the units are dropped in the argument of trigonometric functions, and the maximum values for the velocity and acceleration are placed on the graphs.

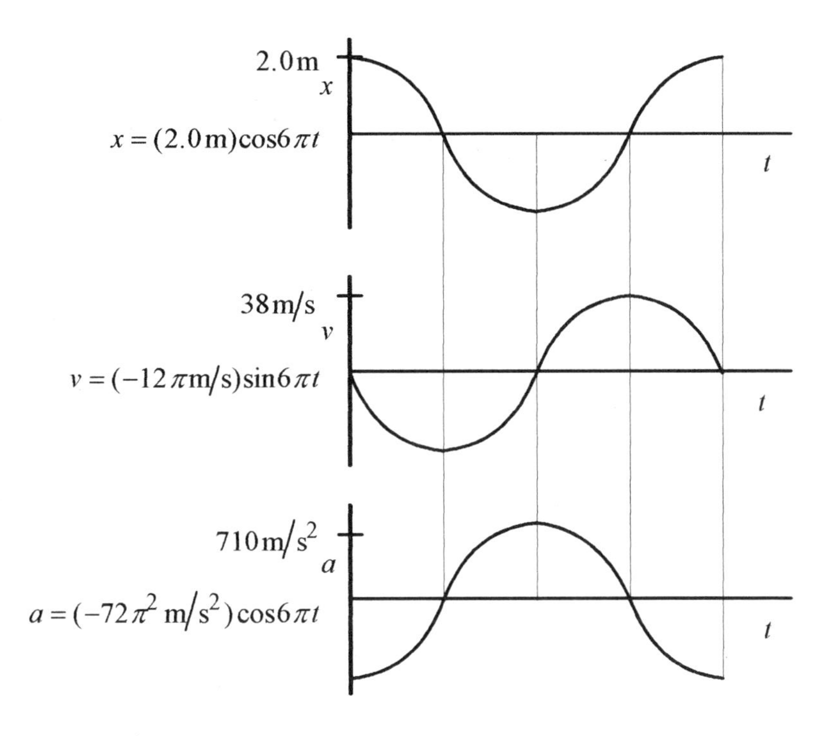

Fig. 15-5

The conservation of energy statement is $\frac{mv^2}{2}+\frac{kx^2}{2}=E$. For a $2.0\,\text{kg}$ mass the total energy can be evaluated when x is zero and the velocity maximum.

$$E=\frac{mv_{max}^2}{2}=\frac{2.0\,\text{kg}(12\pi\,\text{m/s})^2}{2}=1420\,\text{J}$$

15-8 For a $3.0\,\text{kg}$ mass-spring system executing simple harmonic motion according to $y=(4.0\,\text{cm})\cos(\pi t/4-\pi/6)$, what is the total energy and the y position for equal division of the energy between kinetic and potential?

Solution: The general expression for displacement is $y=A\cos(\omega t+\vartheta)$. Looking at the two expressions, $\omega=\sqrt{\frac{k}{m}}=\frac{\pi}{4}\frac{1}{\text{s}}$ so $k=\omega^2 m=\frac{\pi^2}{4^2}\frac{1}{\text{s}^2}3.0\,\text{kg}\left(\frac{\text{m}}{\text{m}}\right)=1.85\frac{\text{N}}{\text{m}}$.

Knowing k, the total energy is $E=\frac{kA^2}{2}=\frac{1.85\,\text{N}}{\text{m}}(16\times10^{-4}\,\text{m}^2)=29.6\times10^{-4}\,\text{J}$

The potential energy due to position is $ky^2/2$ and $y = A\cos(\omega t + \vartheta)$ so that the energy due to position is $kA^2\cos^2(\omega t + \vartheta)/2$. The energy due to position is 1/2 the total when the factor $\cos^2(\omega t + \vartheta) = 1/2$ or $\cos(\omega t + \vartheta) = 1/\sqrt{2}$.

For this particular system, the time for equal division of the energy is when

$$\cos\left(\frac{\pi}{4}t - \frac{\pi}{6}\right) = \frac{1}{\sqrt{2}}, \quad \frac{\pi}{4}t - \frac{\pi}{6} = \cos^{-1}\frac{1}{\sqrt{2}}, \quad \frac{\pi}{4}t - \frac{\pi}{6} = \frac{\pi}{4}, \quad \frac{\pi}{4}t = \frac{5\pi}{12}, \quad \text{or } t = \frac{5}{3}\text{s}$$

The position of the mass for equal division of the energy is

$$y|_{E/2} = (4.0\,\text{cm})\cos\left(\frac{\pi}{4}\frac{5}{3} - \frac{\pi}{6}\right) = (4.0\,\text{cm})\cos\left(\frac{3\pi}{12}\right) = 2.83\,\text{cm}$$

As an exercise, calculate the velocity at this time.

15-9 A $0.20\,\text{kg}$ block traveling at $20\,\text{m/s}$ slides into and sticks to an $0.80\,\text{kg}$ block resting on a frictionless surface and connected to a spring with force constant $80\,\text{N/m}$. What is the angular frequency, frequency, and displacement as a function of time? Also what fraction of the original energy in the moving block appears in the system?

Solution: The collision is inelastic; the blocks stick together, so applying conservation of momentum $0.20\,\text{kg}(20\,\text{m/s}) = (1.0\,\text{kg})V$. The velocity of the $1.0\,\text{kg}$ combination is $V = 4.0\,\text{m/s}$. This is the velocity of the system at zero displacement. The analysis of this collision is similar to the ballistic pendulum problem in the chapter on collisions.

The total energy of the oscillating system is $E = \dfrac{mV^2}{2} = \dfrac{1.0\,\text{kg}}{2}\dfrac{16\,\text{m}^2}{\text{s}^2} = 8.0\,\text{J}$.

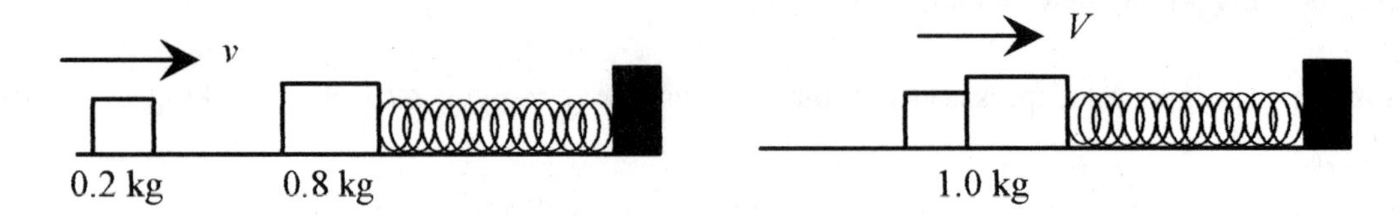

Fig. 15-6

The angular frequency is $\omega = \sqrt{\dfrac{k}{m}} = \sqrt{\dfrac{80\,\text{N/m}}{1.0\,\text{kg}}} = 8.9\dfrac{1}{\text{s}}$

The frequency is from $\omega = 2\pi f$ or $f = \omega / 2\pi = 1.4\,(1/\text{s})$

The general expression for the displacement is $y = A \sin \omega t$. The sine function without a phase angle completely describes the motion because it starts ($t = 0$) at the equilibrium position with maximum velocity. The amplitude of the motion is obtained from

$$kA^2 / 2 = E \qquad \text{or} \qquad A = \sqrt{\frac{2E}{k}} = \sqrt{\frac{2 \cdot 8.0\,\text{J}}{80\,\text{N/m}}} = 0.45\,\text{m}$$

so the specific expression for the displacement is $y = (45\,\text{cm}) \sin(1.4t/\text{s})$

The initial energy is all in the small block $E_i = \frac{mv^2}{2} = \frac{0.20\,\text{kg}(20\,\text{m/s})^2}{2} = 40\,\text{J}$

The energy in the oscillating system is $8.0\,\text{J}$, so the fraction of the original energy that appears in the oscillating system is $F = 8.0\,\text{J} / 40\,\text{J} = 0.20$

Applications

The oscillating simple pendulum can be analyzed in terms of force. For small oscillations the tension in the cable can be written in components; one vertical and the other the "restoring force," the force that returns the mass to equilibrium. For small θ, $\cos\theta \approx 1$ and $\sin\theta \approx x/\ell$, with x being the displacement, approximately equal to the arc length.

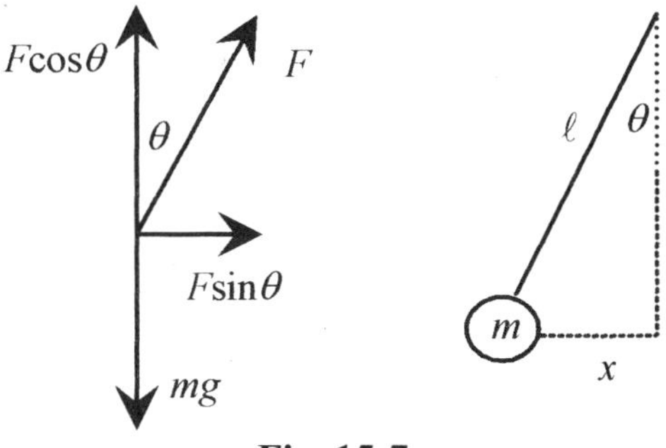

Fig. 15-7

For small angle oscillations $F\cos\theta \approx F$ and $F\sin\theta \approx -mgx/\ell = -(mg/\ell)x$.

Notice that in this equation mg/ℓ plays the same role as k in the mass-spring systems.

By analog then $\omega = \sqrt{\frac{mg/\ell}{m}} = \sqrt{\frac{g}{\ell}}$ and $T = 2\pi\sqrt{\frac{\ell}{g}}$

The most important point to note is that the period is independent of the mass. Historically this was one of the first instruments to measure "g," the gravitational constant.

C The torsional pendulum, a disk suspended by a thin rod, can be analyzed using torque. The disk is rotated by applying a torque. Torque is proportional to the angle in the same manner as force is proportional to compression (or elongation) in a spring. The torque is written in terms of the moment of inertia and the angular acceleration $[T = I\alpha = I(d^2\theta/dt^2)]$ which is set equal to the total torque, $-\kappa\theta$ the torsional constant times the angle, yielding the equation

$I\dfrac{d^2\theta}{dt^2} = -\kappa\theta$, which is the same form as the force equation for the oscillating mass-spring system. By analog then $\theta = \theta_{max}\sin\omega t$ where $\omega = \sqrt{\dfrac{\kappa}{I}}$ and $T = 2\pi\sqrt{\dfrac{I}{\kappa}}$

Fig. 15-8

15-10 What is the restoring constant (torsional constant) for a disk of mass $3.0\,\mathrm{kg}$ and radius $0.20\,\mathrm{m}$ oscillating with a period of $2.0\,\mathrm{s}$?

Solution: The moment of inertia for a disk is

$$I = \frac{mr^2}{2} = \frac{3.0\,\mathrm{kg}(0.20\,\mathrm{m})^2}{2} = 0.060\,\mathrm{kg\cdot m^2}$$

The restoring constant is $\kappa = \dfrac{4\pi^2 I}{T^2} = \dfrac{4\pi^2 \cdot 0.060\,\mathrm{kg\cdot m^2}}{4.0\,\mathrm{s}^2} = 0.59\,\mathrm{N\cdot m}$

Note that these are the correct units since torque has the dimensions $\mathrm{N\cdot m}$, and in $\text{Torque} = -\kappa\theta$ the angle is in (dimensionless) radians.

Damped Mass-Spring System

(C) The system is usually depicted as a mass-spring with a dashpot, a piston in a thick liquid. In the case of the spring, the restoring force is proportional to displacement, compression or elongation of the spring. With the piston in the fluid, however, the retarding force is proportional to velocity, not position. Move your hand through water and notice that the retarding force (the force on your hand) is proportional to velocity! The velocity-dependent force component of a typical automobile suspension is called a shock absorber. It is usually a piston in a cylinder of oil. Since the retardation force is velocity dependent the differential equation governing the motion is

$$m\frac{d^2x}{dt^2} = -kx - b\frac{dx}{dt} \tag{15-5}$$

Notice that the velocity-dependent term has the same sign as the displacement-dependent term. It too opposes the motion.

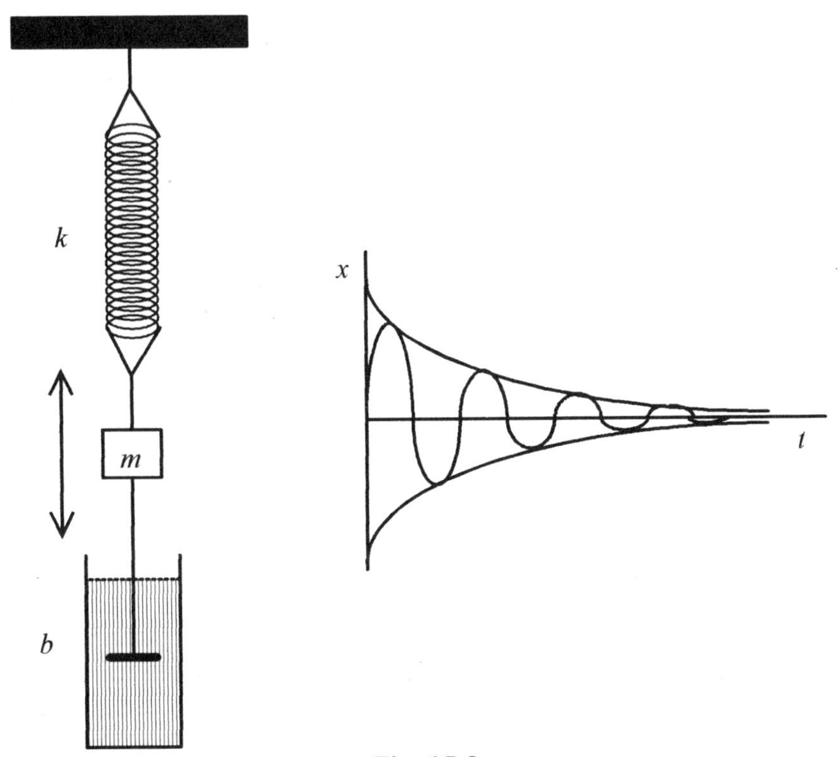

Fig. 15-9

This differential equation is solved with sophisticated techniques. Here we give an intuitive solution.

This second order differential equation is solved as a product of solutions, the position-dependent part and the velocity-dependent part. The position-dependent part is a sine or cosine function with a phase angle. The velocity-dependent part is from the velocity-dependent part of the equation

$$m\frac{d^2x}{dt^2} = -b\frac{dx}{dt}$$

The solution to this part of the equation is one which when differentiated once produces the negative of itself. The function that does this is an exponent to the negative power. The particular solution that satisfies this differential equation is

$$x = Ae^{-bt/2m}\cos(\omega' t + \vartheta) \qquad \textbf{(15-6)}$$

where ω' is slightly different from ω. The damping of the exponential does exactly as we know from experience. Oscillations subject to velocity damping (a velocity-dependent restoring force) eventually die out.

CHAPTER 16

FLUIDS

There are several definitions and concepts that are unique to the study of fluids. Let's review these basic definitions and concepts working some problems along the way.

Density is simply mass per unit volume, or symbolically $\rho = m/V$. The density that is most useful in problems is the density of water, which is $1\,\mathrm{g/cm^3}$ or $1000\,\mathrm{kg/m^3}$.

The ratio called **specific gravity** is better called the specific density since it is the ratio of the density of another substance to the density of water. Aluminum has a specific gravity of 2.7, which means that for the same volumes, aluminum has 2.7 times the mass of water. Relative density or specific gravity is used to measure the relative amount of sulfuric acid (compared to water) in a storage battery or the relative amount of ethylene glycol (compared to water) in an automotive cooling system.

Pressure

The **pressure** in a fluid is defined as the force per unit area, $p = F/A$. And the units of pressure are $\mathrm{N/m^2}$, also called a Pascal or Pa. Atmospheric pressure, the force per unit area exerted by the air, is $1.0\times10^5\,\mathrm{Pa} = 15\,\mathrm{lb/in^2} = 1\,\mathrm{atm}$.

To determine the pressure at any depth in a liquid consider the differential change in force due to a column of liquid of cross section A. The increase in force between the top and bottom of this differential piece is equal to the weight of the piece.

It is convenient to write the mass of this little piece as $\rho\Delta V$ so that the force (difference) is

$$\Delta F = \rho g \Delta V \qquad \textbf{(16-1)}$$

Now ΔV for this piece is $A\,\Delta y$ so $\Delta F = \rho g A\,\Delta y$. The pressure is the force per unit area so rewriting

$$\Delta p = \Delta F/A = \rho g \Delta y \qquad \textbf{(16-2)}$$

Dropping the differential notation

$$p_{bot} - p_{top} = \rho g (y_{bot} - y_{top}) \qquad \textbf{(16-3)}$$

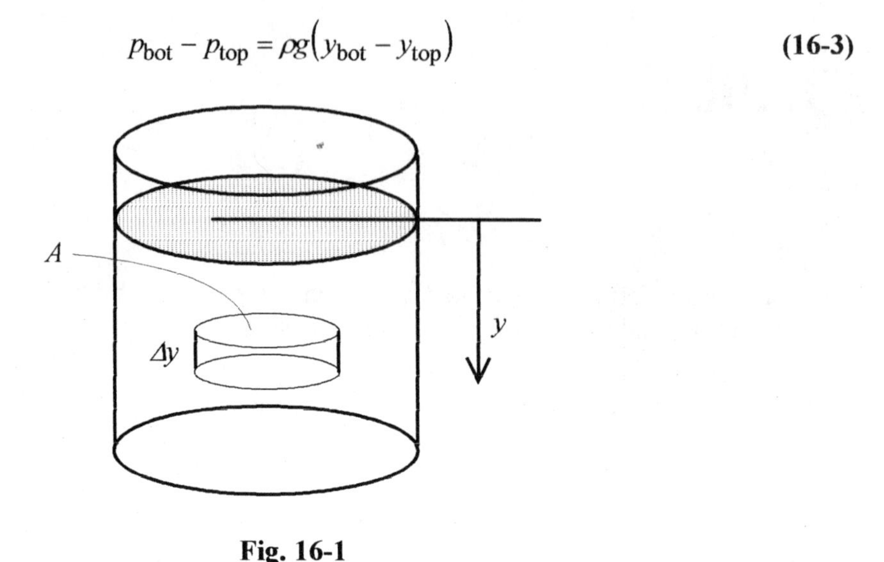

Fig. 16-1

If y is measured from the bottom of the liquid then insert a negative sign on the right side of the equation. This discussion of pressure leading up to equation 16-3 is equally valid whether the delta or differential notation is used. This is true of most of the discussion of fluids. We start with delta notation and switch to differential notation in some later development.

16-1 Calculate the pressure at the bottom of a 3.0 m deep pool.

Solution: $\Delta p = \rho g h = 1000\,\mathrm{kg/m^3}(9.8\,\mathrm{m/s^2})3.0\,\mathrm{m} = 2.9 \times 10^4\ \mathrm{Pa}$

Notice that this pressure is the difference between the top and the bottom. This difference is known as the **gauge pressure**. Adding the atmospheric pressure at the top surface of the pool gives the absolute pressure at the bottom of the pool.

The typical automobile tire pressure is the gauge pressure. With the tire inflated the pressure might read $30\,\mathrm{lb/in^2}$. If the tire were "flat" the pressure would read zero. The gauge pressure for the "flat" tire would be zero but the absolute pressure would be $15\,\mathrm{lb/in^2}$. Likewise the tire with gauge pressure of $30\,\mathrm{lb/in^2}$ would have an absolute pressure of $(30+15)\,\mathrm{lb/in^2}$. The gauge that measures the tire pressure is measuring the difference in pressure between inside and outside the tire.

Pressure exerted on a liquid in a closed container is transmitted throughout the liquid. The pressure is the same in every direction varying only with height as described by equation 16-3. In most devices using fluids to transfer forces, the fluid is nearly incompressible and height variations are insignificant.

16-2 Consider the classic problem of a hydraulic lift. In a typical service station 12 atm is applied to the small area with the lifting column 9.0 cm in radius. Find the force transferred to the large area and the mass of vehicle that can be lifted.

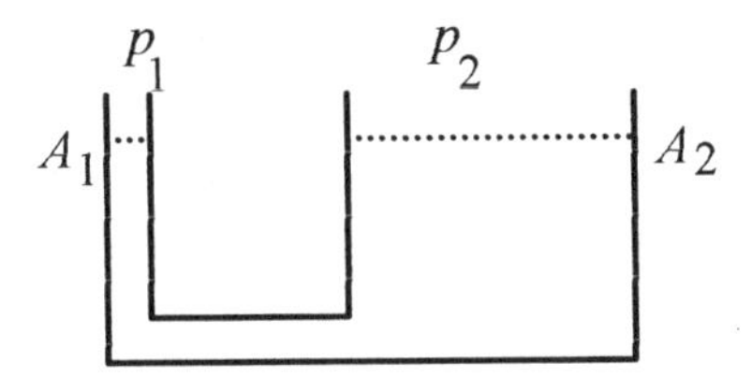

Fig. 16-2

Solution: The hydraulic fluid is at the same level so $p_1 = p_2$ or $\dfrac{F_1}{A_1} = \dfrac{F_2}{A_2}$

A force F_1 applied at A_1 is multiplied by the ratio of the areas so $F_2 = (A_2/A_1)F_1$

The lifting force F_2 can also be rewritten as $F_2 = A_2(F_1/A_1) = A_2 p_1$, and putting in the numbers

$$F_2 = \pi(9.0\,\text{cm})^2\,12\,\text{atm}\frac{1.0\times10^5\,\text{N/m}^2}{\text{atm}}\left(\frac{\text{m}}{100\,\text{cm}}\right)^2 = 3.0\times10^4\,\text{N}$$

And the equivalent mass (from $F = mg$) is $3.1\times10^3\,\text{kg}$. This number is somewhat larger than the mass of typical cars.

Barometer and Manometer

A simple atmospheric air pressure gauge is the mercury barometer. A closed tube is filled with mercury and then inverted in a beaker of mercury. The pressure at the top of the mercury column is zero. And the column is held up by the (atmospheric) pressure on the surface of the mercury. Looking to Fig. 16-3, the pressure at the bottom of the column is ρgh. This experiment was first done by Torricelli, and the pressure was given as the pressure associated with the column of mercury. Pressures are often measured in mm of Hg or Torr. and refer back to this experiment.

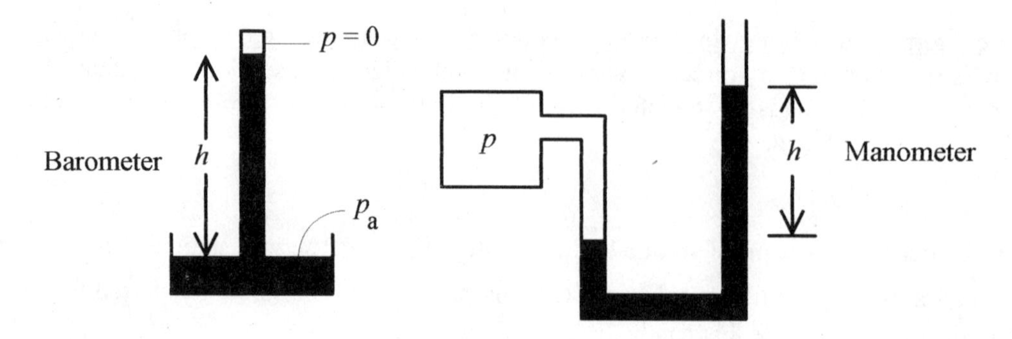

Fig. 16-3

An open tube manometer as depicted in Fig. 16-3 can be used to measure pressure. The force exerted by the gas is pA, and this force raises the column of (usually) mercury by exerting a force equal to ρgV, the weight of the column. The volume of the column, V is Ah so equating forces.

$$pA = \rho gAh \qquad \text{or} \qquad p = \rho gh \tag{16-4}$$

The manometer measures the difference in pressure between two points and as such measures gauge pressure. The open tube manometer is equally valuable as a vacuum gauge. The difference in column height (the height of the mercury in the open tube is now lower than in the tube connected to the vacuum chamber) is a measure of the vacuum.

Buoyancy

Material objects appear to weigh less when partially or completely submerged in liquids. This apparent loss in weight is due to the buoyant force of the liquid. This buoyant force is expressed by **Archimedes' principle:** <u>A body partially or completely immersed in a liquid is buoyed up by a force equal to the weight of the liquid displaced</u>.

16-3 The classic problem in buoyancy is to determine if a "gold" statue is really gold by measuring its weight in and out of water. A certain "gold" statue weighs $70\,\text{N}$ when out of water and $64\,\text{N}$ when immersed in water.

Solution: With the statue suspended from a scale the weight is mg, or density times g, times the volume of the statue. In equation form $mg = \rho_? gV = 70\,\text{N}$.
In water the statue weighs less because of the buoyant force: $\rho_? gV - \rho_W gV = 64\,\text{N}$

Subtracting one equation from the other we can determine the volume from

$$\rho_W gV = 6\,\text{N}, \qquad 1000\,\text{kg/m}^3 (9.8\,\text{m/s}^2) V = 6\,\text{N}, \qquad V = 6.1 \times 10^{-4}\,\text{m}^3$$

With this value for V, calculate the unknown density from the first equation

$$\rho_? gV = 70\,\text{N}, \qquad \rho_? = \frac{70\,\text{N}}{6.1\times 10^{-4}\,\text{m}^3(9.8\,\text{m/s}^2)} = 12\times 10^3\,\frac{\text{kg}}{\text{m}^3}$$

This density is much less than the density of gold ($\rho_{\text{gold}} = 19\times 10^3\,\text{kg/m}^3$) and closer to the density of lead ($\rho_{\text{lead}} = 11\times 10^3\,\text{kg/m}^3$) suggesting that the statue is lead with a thin covering of gold.

This same experiment can be conducted by lowering the statue into a beaker of water on a scale. The volume is determined by the increase in the level of water in the beaker. The drop in tension in the cord is offset by the increase in the scale reading when the statue is lowered into the beaker and the calculations proceed as before.

16-4 A U-tube originally containing mercury has water added to one arm to a depth of $20\,\text{cm}$. What is the pressure at the water-mercury interface? What is the height of the mercury column as measured from the water-mercury level?

Solution: Assume that the difference in air pressure can be neglected. The pressure at the water-mercury interface is due to the column of water plus the atmospheric pressure p_o, or $p = p_o + \rho_W g\cdot 20\,\text{cm}$. Neglecting p_o gives the gauge pressure as

$$p = 1000\,\text{kg/m}^3(9.8\,\text{m/s}^2)20\times 10^{-2}\,\text{m} = 1960\,\text{Pa}$$

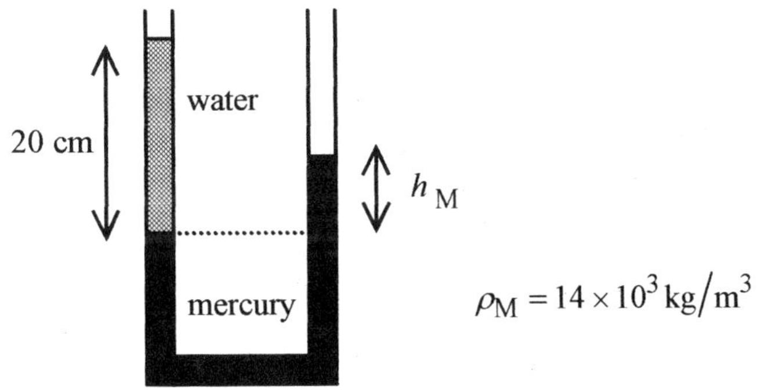

Fig. 16-4

At this level (the level of the interface) the pressure on the mercury arm must be the same. (If it were not the liquids would move!) The height of the mercury level is

$$1960\,\text{Pa} = \rho_M g h_M \qquad \text{or} \qquad h_M = \frac{1960\,\text{N/m}^2}{14\times 10^3\,\text{kg/m}^3(9.8\,\text{m/s}^2)} = 1.4\,\text{cm}$$

As we expect, the same pressure is produced by a much shorter column of mercury.

Second Solution: Analytically the problem can be approached by writing expressions for the pressure at the level of the water-mercury interface.

In the left arm $p_W = p_o + \rho_W g h_W$, where the subscript W refers to the water column.

In the right arm the pressure is $p_M = p_o + \rho_M g h_M$, where the subscript M refers to the mercury column.

Since these pressures must be equal at the water-mercury interface

$$\rho_W g h_W + p_o = \rho_M g h_M + p_o \qquad \text{or} \qquad \rho_W h_W = \rho_M h_M$$

For this problem $h_M = h_W \dfrac{\rho_W}{\rho_M} = 20\,\text{cm}\dfrac{1}{14} = 1.4\,\text{cm}$

Practice variations of this problem with different liquids (different densities).

16-5 A rectangular-shaped, open-top steel barge is $10\,\text{m}$ by $3.0\,\text{m}$ and has sides $1.0\,\text{m}$ high. The mass of the barge is $8.0 \times 10^3\,\text{kg}$. How much mass can the barge hold if it can safely sink $0.75\,\text{m}$?

Solution: The volume of water displaced is length times width times allowed depth, $10\,\text{m} \times 3.0\,\text{m} \times 0.75\,\text{m} = 22.5\,\text{m}^3$. The buoyant force is the weight of the water displaced (Archimedes' principle)

$$F_B = \rho g V = 1000\,\text{kg}/\text{m}^3 (9.8\,\text{m}/\text{s}^2) 22.5\,\text{m}^3 = 2.2 \times 10^5\,\text{N}$$

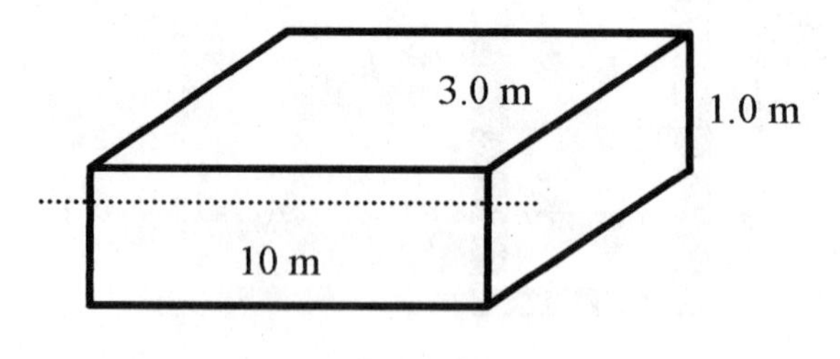

Fig. 16-5

This buoyant force must support the barge (m_B) plus the load (m_L).

$$2.2 \times 10^5\,\text{N} = g(m_B + m_L), \qquad m_B + m_L = 22 \times 10^3\,\text{kg}, \qquad \text{and} \qquad m_L = 14 \times 10^3\,\text{kg}$$

As an additional exercise calculate how deep the unloaded barge will sink in the water.

16-6 An iceberg floats in seawater . What percentage of the iceberg is submerged? $(\rho_{\text{ice}} = 0.92 \times 10^{-3}\,\text{kg/m}^3)$ $(\rho_{\text{SW}} = 1.0 \times 10^3\,\text{kg/m}^3)$

Solution: Assume a cubical iceberg of side s. You can assume any shape you wish. Cubical, however, seems convenient. The mass of the cubical iceberg is $\rho_{\text{ice}} s^3$, and the weight is $\rho_{\text{ice}} s^3 g$. Since the density of the iceberg is less than the density of the seawater the iceberg will float. The buoyant force of the seawater is $\rho_{\text{SW}} g$ times the volume submerged. Since the iceberg is cubical the submerged volume must be s^2, the area of the bottom of the iceberg, times a fraction of the side K times s.

This buoyant force must equal the weight of the iceberg. $\rho_{\text{ice}} s^3 g = \rho_{\text{SW}} g s^2 (Ks)$.

So K, the fraction of the iceberg submerged is $K = \dfrac{\rho_{\text{ice}}}{\rho_{\text{SW}}} = \dfrac{0.92}{1.0} = 0.92$.

Thus 92% of the iceberg is submerged.

Fluid Flow

The last, and possibly most important, of the fluid laws are associated with the flow of liquids through tubes of varying cross section. The best place to start on this is with a simple **statement of mass flow** in a pipe of variable thickness.

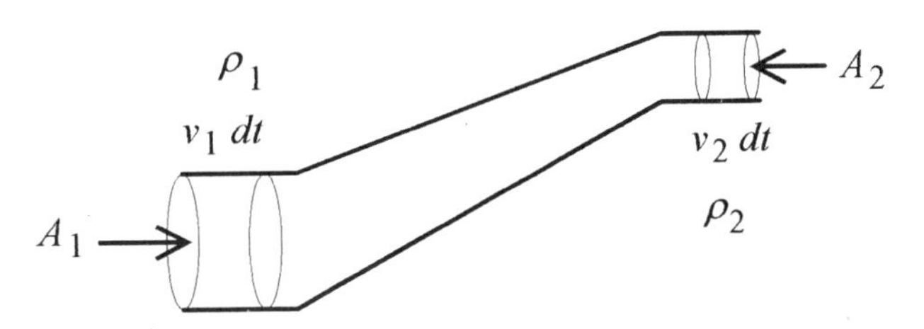

Fig. 16-6

Mass flow requires $\rho_1 A_1 v_1 dt = \rho_2 A_2 v_2 dt$

Density is mass per unit volume, and $Avdt = Adx$ is a differential volume, so that the above equation is a statement that there is the same mass of fluid in the two differential volumes shown. Since in most fluids $\rho_1 = \rho_2$, the **mass flow** (also called the **continuity equation**) statement reduces to

$$A_1 v_1 = A_2 v_2 \qquad \textbf{(16-5)}$$

Now look to the **Bernoulli equation,** which takes account of changes in the gravitational potential and pressure. The law is understood from a work-energy point of view. If liquid is flowing through a tube as shown in Fig. 16-7, then the work done on the fluid in pushing it through the tube must be manifest as increased potential and kinetic energy.

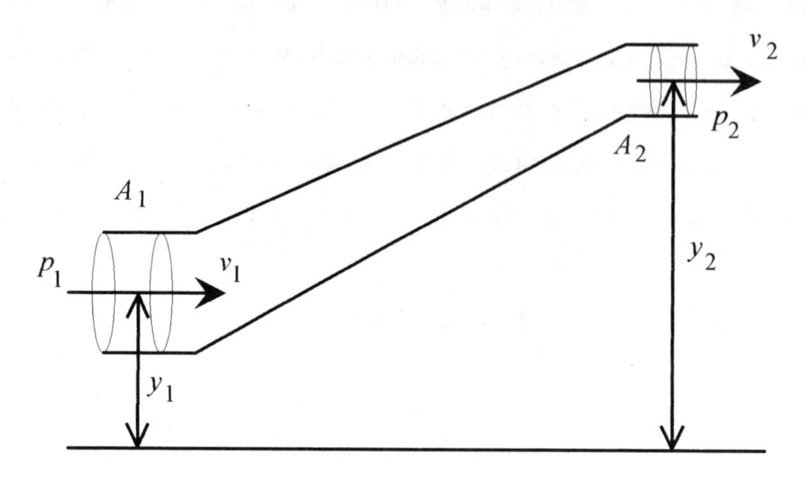

Fig. 16-7

$$dW = dK + dU$$

Consider a mass of fluid moving through this tube. The volume of this mass of fluid is

$$dV = A_1 ds_1 = A_2 ds_2$$

where the *ds*'s are the respective distances along the tube necessary to make up the same mass of fluid. The work performed on this mass of fluid is (note that the work is force, *pA*, times distance, *ds*).

$$dW = p_1 A_1 ds_1 - p_2 A_2 ds_2 = (p_1 - p_2) dV$$

The change in kinetic energy, basically $(1/2)mv^2$, is $dK = (1/2)\rho dV\left(v_2^2 - v_1^2\right)$.

And the change in potential energy (analogous to *mgh*) is $dU = \rho g dV(y_2 - y_1)$.

Equating the work performed to the energy change yields the **Bernoulli equation**

$$(p_1 - p_2)dV = \frac{1}{2}\rho dV\left(v_2^2 - v_1^2\right) + \rho g dV(y_2 - y_1)$$

or

$$p_1 + \rho g y_1 + \frac{1}{2}\rho v_1^2 = p_2 + \rho g y_2 + \frac{1}{2}\rho v_2^2 \qquad \textbf{(16-6)}$$

This is all the information necessary to do fluid problems as found in most physics texts. Buoyancy and pressure problems do not usually present a problem. Calculating pressures is sometimes encountered in the context of a flow problem. In flow problems the usual procedure is to first apply the mass flow statement, equation 16-5. Then apply the Bernoulli equation. In the application of this equation it is often necessary to take some of the terms as zero. The ability to know when to take a term as zero comes from experience.

Venturi Tube

The venturi tube is used as a siphon or speed indicator. The tube is constructed as shown in Fig. 16-8. Apply the Bernoulli equation without the gravitational term (most venturi tubes are mounted horizontally).

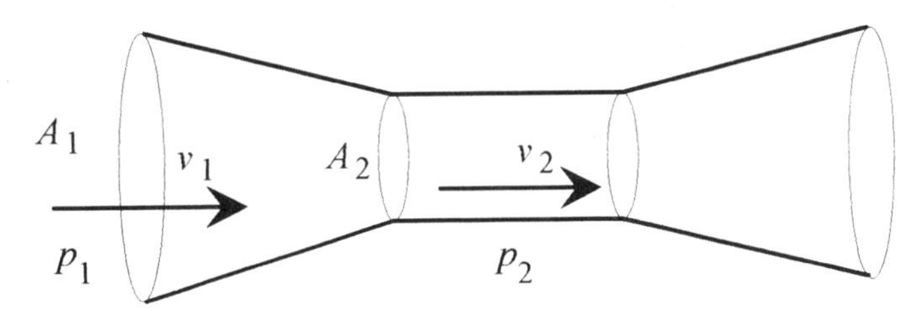

Fig. 16-8

$$p_1 + \frac{1}{2}\rho v_1^2 = p_2 + \frac{1}{2}\rho v_2^2 \quad \text{or} \quad p_1 - p_2 = \frac{1}{2}\rho\left(v_2^2 - v_1^2\right)$$

From the mass flow equation, equation 16-5, $A_1 v_1 = A_2 v_2$ so

$$p_1 - p_2 = \frac{1}{2}\rho\left[\left(\frac{A_1}{A_2}\right)^2 - 1\right]v_1^2 \qquad \textbf{(16-7)}$$

Since A_1 is larger than A_2, $p_1 - p_2$ is positive. The pressure is reduced in the smaller section, or throat, of the venturi. A side tube connected to the narrow region will siphon liquid or gas. The side tube also can be connected to a pressure gauge to measure velocity, which is proportional to the square root of pressure.

16-7 Calculate the speed at which liquid flows out a small hole in the bottom of a large tank containing liquid to a depth of $1.0\,\mathrm{m}$.

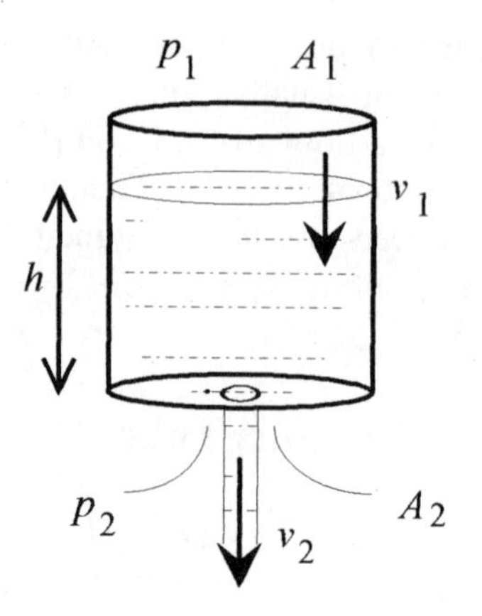

Fig. 16-9

Solution: Apply the Bernoulli equation $p_1 + \rho g y_1 + \frac{1}{2}\rho v_1^2 = p_2 + \rho g y_2 + \frac{1}{2}\rho v_2^2$

Note that $p_1 = p_2$, so remove these terms from the equation. Because $A_1 >> A_2$, v_1 is very small and is taken as zero. Take $y_1 = 0$, so y_2 is the depth of the liquid. The Bernoulli equation for this case reduces to

$$\rho g h = \frac{1}{2}\rho v_2^2 \quad \text{or} \quad v_2 = \sqrt{2gh}$$

The velocity of liquid from a small hole in a container depends only on the height of the column of liquid.

For a liquid height of 1.0 m, $v_2 = \sqrt{2(9.8\,\mathrm{m/s^2})1.0\,\mathrm{m}} = 4.4\,\mathrm{m/s}$.

Go back through this problem and note how the terms of the Bernoulli equation were handled, the pressure being the same, and the logic leading to the v_1 term being taken as zero. This is the hard part of this problem. If, in doing a problem like this, you do not see that v_1 can be taken as zero you will be stuck. You can move no further on the problem because there is not enough data to calculate v_1.

(C) **16-8** For the situation of problem 16-7, find the volume flow rate, that is *dV/dt*, for a hole of radius 3.0 mm.

Solution: The flow rate is the instantaneous rate of liquid flowing out of the container. This is most easily obtained from the statement of the differential volume.

$$dV = A_1 dl = A_2 v_2 dt \quad \text{so that} \quad \frac{dV}{dt} = A_2 v_2 = A_2\sqrt{2gh}$$

For a hole of radius 3.0 mm and liquid depth 1.0 m

$$\frac{dV}{dt} = \pi\left(3.0\times10^{-3}\,\text{m}\right)^2 4.4\frac{\text{m}}{\text{s}} = 125\times10^{-6}\frac{\text{m}^3}{\text{s}} = 1.25\times10^{-4}\frac{\text{m}^3}{\text{s}}$$

16-9 A cylindrical container of radius 20 cm contains water to a height of 3.0 m. Two meters down from the water level there is a hole 0.50 cm in radius. Find the velocity and volume flow rate of water leaving the hole. What is the shape of the stream, and where does it strike the ground?

Solution: Apply the Bernoulli equation $p_1 + \rho g y_1 + \frac{1}{2}\rho v_1^2 = p_2 + \rho g y_2 + \frac{1}{2}\rho v_2^2$

The pressure is the same: $p_1 = p_2$. The velocity v_1 is so small compared to v_2 that it can be neglected: set $v_1 = 0$. The distance from the top of the water to the hole is 2.0 m.

$$gh = \frac{1}{2}v_2^2, \qquad v_2 = \sqrt{2gh} = \sqrt{2(9.8\,\text{m/s}^2)2.0\,\text{m}} = 6.3\,\text{m/s}$$

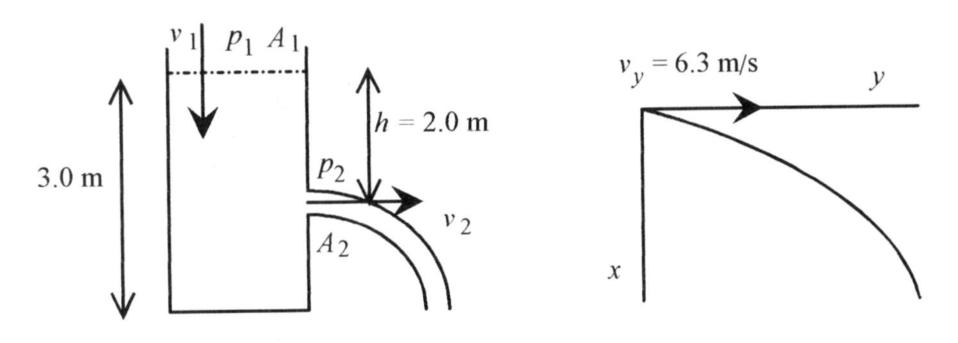

Fig. 16-10

(C) The differential volume of liquid flowing out of the hole is $dV = A_2 v_2 dt$

$$\frac{dV}{dt} = A_2 v_2 = A_2\sqrt{2gh} \quad \text{and} \quad \frac{dV}{dt} = \pi\left(0.5\times10^{-2}\,\text{m}\right)^2 6.3\frac{\text{m}}{\text{s}} = 4.9\times10^{-4}\frac{\text{m}^3}{\text{s}}$$

To analyze the motion look at a differential piece of the stream. Set up a right-handed coordinate system with origin at the exit hole. This coordinate system is similar to ones used in some projectile problems. The differential piece of liquid has initial velocity in the $+y$ direction of $6.3\,\text{m/s}$ and is acted upon by gravity in the $+x$ direction.

The six equations of motion for acceleration, velocity, and position in the x and y directions are

$$a_x = 9.8\,\text{m/s}^2 \qquad a_y = 0$$

$$v_x = (9.8\,m/s^2)\,t \qquad v_y = 6.3\,\text{m/s}$$

$$x = (4.9\,\text{m/s}^2)\,t^2 \qquad y = (6.3\,\text{m/s})\,t$$

These are the equations of a parabola or half of a parabola in this case.

In this problem $x = 1.0\,\text{m}$ so set $1.0\,\text{m} = (4.9\,\text{m/s}^2)\,t^2$ for $t = 0.45\,\text{s}$, the time for the differential piece of the stream to reach the ground.

Then y at $t = 0.45\,\text{s}$ is $y|_{t=0.45} = 2.8\,\text{m}$. The stream strikes the ground at $2.8\,\text{m}$ from the tank.

The approximation of v_1 as zero is justified with the continuity equation, $A_1 v_1 = A_2 v_2$.

$$\pi(20\,\text{cm})^2 v_1 = \pi(0.50\,\text{cm})^2 v_2 \quad \text{or} \quad v_1 = \left(\frac{0.50}{20}\right)^2 v_2 = \left(\frac{1}{40}\right)^2 v_2 = \frac{1}{1600} v_2$$

The v_2 is 1600 times v_1!

As a side exercise solve the equation $y = 6.3t$ for t and substitute into the equation for x, obtaining an equation in the form $x = a y^2$, the equation of a parabola. Then substitute $x = 1.0\,\text{m}$ to obtain y.

CHAPTER 17

TEMPERATURE AND CALORIMETRY

Temperature and variations in temperature are measured by a variety of physical properties: volume of a liquid, length of a rod, resistance of a metal, pressure of a gas at constant volume, or volume of a gas at constant pressure. Thermometric devices (thermometers) are not all linear, though some are linear over a fairly wide range, and it is necessary in selecting a thermometer to fit the linearity of the thermometer to the range of temperatures to be measured.

The standard in temperature measurement is the constant volume gas thermometer. An ideal gas at low density, usually helium, confined to a constant volume has a linear relation between temperature and pressure as shown in Fig. 17-1. The dotted portion of the line is an extension of the linear relationship. Gas thermometers stop being linear at low temperatures because the gas stops being a gas and becomes liquid or solid. Extrapolating the straight line relationship gives an intercept of $-273.15^{o}\mathrm{C}$ corresponding to zero pressure. Figure 17-1 shows graphs of pressure versus temperature for several different gases. Other gases also show this linear relationship between pressure and temperature; and while the slopes of the lines are different, they all extrapolate to the same temperature.

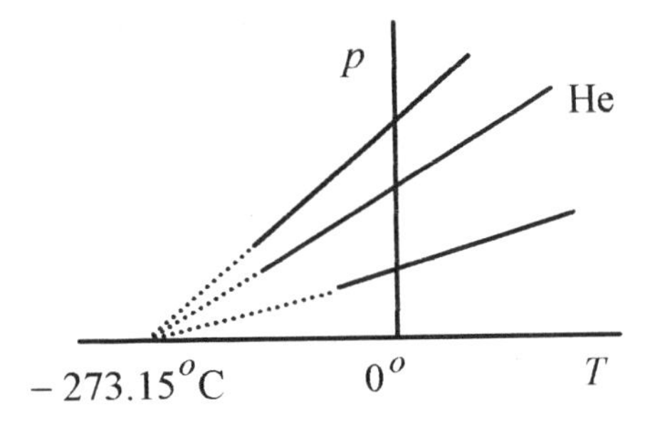

Fig. 17-1

The important point is that while the slopes are different, the relation between pressure and temperature is a straight line over a wide range of temperatures, and the lines all extrapolate to one specific temperature, which is taken as the absolute zero of temperature. This temperature is $0^{o}\mathrm{K}$ (Kelvin) or $-273.15^{o}\mathrm{C}$. (The international standard is not to use the degree symbol with the Kelvin scale. Popular practice is either way.) The reason for this rather odd number (−273.15) is that when the zero on the Celsius scale is calibrated with the freezing point of water at standard atmospheric pressure, the absolute zero is 273.15 degree units below (this zero). The most accurate calibration point for a thermometer is the triple point of water, where the conditions of pressure and temperature are such that water, ice, and vapor all coexist at the same time.

The Kelvin and Celsius scales have the same size degree. The Fahrenheit scale uses the same end points for calibration (freezing and vapor point of water) but the scale starts at 32 and ends at 212, giving a much smaller degree. Temperature is written as 20^oC or 50^oC while temperature difference is written as 30C^o.

17-1 Consider a linear temperature device such as the resistance of a piece of metal. If the resistance of the device is 800Ω at 20^oC and 900Ω at 60^oC, what is the temperature when the device reads 835Ω?

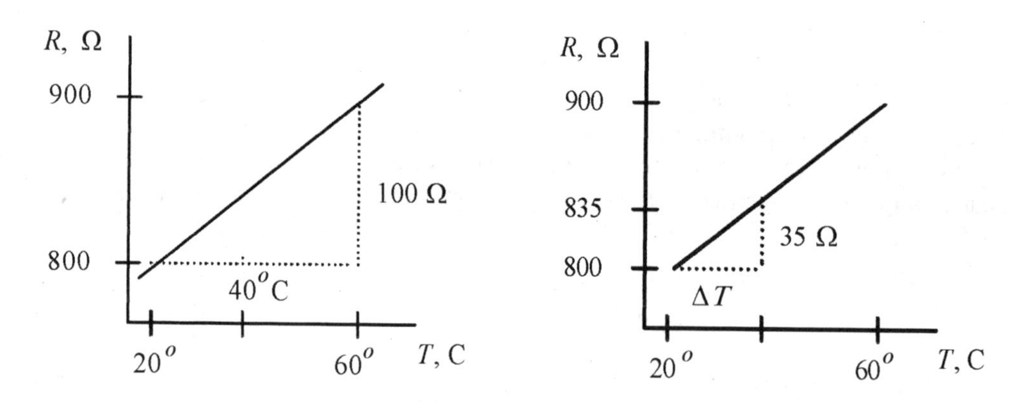

Fig. 17-2

Solution: Figure 17-2 shows the linear relationship of the device. Since the slope is a constant, the slope of the graph on the right is equal to the slope of the graph on the left (similar triangles).

So $\dfrac{100\Omega}{40^o\text{C}} = \dfrac{35\Omega}{\Delta T}$ or $\Delta T = 14^o\text{C}$ and the measured temperature is 34^oC.

C Hot objects placed in contact with cold objects cool and warm (respectively) according to a constant that depends on the conditions of contact and the difference in temperature. This heating or cooling is most conveniently expressed as a differential equation

$$\frac{d\Delta T}{dt} = -k\Delta T \tag{17-1}$$

In words this differential equation states that the change over time of the difference in temperature is proportional to the difference in temperature. The rate at which something cools is directly proportional to the difference in temperature with its surroundings. The negative sign is a reminder that the change in ΔT is such as to reduce ΔT. The equation is solved by separating

$$\frac{d\Delta T}{\Delta T} = -kdt \text{ and integrating } \ln \Delta T = -kt + \text{const}$$

The constant can be evaluated by taking an initial temperature difference ΔT_O. Substituting into this equation for $t = 0$ and ΔT_o yields $\ln \Delta T_o = \text{const}$ and the equation reads $\ln \Delta T = \ln \Delta T_o - kt$ or

$\ln \frac{\Delta T}{\Delta T_o} = -kt$ and switching to exponential format

$$\Delta T = \Delta T_o e^{-kt} \tag{17-2}$$

This equation states that the difference in temperature between a hot body and a cold body in contact is proportional to the initial difference and a decaying exponential. This is very much what we expect from experience. Note that this equation is in the same form as other decay equations that describe, for example, the rate of radioactive decay as proportional to the amount of decaying product remaining.

17-2 Two objects have an initial temperature difference of 20C^o. In two minutes the temperature difference is 18C^o. Calculate the time for the temperature difference to be 10C^o.

Solution: This is a classic type of problem where the physical law is known, data is given to calculate the specific relationship, and this relationship is used to predict something. Keep this process in mind as you go through this problem. It is used in the solution of many problems in physics.

The basic law is $\Delta T = \Delta T_o e^{-kt}$

Now, use the given data to determine *k*. Units are left off in the exponent to avoid clutter, but keep in mind that the 2.0 is in minutes, so *k* will be in reciprocal minutes.

$$18\text{C}^o = (20\text{C}^o)e^{-2.0k} \quad \text{so} \quad \ln\frac{18}{20} = -2.0k \quad \text{or} \quad k = -\frac{1}{2}\ln 0.90 = 0.053\left(\frac{1}{\text{min}}\right)$$

This gives *k* for this specific situation, and the law for this specific situation is

$$\Delta T = \Delta T_o e^{-0.053t}$$

Now ask, what time is required for the temperature difference to decline to 10^oC, $\Delta T = 10^o\text{C}$ for an initial difference of 20^oC.

$10 = 20e^{-0.053t}$ and switching to logarithms $\ln\frac{1}{2} = -0.053t$ or $t = -\frac{1}{0.053}\ln\frac{1}{2} = 13\,\text{min}$

Go back over this problem and be sure you understand the procedure and how to manipulate the exponents and logarithms. Practice performing these operations on your calculator so you will be able to do them rapidly and accurately on a test.

Solids expand and contract with their change in length proportional to their length and change in temperature

$$\Delta\ell = \alpha\ell\Delta T \quad \text{or} \quad \frac{\Delta\ell}{\ell} = \alpha\Delta T \qquad \textbf{(17-3)}$$

The α's of various materials are tabulated in most physics texts.

17-3 A steel bridge is set in concrete pillars on opposite sides of a river. The bridge is $300\,\text{m}$ long, and the work is performed on a day when the temperature is 18°C. What clearance is required for the bridge not to buckle at 45°C? ($\alpha_{\text{steel}} = 11\times10^{-6}/\text{C}^{o}$)

Solution: The change in length is

$$\Delta\ell = \alpha\ell\Delta T = (11\times10^{-6}\,1/\text{C}^{o})300\,\text{m}(27\,\text{C}^{o}) = 8.9\times10^{-2}\,\text{m}$$

A minimum of $8.9\,\text{cm}$ would be required for expansion.

For flat objects the change in length is isotropic (the same in all directions) so that the diagonal of a plate, the diameter or circumference of a hole punched in the plate, or any length changes according to $\Delta\ell = \alpha\ell\Delta T$; so that for areas

$$\Delta A = 2\alpha A\Delta T \quad \text{and for volumes} \quad \Delta V = 3\alpha V\Delta T \qquad \textbf{(17-4)}$$

17-4 Calculate the change in volume of a brass sphere of $8.0\,\text{cm}$ in radius on going from 0°C to 100°C. ($\alpha_{\text{brass}} = 19\times10^{-6}/\text{C}^{o}$)

Solution: The change in volume is

$$\Delta V = 3\alpha V\Delta T = 3(19\times10^{-6}\,1/\text{C}^{o})(4/3)\,\pi(8.0\,\text{cm})^3\,100\,\text{C}^{o} = 12.2\,\text{cm}^3$$

The fractional change in volume $\Delta V / V$ is

$$\frac{\Delta V}{V} = \frac{3\alpha V \Delta T}{V} = 3\alpha\Delta T = 3(19\times10^{-6}\,1/\mathrm{C}^o)100\mathrm{C}^o = 0.0057$$

The percentage change in volume is 0.57%.

(C) Consider the effect of temperature on the period of a pendulum clock. The period of a pendulum is

$$t = 2\pi\sqrt{\frac{\ell}{g}} = 2\pi\sqrt{\frac{1}{g}}\,\ell^{1/2}$$

Taking the differential with respect to ℓ, the variable that depends on temperature

$$dt = 2\pi\sqrt{\frac{1}{g}}\frac{1}{2}\ell^{-1/2}d\ell \qquad \text{or in } \Delta \text{ form} \qquad \Delta t = \frac{\pi}{\sqrt{g\ell}}\Delta\ell$$

Now the $\Delta\ell$ can be written in terms of change in temperature as $\Delta\ell = \alpha\ell\Delta T$, so

$$\Delta t = \pi\alpha\sqrt{\frac{\ell}{g}}\Delta T$$

A more interesting formula is one for the fractional change in period $\Delta t / t$

$$\frac{\Delta t}{t} = \alpha\frac{\Delta T}{2}$$

17-5 Calculate the fractional change in period for a brass pendulum going through a $5\mathrm{C}^o$ change from summer to winter (not unusual).

Solution: $\dfrac{\Delta t}{t} = \alpha\dfrac{\Delta T}{2} = 19\times10^{-6}\dfrac{1}{\mathrm{C}^o}\dfrac{5\mathrm{C}^o}{2} = 47.5\times10^{-6} = 0.0000475$ or 0.00475%.

As the temperature increases, the length of the pendulum gets longer, and the time for one oscillation gets longer. Pendulum clocks run slower in the summer.

Calorimetry

Heat is a form of energy. The unit of heat is the **calorie**, the heat to raise $1.0\mathrm{g}$ of water from $14.5^o\mathrm{C}$ to $15.5^o\mathrm{C}$. The kilocalorie or kcal is also popular. Food calories are actually kilocalories.

The British heat unit is the **British Thermal Unit (BTU)**, the heat to raise 1.0 pound of water from $63^o\,\mathrm{F}$ to $64^o\,\mathrm{F}$. The joule equivalence is $1\,\mathrm{cal} = 4.2\,\mathrm{J}$.

The quantity of heat necessary to raise the temperature of a mass of material is

$$Q = mc\Delta T \qquad \textbf{(17-5)}$$

where c, the **specific heat**, is the amount of heat per $\mathrm{g \cdot C^o}$ for the material. The specific heat for water is $1.0\,\mathrm{cal/g \cdot C^o}$. Tables of specific heats are in most physics texts.

17-6 How much heat is required to raise a $1.8\,\mathrm{kg}$ copper tea kettle containing $2.0\,\mathrm{kg}$ of water from $20^o\,\mathrm{C}$ to $100^o\,\mathrm{C}$? The specific heat of copper is $0.092\,\mathrm{cal/g \cdot C^o}$.

Solution: The total heat is $Q = \left[m_{\mathrm{cu}}c_{\mathrm{cu}} + m_{\mathrm{wat}}c_{\mathrm{wat}}\right]\Delta T$

$$Q = \left[1800\,\mathrm{g}(0.092\,\mathrm{cal/g \cdot C^o}) + 2000\,\mathrm{g}(1.0\,\mathrm{cal/g \cdot C^o})\right]80^o\,\mathrm{C} = 1.73 \times 10^5\,\mathrm{cal}$$

The specific heat is sometimes expressed in $\mathrm{J/kg \cdot K}$ rather than $\mathrm{cal/g \cdot C^o}$. For the previous problem the calculations would read

$$Q = \left[1.8\,\mathrm{kg}(390\,\mathrm{J/kg \cdot K}) + 2.0\,\mathrm{kg}(4200\,\mathrm{J/kg \cdot K})\right]80\,\mathrm{K} = 728000\,\mathrm{J}$$

Use the calorie to joule equivalence to verify the numbers in this problem.

The quantity of material is sometimes given in moles, n. The total mass is related to the number of moles via $m = nM$ where M is the mass per mole of the material. Measuring the mass in number of moles, heat capacity would be calculated with the molar heat capacity, the heat in joules per $\mathrm{mole \cdot C^o}$ or $\mathrm{mole \cdot K}$. (Remember, the Celsius and Kelvin degrees are the same size.)

The equilibrium temperature of a mixture can be determined by knowing the quantity of heat necessary to affect a change in temperature

17-7 A $150\,\mathrm{g}$ cup of coffee (water) at $80^o\,\mathrm{C}$ has added to it $20\,\mathrm{g}$ of sugar (carbon) at $25^o\,\mathrm{C}$. What is the final temperature of the insulated mixture? ($c_{\mathrm{c}} = 0.12\,\mathrm{cal/g \cdot C^o}$)

Solution: In this problem heat from the water raises the temperature of the carbon until they are both at the same temperature. Saying it simply, this is a **Goesinto** problem. The heat in the water **Goesinto** heating the carbon until they both reach an equilibrium temperature, T.

The heat leaving the water is $150\text{g}(1.0\,\text{cal}/\text{g}\cdot\text{C}^o)(80-T)^o\text{C}$.

The final temperature is between 80 and 25, so writing the temperature difference as $80-T$ produces a positive number (of calories).

This heat goes to raise the temperature of the carbon $20\text{g}(0.12\,\text{cal}/\text{g}\cdot\text{C}^o)(T-25)^o\text{C}$.

Notice that the temperature difference in parentheses is written so as to produce a positive number.

The equation will read: The heat from the water **Goesinto** heating the carbon.

$$150\text{g}(1.0\,\text{cal}/\text{g}\cdot\text{C}^o)(80-T)^o\text{C} = 20\text{g}(0.12\,\text{cal}/\text{g}\cdot\text{C}^o)(T-25)^o\text{C}$$

or

$$12000 - 150T = 2.4T - 60 \qquad \text{or} \qquad T = 79.1^o\text{C}$$

These problems can be complicated with three or more components at different temperatures. If you guess wrong as to where the equilibrium temperature lies, the calculations will produce a temperature higher or lower than the original temperatures. If this occurs you know that the initial assumption was wrong, so go back with another (better) assumption of the equilibrium temperature and do the problem over again. The main problem with calorimetry problems is keeping the algebraic signs correct. These problems can be done with a positive heat minus negative heat approach, but the **Goesinto** approach cuts down on the number of negative signs and reduces the possible number of places where you can get a sign wrong.

Phase Change

If water is heated from room temperature to the vapor point it takes 1 calorie of heat (energy) to raise each gram 1K degree. When the vapor point is reached a discrete amount of energy is required to convert each gram (or kilogram) of water at 100^oC to vapor without raising its temperature. This amount of energy is called the **heat of vaporization** and for water is $2.3\times10^6\,\text{J}/\text{kg}$.

A similar phenomenon is observed when a solid is taken to the liquid phase. The **heat of fusion**, the heat to change $1.0\,\text{kg}$ of solid (water) to liquid at 0^oC is $3.3\times10^5\,\text{J}/\text{kg}$. Tables of heats of fusion and vaporization are in most physics texts. Problems involving heat of fusion or heat of vaporization must be done carefully, because the energy requirements for these changes of state are very large. Converting a small amount of ice to water consumes a large amount of energy compared to raising the temperature of the water. It takes $4.2\times10^3\,\text{J}$ per kg for each degree of temperature while it takes $3.3\times10^5\,\text{J}$ per kg to convert ice to water with no temperature rise!

17-8 An insulated container holds $3.0\,\text{kg}$ of water at 25^oC. One kilogram of ice at 0^oC is added to the mixture. What is the final temperature and composition (ice and water) of the mixture?

Solution: First calculate the energy to bring all the water to 0^oC.

$$Q = 3.0\,\text{kg}(4200\,\text{J/kg}\cdot\text{K})25\,\text{K} = 315000\,\text{J}$$

Now ask how much ice could be melted with this amount of energy.

$$315000\,\text{J} = (3.3\times 10^5\,\text{J/kg})m \qquad \text{or} \qquad m = 0.95\,\text{kg}$$

All the ice will not be melted so the final mixture will be $3.95\,\text{kg}$ of water and $0.05\,\text{kg}$ of ice all at 0^oC. There will be no further heat transfer because both the water and the ice are at the same temperature.

17-9 If in the previous problem the amount of ice is reduced to $0.50\,\text{kg}$ what is the final temperature of the composition?

Solution: Based on the previous calculation there is not enough ice to lower the temperature of the water to 0^oC. One way of doing the problem is to calculate the drop in temperature of the $3.0\,\text{kg}$ of water at 25^oC due to the melting ice, then treat the problem as a mixture of two amounts of water at different temperatures.

However, since we already know that all the ice melts and the final mixture is between 0^oC and 25^oC, the equation for the final temperature can be set up with the statement: The energy in the 3.0kg of water at 25^oC **Goesinto** melting the ice and raising the temperature of the 0^oC water.

$$3.0\,\text{kg}(4200\,\text{J/kg}\cdot\text{C}^o)(25-T)^o\text{C} = 0.50\,\text{kg}(3.3\times 10^5\,\text{J/kg}) + 0.50\,\text{kg}(4200\,\text{J/kg}\cdot\text{C}^o)(T-0)^o\text{C}$$

$$315000 - 12600T = 165000 + 2100T \qquad \text{or} \qquad T = 10^o\text{C}$$

Heat Flow

Conduction of heat or heat flow depends on the temperature difference, cross section, separation, and a constant of the material according to

$$H = \frac{\Delta Q}{\Delta t} = \frac{kA\,\Delta T}{L} \qquad \textbf{(17-6)}$$

The units of H are energy per time or Joule/sec which is a watt. The thermal conductivity, k, has units of $\mathrm{W/m \cdot K}$.

17-10 Calculate the heat flow through a wooden door of area $2.0\,\mathrm{m^2}$ and thickness $5.0\,\mathrm{cm}$ when the temperature difference is $20\,\mathrm{K}$. The k for wood is $0.10\,\mathrm{W/m \cdot K}$.

Solution: The heat flow is $H = \dfrac{kA\,\Delta T}{L} = \dfrac{0.10\,\mathrm{W/m \cdot K}(2.0\,\mathrm{m^2})20\,\mathrm{K}}{5.0 \times 10^{-2}\,\mathrm{m}} = 80\,\mathrm{W}$

17-11 Two rods with cross sections of $4.0\,\mathrm{cm^2}$, one copper of $1.2\,\mathrm{m}$ length, and the other steel of $1.0\,\mathrm{m}$ length, are connected together with their opposite ends held at $100^o\mathrm{C}$ and $0^o\mathrm{C}$ as shown. What is the heat flow and the temperature of the junction at equilibrium? ($k_{cu} = 380\,\mathrm{W/m \cdot C^o}$ and $k_{st} = 50\,\mathrm{W/m \cdot C^o}$)

Solution: The key to this problem is that the heat flow in the two rods must be the same. If the heat flow in the copper is greater than the heat flow in the steel then the joint will get hot. This does not happen so the heat flow must be the same.

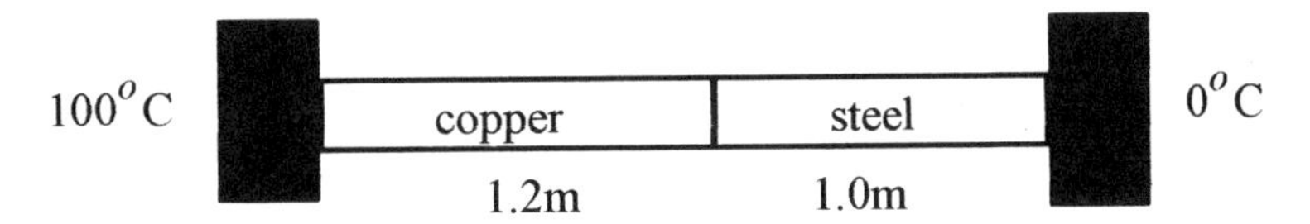

Fig. 17-3

Take the temperature of the junction as T_j, and set the heat flows equal

$$\frac{k_{cu}A(100 - T_j)^o\mathrm{C}}{1.2\,\mathrm{m}} = \frac{k_{st}A(T_j - 0)^o\mathrm{C}}{1.0\,\mathrm{m}}$$

Putting in the numbers

$$1.0\,\mathrm{m}(380\,\mathrm{W/m \cdot C^o})(100 - T_j)^o\mathrm{C} = 1.2\,\mathrm{m}(50\,\mathrm{W/m \cdot C^o})(T_j - 0)^o\mathrm{C}$$

$$38000 - 380T_j = 60T_j \quad \text{or} \quad T_j = 86^o C$$

The heat flow is from either formula

$$H_{cu} = \frac{380\,\mathrm{W/m \cdot C^o}\,4.0 \times 10^{-4}\,\mathrm{m^2}}{1.2\,\mathrm{m}}(100 - 86)^o\mathrm{C} = 1.8\,\mathrm{W}$$

Calorimetry combined with other energy analysis techniques can be used to solve other types of problems.

17-12 A block of ice initially at $50\,\text{kg}$ and traveling at $5.4\,\text{m/s}$ slides along a horizontal surface until it comes to rest. Assume all the heat generated due to friction is used to melt the ice. How much ice is melted?

Solution: All the kinetic energy in the block **Goesinto** work expended to overcome friction, and all this energy **Goesinto** melting the ice. The initial kinetic energy is

$$KE = (1/2)mv^2 = (1/2)50\,\text{kg}(5.4\,\text{m/s})^2 = 730\text{ J}$$

All this energy is used to melt the ice.

$$730\text{ J} = (3.3\times 10^5\text{ J/kg})m \qquad \text{or} \qquad m = 0.0022\,\text{kg} = 2.2\,\text{g}$$

17-13 A $5.0\,\text{kg}$ bullet traveling at $800\,\text{m/s}$ passes through a $5.0\,\text{kg}$ copper block at rest on a frictionless table. Twenty percent of the kinetic energy of the bullet goes into heating the block. Thirty percent goes into kinetic energy of the block. Fifty percent remains with the bullet. The block and bullet are initially at 25^oC. What is the temperature and speed of the block and the speed of the bullet? ($c_{cu} = 390\text{ J / kg}\cdot\text{C}^o$)

Solution: First find the initial kinetic energy of the bullet.

$$KE = (1/2)mv^2 = (1/2)5.0\times 10^{-3}\,\text{kg}(800\,\text{m/s})^2 = 1600\text{J}$$

Twenty percent of this, or 320J, goes to heating the copper block.

$$320\text{J} = 5.0\text{kg}(390\text{J/kg}\cdot\text{C}^o)\Delta T \qquad \text{or} \qquad \Delta T = 0.16^o\text{C}$$

Thirty percent, or 480J, goes to kinetic energy of the block.

$$480\text{J} = (1/2)(5.0\text{kg})v^2 \qquad \text{or} \qquad v = 14\,\text{m/s}$$

Fifty percent, or 800J, stays with the bullet.

$$800\text{J} = (1/2)(5.0\times 10^{-3}\,\text{kg})v^2 \qquad \text{or} \qquad v = 570\text{m/s}$$

CHAPTER 18

KINETICS AND THE GAS LAWS

A complete description of a gas is given by the pressure, volume, temperature, and amount (total mass or number of moles) of the gas. What is called the **equation of state** (equation 18-1) describes the relation between p, V, T and m or n. These four properties completely describe a gas just as position, velocity, and acceleration completely describe the motion of a particle.

$$pV = nRT \tag{18-1}$$

Early experiments on gases produced **Boyle's law,** $pV = \text{constant}$ (at constant temperature), **Charles' law,** $p = \text{constant} \cdot T$ (at constant volume), and finally, the ideal gas **equation of state** $pV = nRT$ with n the number of moles and R the gas constant equal to $2.0\,\text{cal/mol}\cdot\text{K}$ or $8.3\,\text{J/mol}\cdot\text{K}$. A mole of a gas contains a specific number of molecules. The number of molecules per mole is **Avogadro's number,** $N_A = 6.0 \times 10^{23}\,\text{molecule/mole}$. The mass of a mole of gas depends on the atomic weight of the gas. The mole and Avogadro's number are defined as the number of carbon 12 atoms in $12\,\text{g}$ of carbon 12. For helium (atomic weight 4.0) one mole of the gas contains Avogadro's number of helium atoms and has mass of $4.0\,\text{g}$.

The ideal gas law not only describes the p, V, T relationship but allows that for the same amount of gas

$$\frac{p_1 V_1}{T_1} = \frac{p_2 V_2}{T_2} = nR \tag{18-2}$$

allowing a large number of problems to be solved. In applying this formula, Kelvin temperatures must be used throughout, and the pressure and volume must be written in the same units.

18-1 What is the volume of a mole of ideal gas at standard temperature and pressure (STP)?

Solution: Standard temperature is $0^o\text{C} = 273\,\text{K}$. Standard pressure is $1\,\text{atm} = 1.0 \times 10^5\,\text{Pa}$. From the ideal gas law

$$V = \frac{nRT}{p} = \frac{1\text{mol}(8.3\,\text{J}/\text{mol}\cdot\text{K})273\text{K}}{1.0\times10^5\,\text{Pa}} = 0.023\text{m}^3$$

This is a 23 liter container or a box 0.28m on a side. Remember that a Pascal is a N/m^2.

18-2 An air bubble at the bottom of a lake has a volume of $20\,\text{cm}^3$, pressure of $4.9\,\text{Pa}$, and temperature 4^oC. The bubble rises to the surface where the temperature is 20^oC and the pressure $1.0\,\text{Pa}$. Find the volume as the bubble reaches the surface. Also find the number of moles of the gas.

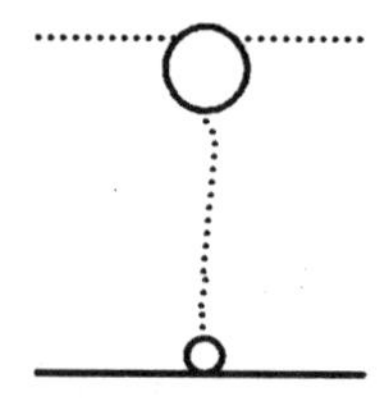

Fig. 18-1

Solution: Use $\frac{p_1V_1}{T_1} = \frac{p_2V_2}{T_2}$ remembering that T is written in kelvin.

$$\frac{(4.9\times10^5\,\text{N}/\text{m}^2)20\text{cm}^3}{277\text{K}} = \frac{(1.0\times10^5\,\text{N}/\text{m}^2)}{293\text{K}}V_2$$

or

$$V_2 = \frac{293\text{K}}{277\text{K}}\frac{4.9\times10^5\,\text{N}/\text{m}^2}{1.0\times10^5\,\text{N}/\text{m}^2}20\text{cm}^3 = 104\,\text{cm}^3$$

To find the number of moles use $pV/T = nR$ and the given conditions at the bottom of the lake.

$$n = \frac{pV}{RT} = \frac{(4.9\times10^5\,\text{N}/\text{m}^2)20\text{cm}^3}{(8.3\,\text{J}/\text{mol}\cdot\text{K})(277\text{K})}\left(\frac{\text{m}}{10^2\,\text{cm}}\right)^3 = 0.0043\text{mol}$$

18-3 A weather balloon is inflated with helium at just over atmospheric pressure of $1.2\times10^5\,\text{Pa}$, temperature 20^oC, and volume 2.5m^3. When the balloon rises to where the pressure is $0.50\times10^5\,\text{Pa}$ and temperature is -58^oC, what is the volume?

Solution: Using $\frac{p_1V_1}{T_1} = \frac{p_2V_2}{T_2}$, $\frac{1.2\times10^5\,\mathrm{N/m^2}\,2.5\mathrm{m}^3}{293\mathrm{K}} = \frac{0.5\times10^5\,\mathrm{N/m^2}\,V_2}{215\mathrm{K}}$

or $$V_2 = \frac{215\mathrm{K}}{293\mathrm{K}}\frac{1.2\times10^5\,\mathrm{N/m^2}}{0.5\times10^5\,\mathrm{N/m^2}}2.5\mathrm{m}^3 = 4.4\,\mathrm{m}^3$$

As an exercise find the number of moles of this gas.

Variation of Pressure with Elevation

(C) Consider a slab of air in a column of air and note that the increase in force between the top and bottom of the slab is due to the weight of the slab. Envision this as holding your hand horizontally at two different heights. Higher up in the column of air there is less air above your hand and thus a lower pressure.

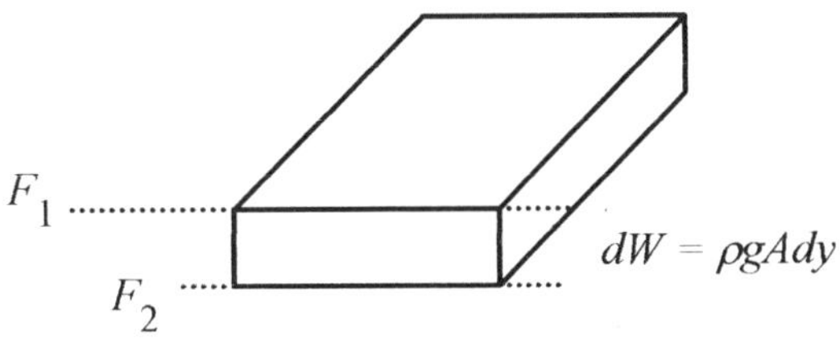

Fig. 18-2

Writing this difference with differentials, $F_2 = F_1 + dW$

Since $p = F/A$ and $dW = \rho gAdy = \rho gdV$, this statement can be written as

$$p_2A = p_1A + \rho gAdy \quad \text{or} \quad dp = -\rho gdy \tag{18-3}$$

where dp is the difference, top to bottom, of the slab. The minus sign serves to remind us that as we go up in the atmosphere (column) the pressure goes down (there is less air in the column).

Going back to $pV = nRT$ and the relation mass equals moles times molecular mass ($m = nM$) (remember that each mole of Helium is $4.0\mathrm{g}$)

$pV = \frac{m}{M}RT$, or using density $\rho = \frac{m}{V} = \frac{pM}{RT}$, so equation 18-3 can be written as

$$\frac{dp}{dy} = -\rho g = -\frac{pMg}{RT} \quad \text{or} \quad \frac{dp}{p} = -\frac{Mg}{RT}dy \tag{18-4}$$

which integrates to $\ln p = -\frac{Mg}{RT}y + \ln A$ with $\ln A$ the constant of integration.

At $y = 0$ (on the ground), $p = p_o$, $\ln p_o = \ln A$, and

$$\ln\frac{p}{p_o} = -\frac{Mg}{RT}y \qquad \text{or switching to exponents} \qquad p = p_o e^{-\frac{Mg}{RT}y} \tag{18-5}$$

Note that in this derivation the temperature is assumed constant. The density of the air is not assumed constant because V in the density relation $\rho = (m/V)$ was replaced with nRT/p.

18-4 What is the pressure at 3000 m elevation above the earth?

Solution: Assume $T = 0^oC = 273K$. T decreases with height because the air is heated, primarily by radiation from the earth, so consider this an average temperature over the 3000 m. Air is mostly oxygen and nitrogen and has an average molecular weight of 28.8 g / mole or 28.8×10^{-3} kg / mole. The gravitational constant varies only slightly over this distance.

First calculate the exponent $\frac{Mg}{RT}y = \frac{28.8 \times 10^{-3}\,\text{kg/mol}(9.8\,\text{m/s}^2)}{8.3\,\text{J/mol}\cdot\text{K}(273\text{K})}3000\,\text{m} = 0.374$

then the pressure $p = p_o e^{-0.374} = 1.0\,\text{atm}(0.69) = 0.69\,\text{atm}$

pV Diagrams

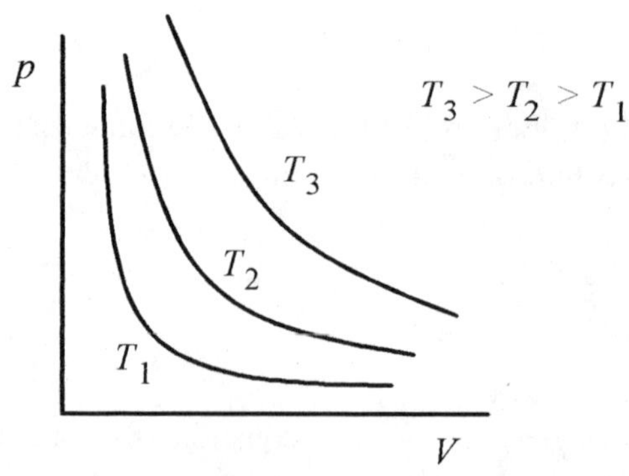

Fig. 18-3

The $pV/T = \text{constant}$ relationship for an ideal gas is depicted in Fig. 18-3 as a collection of $pV = \text{constant}$ curves for different temperatures. Each line of this family of hyperbolas is called an **isotherm**, a curve of constant temperature. A gas at constant temperature has its pV relation defined by these curves. For example, if a gas is expanded at constant temperature, the pressure follows the pV isotherm for that temperature.

Kinetic Theory

Thermodynamics is a very neat and compact subject in that the equation of state $pV = nRT$ relation can be derived with some simple assumptions and the application of basic force, momentum, and energy concepts. The kinetic theory of gases leads directly to the equation of state for a gas.

Start with N identical molecules in a cube of volume V and sides d, and make the following assumptions and restrictions on their behavior.

1. The molecules are small compared to the dimensions of the container and the distance between molecules.

2. Only collisions with the walls are considered; collisions between molecules are ignored.

3. Collisions with the walls are elastic, and the momentum transferred is manifest as pressure on the walls.

4. The molecules and the walls are in thermal equilibrium. There is no energy transfer between the molecules and the walls.

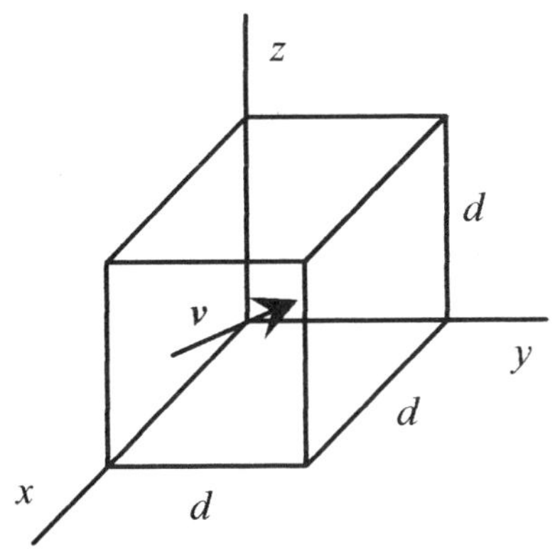

Fig. 18-4

A molecule with velocity $\boldsymbol{v}$ and components v_x, v_y, v_z makes an elastic collision (momentum conserved) with the x-wall. The change in momentum of the molecule is numerically equal to the momentum imparted to the wall $(\Delta p)_{mom} = 2mv_x$.

This is the same expression you would write for a tennis ball bouncing off a wall. The subscript identifies this as momentum and not pressure. The time between collisions with the same wall is the time for the molecule to go to the opposite wall and return, or $2d/v_x = \Delta t$. Therefore the average force exerted on the wall by this molecule is

$$F = \frac{(\Delta p)_{mom}}{\Delta t} = \frac{2mv_x^2}{2d} = \frac{mv_x^2}{d}$$

To find the force over the entire area of the wall, we need to sum all the molecular velocities.

The symbolism for this summation is $\sum v_{xi}^2$. The $\sum$ means sum the velocities, and the index, i, means to take that sum over all the molecules. The average value of the velocity in the x-direction then is

$$(v_x^2)_{avg} = \frac{\sum v_{xi}^2}{N}$$

The total force on the wall due to all the molecules is $F = \dfrac{m}{d} N(v_x^2)_{avg}$.

The velocity is the same in all three directions so $(v_x^2)_{avg}$ is one-third $(v^2)_{avg}$.

Again, the total force on the wall is $F = \dfrac{N}{3}\dfrac{m(v^2)_{avg}}{d}$.

The pressure is the force per unit area, d^2 in this case, so

$$p = \frac{F}{A} = \frac{F}{d^2} = \frac{1}{3}\frac{N}{d^3}m(v^2)_{avg} = \frac{1}{3}\frac{N}{V}\left[m(v^2)_{avg}\right]$$

Adding a factor 2 over 2, the pressure can be written with familiar terms

$$p = \frac{2}{3}\frac{N}{V}\left(\frac{m(v^2)_{avg}}{2}\right) \qquad \textbf{(18-6)}$$

where the expression $m(v^2)_{avg}/2$ is identified as the average translational kinetic energy of the molecule. Writing equation 18-6 another way

$$pV = \frac{2}{3}(KE)_{tr} \qquad \textbf{(18-7)}$$

where $(KE)_{tr}$ is the total translational kinetic energy of all N molecules.

Equation 18-7 can be compared to $pV = nRT$, the equation of state for the gas. However, it is more instructive to rework the equation of state. The expression nR is the product of the number of moles of the gas and the gas constant, in joule per mole-kelvin. The number of moles times the gas constant per mole is the same as number of molecules times a gas constant per molecule. Specifically $nR = Nk$, where k is the gas constant per molecule as opposed to R, which is the gas constant per mole. For one mole N is N_A in $nR = Nk$ and k is easily calculated. The k is the **Boltzmann constant** of value $1.4 \times 10^{-23}\ \text{J/K}$. Writing the equation of state with Nk rather than nR produces

$$pV = NkT \tag{18-8}$$

Comparing equations 18-6 and 18-8

$$kT = \frac{2}{3}\left(\frac{m(v^2)_{avg}}{2}\right) \tag{18-9}$$

This identification shows that the temperature of the gas is directly proportional to the kinetic energy of the molecules. If a container of gas is heated, the pressure increases due to the molecules moving faster!

Rearrange this equation to read $\dfrac{m(v^2)_{avg}}{2} = \dfrac{3}{2}kT$

A gas molecule acting as a mass point confined to a box is viewed as being able to move in three mutually perpendicular directions. In the language of kinetic theory the molecule has three degrees of freedom. Looking at this equation then $(1/2)kT$ of energy is associated with each degree of freedom.

The root-mean-square speed is defined as the square root of $(v^2)_{avg}$, which is

$$v_{rms} = \sqrt{(v^2)_{avg}} = \sqrt{\frac{3kT}{m}} = \sqrt{\frac{3RT}{M}} \tag{18-10}$$

Look back to equation 18-6 and note that for the conditions imposed here, specifically that the gas and the walls are in thermal equilibrium, pV is a constant. This is Boyle's law.

Again looking to equation 18-6, the pressure for constant volume is proportional to $(v^2)_{avg}$ and in equation 18-9 $(v^2)_{avg}$ is proportional to temperature. Pressure being proportional to temperature at constant volume is Charles' law.

18-5 Two moles of helium are in a tank at 25^oC. Find the total translational kinetic energy, the kinetic energy per molecule, and the rms speed of the atoms.

Solution: The total translational kinetic energy is

$$E = \frac{3}{2}nRT = \frac{3}{2}2.0\,\text{mole}\frac{8.3\,\text{J}}{\text{mole}\cdot\text{K}}298\text{K} = 7420\text{J}$$

The energy per molecule is $\frac{m(v^2)_{avg}}{2} = \frac{3}{2}kT = \frac{3}{2}\left[\frac{1.4\times10^{-23}\,\text{J}}{\text{K}}\right]298\,\text{K} = 6.26\times10^{-21}\,\text{J}$

The rms speed is $v_{rms} = \sqrt{\frac{3RT}{M}} = \left[\frac{3(8.3\text{J}/\text{mol}\cdot\text{K})298\text{K}}{2.0\times10^{-3}\,\text{kg}/\text{mol}}\right]^{1/2} = 1930\frac{\text{m}}{\text{s}}$

18-6 Isotopes of uranium are sometimes separated by gaseous diffusion. The lighter isotope, having a higher speed diffuses more rapidly than the heavier isotope. What is the ratio of the rms speeds of U 235 and U 238?

Solution: The rms speed is $v_{rms} = \sqrt{3kT/m}$. The ratio of these speeds is

$$\frac{(v_{rms})_{235}}{(v_{rms})_{238}} = \frac{\sqrt{3kT/235}}{\sqrt{3kT/238}} = \sqrt{\frac{238}{235}} = 1.006$$

Heat Capacity of Gases

In the previous section we took the energy in the gas as the translational motion, with $kT/2$ or $nRT/2$ the amount of energy associated with each degree of freedom. For a container where the gas has three degrees of freedom, then the energy associated with the motion of the molecules is

$$KE_{tr} = (3/2)nRT$$

and the change in energy with temperature is $\Delta KE_{tr} = (3/2)nR\Delta T$

The quantity of heat associated with a rise in temperature for n moles of a gas is

$$\Delta Q = nC_V\Delta T$$

where C_V is the molar heat capacity, the heat to raise one mole of the gas one kelvin. The units of C_V are $\text{J}/\text{mole}\cdot\text{K}$. It is important to differentiate between the heat capacity for constant volume and for constant pressure. More heat is required to raise the temperature of a gas one kelvin at constant pressure than at constant volume because the change in volume requires (additional) work to be performed in the expansion. Associating ΔKE_{tr} with ΔQ, C_V should equal $3R/2$. Measurements on monatomic gases verify this relation.

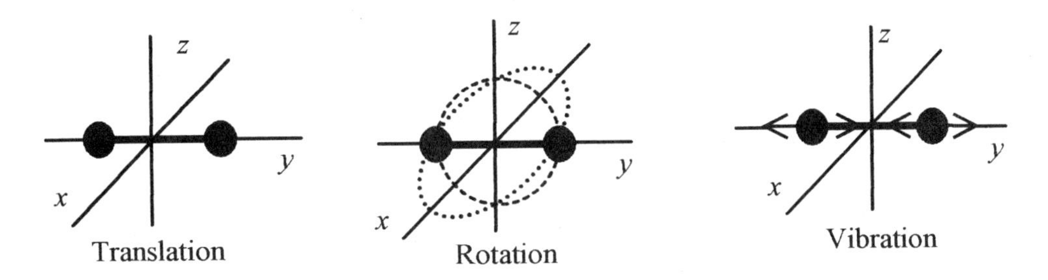

Fig. 18-5

The diatomic gases have additional degrees of freedom. Figure 18-5 illustrates a diatomic molecule in translation, then in rotation, first in the *x-y* plane and then in the *x-z* plane, and finally with the molecule in vibration. The rotation adds two new degrees of freedom and the vibration adds another two degrees of freedom.

Experimental measurements on diatomic molecules show that for low temperatures (three translational degrees of freedom only) $C_V = 3R/2$. As the temperature of the gas is increased the diatomic molecule begins to rotate (adding two more degrees of freedom) and the C_V climbs to $5R/2$. At even higher temperatures the molecule begins to vibrate (adding another two degrees of freedom) and the C_V reaches $7R/2$. This is shown in Fig. 18-6.

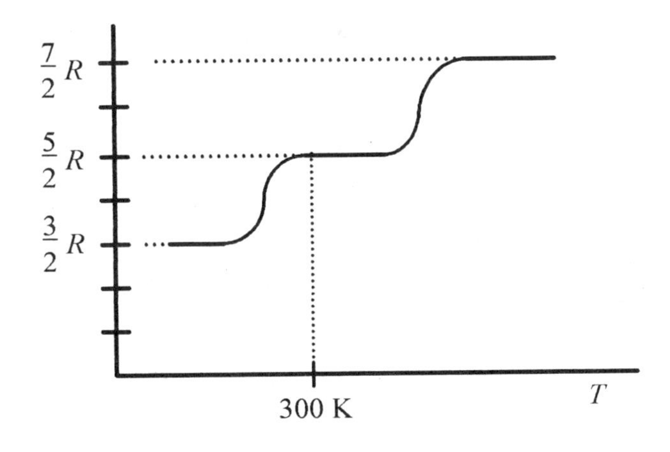

Fig. 18-6

At sufficiently high temperatures diatomic molecules require more heat to raise their temperatures one kelvin than do monatomic molecules because diatomic molecules are not only translators but rotators and possibly vibrators. Most diatomic molecules are rotators at room temperature.

18-7 How much heat is required to raise the temperature of 2.0 mole of monatomic gas 25K starting at room temperature.

Solution: Use $\Delta Q = nC_V\Delta T$ with C_V for a diatomic gas as $3R/2$ so

$$\Delta Q = n\frac{3R}{2}\Delta T = 2.0\,\text{mol}\frac{3(8.3\,\text{J/mol}\cdot\text{K})}{2}25\,\text{K} = 620\,\text{J}$$

18-8 For the previous problem find the heat required if the gas were diatomic.

Solution: Assume that at room temperature the diatomic molecule is a translator and rotator (see Fig. 18-6) so that $C_V = 5R/2$. Then

$$\Delta Q = n\frac{5R}{2}\Delta T = 2.0\,\text{mol}\frac{5(8.3\,\text{J/mol}\cdot\text{K})}{2}25\,\text{K} = 1040\,\text{J}$$

CHAPTER 19

FIRST LAW OF THERMODYNAMICS

The **first law of thermodynamics** is a conservation of energy statement for thermodynamic systems that exchange energy with their surroundings. It states that energy cannot be created or destroyed, but may only be changed from one form to another. The discussion is restricted to ideal gases and systems with constant mass. The quantity of heat, Q, added to the system and the work performed by the system, W, are taken as positive.

Work During Volume Change

Place a gas in a cylinder with a moveable piston. If the gas is allowed to expand at constant temperature, there is a force on the piston $F = pA$, and this force acting over a distance Δx is the work performed by the system.

$$\Delta W = F\Delta x = p(A\,\Delta x) = p\Delta V \qquad \textbf{(19-1)}$$

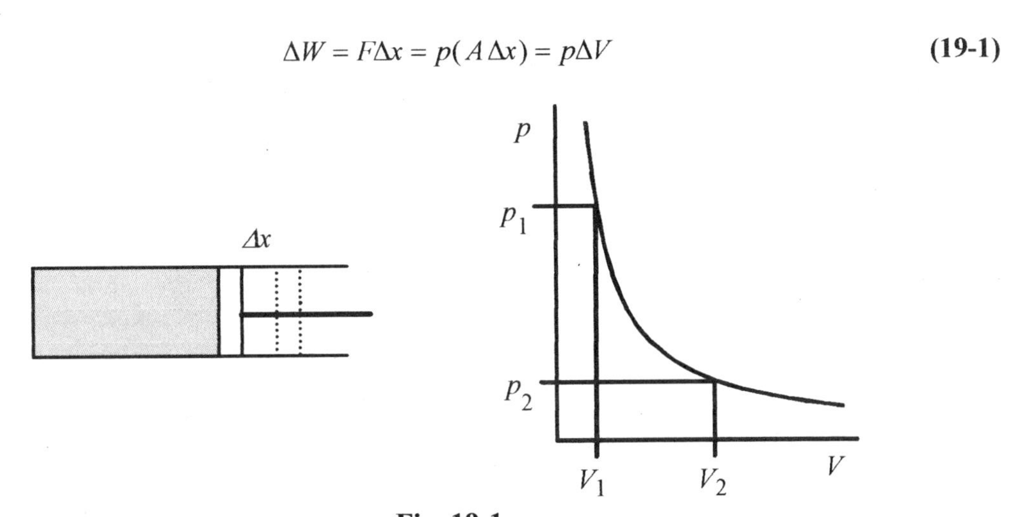

Fig. 19-1

Figure 19-1 shows the cylinder with moveable piston and a p-V diagram. The single (p-V) curve is an isotherm, a constant temperature curve.

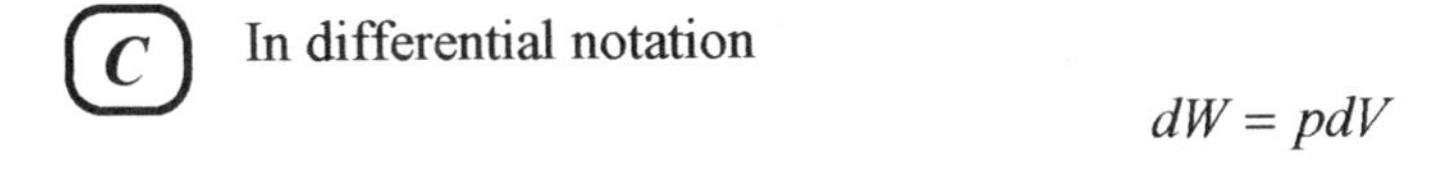

In differential notation

$$dW = pdV$$

The total work in expanding from V_1 to V_2 is the area under the p-V curve from V_1 to V_2. This is analogous to a spring where the work to compress the spring is the area under the curve of F vs x (see Chapter 7, Work and the Definite Integral).

$$W = \int_{V_1}^{V_2} p dV \tag{19-2}$$

The pressure depends on p through $pV = nRT$. At constant temperature the system follows the curve in Fig. 19-1. Writing p in terms of V

$$W = \int_{V_1}^{V_2} nRT \frac{dV}{V} = nRT \int_{V_1}^{V_2} \frac{dV}{V} = nRT \ln|_{V_1}^{V_2} = nRT \ln \frac{V_2}{V_1} \tag{19-3}$$

Since the gas is expanding at constant temperature, it is moving along an isotherm ($T = \text{const}$) line so $p_1 V_1 = p_2 V_2$, and another expression for work is

$$W = nRT \ln[p_1/p_2] \tag{19-4}$$

19-1 Two moles of an ideal gas maintained at 20^oC expand until the pressure is one-half the original. How much work is done by the gas?

Solution: Since the gas remains at constant temperature, it goes from one state to the other along an isotherm (see Fig. 19-1).

$$W = nRT \ln \frac{p_1}{p_2} = 2\,\text{mol} \frac{8.3\,\text{J}}{\text{mol}\cdot\text{K}} 293\,\text{K} \cdot \ln 2 = 3370\,\text{J}$$

19-2 A gas at constant pressure of $4.0 \times 10^5\,\text{Pa}$ is cooled so that its volume decreases from $1.6\,\text{m}^3$ to $1.2\,\text{m}^3$. What work is performed by the gas?

Solution: Since pressure is a constant the work performed is from equation 19-1.

$$\Delta W = p\Delta V = 4.0 \times 10^5\,\text{N}/\text{m}^2 (-0.4\,\text{m}^3) = -1.6 \times 10^5\,\text{J}$$

The negative sign indicates that work was done <u>on</u> the system by an outside agent.

Internal Energy and the First Law

In going from state (p, V, T) to state (another p, V, T) a gas goes through a set of intermediary states. These states are called the **path** the system takes. On the p-V diagram in Fig. 19-2 the point p_1V_1 represents the initial point and p_2V_2 the final point.

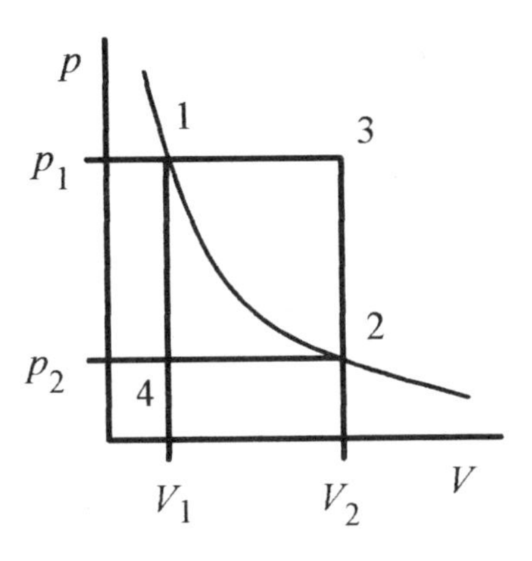

Fig. 19-2

If the gas were taken along the isothermal line, then the work performed would be the area under the p-V curve as shown earlier. Consider, however, two different paths. First, the path from 1-3-2 could be accomplished by expanding the gas at constant pressure (1-3) then reducing the pressure to reach p_2. In this case the work performed would be the area under the 1-3 line.

A second path from 1-4-2 is accomplished by reducing the pressure to p_2 then expanding the gas to V_2. In this case the work performed would be the area under the 4-2 line.

The work performed by the system in going from state 1 to state 2 via the 1-3-2 route is different from the work performed via the 1-4-2 route. Since there are an infinite number of possible routes from state 1 to state 2, the amount of work performed in going from state 1 to state 2 depends on the path.

Now consider an ideal gas in a container with a moveable piston and heat source (Fig. 19-3). The gas at $p_1V_1T_1$ is heated to a state $p_2V_2T_1$. Heat is added from the reservoir to change p and V but not T.

The same amount of gas is placed in another container of volume V_2, but the gas is confined to V_1 with a breakable partition and the space above V_1 is evacuated. If the partition is broken, the gas expands to V_2 without a change in temperature and sufficiently rapidly so there is no heat transferred in the process. This rapid expansion of a gas into a vacuum is called **free expansion**. These two extreme routes are illustrative of the infinite number of possible routes, each requiring a different amount of heat, from $p_1V_1T_1$ to $p_2V_2T_1$. Therefore, the amount of heat necessary to go from one state to another depends on the path.

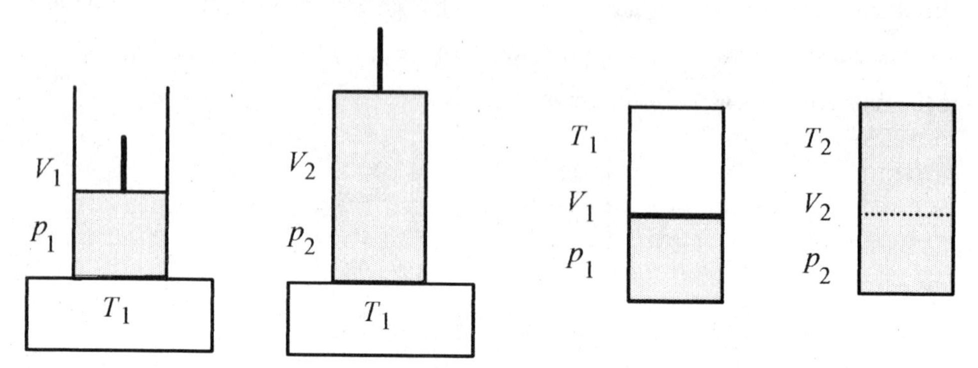

Fig. 19-3

While the amount of heat to go from one state to another depends on the path and the amount of work to go between these two states depends on the path, the difference, heat minus work, is independent of the path and always appears as a change in internal energy of the gas. In equation form this is

$$\Delta U = Q - W \quad \text{or} \quad Q = \Delta U + W \tag{19-5}$$

This last statement is the **first law of thermodynamics** and in words states that heat into a system appears either as an addition to the total internal energy or work performed by the system. Going from one state to another where the temperature of the state is the same, the heat in equals the work performed. Going from one state to another where the temperature of the state is not the same the heat equals the increase in internal energy plus the work performed.

19-3 You eat 100 food calories worth of nuts. How high would you, at $65\,\text{kg}$, have to climb to "work off" these 100 calories?

Solution: The 100 food calories is $100\,\text{kcal}$ into your (thermodynamic) system. The work to reduce your total internal energy back to what it was before eating the nuts is

$$1.0\times 10^5\,\text{cal}(4.2\,\text{J/cal}) = 65\,\text{kg}(9.8\,\text{m/s}^2)h \quad \text{or} \quad h = 660\,\text{m}$$

19-4 In a certain process $1.5\times 10^5\,\text{J}$ of heat is added to an ideal gas to keep the pressure at $2.0\times 10^5\,\text{Pa}$ while the volume expands from $6.3\,\text{m}^3$ to $7.1\,\text{m}^3$. What is the change in internal energy for the gas?

Solution: Apply the first law $\Delta U = \Delta Q - \Delta W$. Since the pressure is constant

$$\Delta W = p\Delta V = 2.0 \times 10^5 \,\mathrm{N/m^2}\,(0.8\,\mathrm{m^3}) = 1.6 \times 10^5\,\mathrm{J}$$

and

$$\Delta U = 1.5 \times 10^5\,\mathrm{J} - 1.6 \times 10^5\,\mathrm{J} = -1.0 \times 10^4\,\mathrm{J}$$

The internal energy of the gas has been decreased by $1.0 \times 10^4\,\mathrm{J}$.

19-5 An ideal gas initially at $3.0 \times 10^5\,\mathrm{Pa}$, volume of $0.030\,\mathrm{m^3}$, and temperature $20^o\mathrm{C}$ is heated at constant pressure to a volume of $0.120\,\mathrm{m^3}$. How much work was performed by the gas? If $80\,\mathrm{kJ}$ of heat energy was supplied to the gas, what was the change in internal energy?

Solution: The work performed is

$$W = p\Delta V = 3.0 \times 10^5\,\mathrm{N/m^2}(0.120 - 0.030)\,\mathrm{m^3} = 2.7 \times 10^4\,\mathrm{J}$$

The increase in internal energy is

$$\Delta U = Q - W = 8.0 \times 10^4\,\mathrm{J} - 2.7 \times 10^4\,\mathrm{J} = 5.3 \times 10^4\,\mathrm{J}$$

There are several different thermodynamic processes that occur often enough to warrant definition.

An **isothermal** process occurs at constant temperature.
An **adiabatic** process is one where no heat is transfered in or out of the system ($\Delta Q = 0$).
An **isobaric** process occurs at constant pressure ($W = p\Delta V$).
An **isochoric** process occurs at constant volume ($W = 0$).

Heat Capacities of Ideal Gases

The heat capacity of an ideal gas depends on whether the measurement is made at constant volume, C_V, or constant pressure, C_p.

At constant volume ($\Delta V = 0$) no work is performed by the system, so according to the first law $\Delta Q = \Delta U$. For n moles requiring C_V amount of heat per mole

$$\Delta Q = \Delta U = nC_V\Delta T$$

At constant pressure work is performed. Defining the heat capacity at constant pressure as the heat to raise one mole of a gas one degree,

$$\Delta Q = nC_p \Delta T$$

Since the gas expands in this process, work, $\Delta W = p\Delta V$ is performed. From the ideal gas law $p\Delta V = nR\Delta T$ so $\Delta W = nR\Delta T$. Substituting in the first law in the form $\Delta Q = \Delta U + \Delta W$, $nC_p \Delta T = \Delta U + nR\Delta T$. Since the internal energy depends only on temperature $\Delta U = nC_V \Delta T$ and

$$nC_p \Delta T = nC_V \Delta T + nR\Delta T \quad \text{or} \quad C_p = C_V + R \qquad \textbf{(19-6)}$$

Measurements on gases confirm this relationship.

19-6 What is the total internal energy of 10 moles of an ideal monatomic gas at $20^o\mathrm{C}$?

Solution: A monatomic gas is composed of single atoms that behave as mass points, and as such, they can only translate. They cannot rotate or vibrate. See problem 18-5 and the associated discussion to confirm that the internal energy is $(3/2)nRT$.

$$U = nRT = (3/2)10\,\mathrm{mol}(8.3\,\mathrm{J/mol\cdot K})293\,\mathrm{K} = 3.6\times 10^4\,\mathrm{J}$$

19-7 The heat necessary to raise the temperature of 10 moles of an ideal gas at a constant pressure of $1.0\times 10^5\,\mathrm{Pa}$ 10 degrees is $2100\,\mathrm{J}$. What is the volume change and the heat capacity at constant pressure and constant volume?

Solution: The work performed in the volume change is $\Delta W = p\Delta V = nR\Delta T$ so

$$\Delta V = \frac{nR\Delta T}{p} = \frac{10\,\mathrm{mol}(8.3\,\mathrm{J/mol\cdot K})10\,\mathrm{K}}{1.0\times 10^5\,\mathrm{N/m^2}} = 8.3\times 10^{-3}\,\mathrm{m^3}$$

Starting with the basic statement $\Delta Q = nC_p \Delta T$ the heat capacity at constant pressure is

$$C_p = \frac{1}{n}\frac{\Delta Q}{\Delta T} = \frac{1}{10\,\mathrm{mol}}\frac{2100\,\mathrm{J}}{10\,\mathrm{K}} = \frac{21\,\mathrm{J}}{\mathrm{mol\cdot K}}$$

The gas constant R is $8.3\,\mathrm{J/mol\cdot K}$ so

$$C_V = C_p - R = 21\,\mathrm{J/mol\cdot K} - 8.3\,\mathrm{J/mol\cdot K} = 12.7\,\mathrm{J/mol\cdot K}$$

19-8 Two moles of an ideal gas at a pressure of $1.0\,\text{atm}$ has $400\,\text{J}$ of heat added in expanding from a volume of $4.7\times10^{-3}\,\text{m}^3$ to $7.0\times10^{-3}\,\text{m}^3$. What is the change in internal energy and temperature? What are the specific heats?

Solution: The work performed is $W = p\Delta V = 1.0\times10^5\,\text{N}/\text{m}^2(2.3\times10^{-3}\,\text{m}^3) = 230\,\text{J}$.

The change in internal energy is from the first law $\Delta U = \Delta Q - W = 400\text{J} - 230\text{J} = 170\text{J}$.

The change in temperature is from $\Delta U = (3/2)nR\Delta T$, or

$$\Delta T = \frac{2\Delta U}{3nR} = \frac{2\cdot 170\,\text{J}}{3(2\,\text{mol})(8.3\,\text{J}/\text{mol}\cdot\text{K})} = 6.8\,\text{K}$$

The heat capacity at constant pressure is from $\Delta Q = nC_p\Delta T$, or

$$C_p = \frac{\Delta Q}{n\Delta T} = \frac{400\text{J}}{2\,\text{mol}\cdot 6.8\,\text{K}} = \frac{29\,\text{J}}{\text{mol}\cdot\text{K}}$$

The heat capacity at constant volume is

$$C_V = C_p - R = 29\,\text{J}/\text{mol}\cdot\text{K} - 8.3\,\text{J}/\text{mol}\cdot\text{K} = 20.7\,\text{J}/\text{mol}\cdot\text{K}$$

Adiabatic Processes

Adiabatic processes are ones where there is no heat transfer between the system and its surroundings. Rapid volume changes where there is no time for heat transfer are characteristic of adiabatic processes.

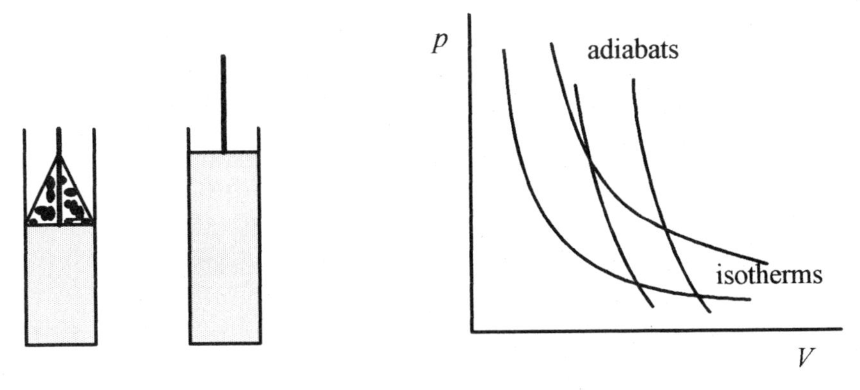

Fig. 19-4

A model adiabatic process is shown in Fig. 19-4. An ideal gas in an insulated container with a moveable weighted piston rapidly expands when the weight is removed. The volume increases, and the pressure and temperature drop with no heat in or out.

From the first law $\Delta U = -W$ and $\Delta U = nC_V\Delta T$ and $W = p\Delta V$. Writing the first law with p replaced from the ideal gas law $pV = nRT$

$$nC_V\Delta T = -p\Delta V = -\frac{nRT}{V}\Delta V$$

C Switching to differential format $\dfrac{dT}{T} = -\dfrac{R}{C_V}\dfrac{dV}{V}$

The factor R/C_V is usually written in the form

$$\frac{R}{C_V} = \frac{C_p - C_V}{C_V} = \frac{C_p}{C_V} - 1 = \gamma - 1 \quad \text{with } \gamma \text{ defined as } \frac{C_p}{C_V}$$

Rewriting, $\dfrac{dT}{T} + (\gamma - 1)\dfrac{dV}{V} = 0$ is integrated directly as a logarithm

$\ln T + (\gamma - 1)\ln V = \ln(\text{const})$, and applying the laws of logarithms $TV^{\gamma-1} = \text{const}$

Again applying the gas law $pV = nRT$, $\dfrac{pV}{nR}V^{\gamma-1} = \text{const}$, or $pV^{\gamma} = (\text{another})\text{const}$.

For a monatomic gas $\gamma = 1.40$. For adiabatic expansions there are two relationships

$$p_1V_1^{\gamma} = p_2V_2^{\gamma} \quad \text{and} \quad T_1V_1^{\gamma-1} = T_2V_2^{\gamma-1} \qquad \textbf{(19-7)}$$

The p-V curves for the $pV^{\gamma} = \text{const}$ are a lot like $pV = \text{const}$ curves. The $pV^{\gamma} = \text{const}$ curves are called adiabats and are shown in Fig. 19-4 along with (for comparison) some isotherms.

19-9 Two moles of an ideal gas expand isothermally at 295K from 0.60m^3 to 0.80m^3. What would be the temperature of the gas if the same expansion were adiabatic?

Solution: The temperature is from equation 19-7.

$$T_2 = T_1(V_1/V_2)^{\gamma-1} = 295\text{K}(0.60/0.80)^{1.40-1} = 263\text{K}$$

19-10 Air and fuel at 300K and 1.0atm is compressed in an automobile engine with a compression ratio of 8 to 1. Take $\gamma = 1.40$ and calculate the final pressure and temperature of the mixture.

Solution: The pressure is from $p_1V_1^\gamma = p_2V_2^\gamma$ or

$$p_2 = p_1(V_1/V_2)^\gamma = 1.0\,\text{atm}(8)^{1.4} = 18.4\,\text{atm}$$

The temperature is from $T_1V_1^{\gamma-1} = T_2V_2^{\gamma-1}$ or

$$T_2 = T_1(V_1/V_2)^{\gamma-1} = 300\,\text{K}(8)^{0.40} = 689\,\text{K}$$

19-11 An ideal gas at 3.0×10^5 Pa and $0.080\,\text{m}^3$ is compressed adiabatically to $0.020\,\text{m}^3$. What is the final pressure and the work performed?

Solution: The final pressure is from $p_1V_1^\gamma = p_2V_2^\gamma$, or

$$p_2 = p_1(V_1/V_2)^\gamma = 3.0\times10^5\,\text{Pa}(4)^{1.4} = 2.1\times10^6\,\text{Pa}$$

The work performed is from $-W = nC_V\Delta T = nC_V(T_2 - T_1)$. Substituting from the ideal gas law $pV = nRT$ and $R/C_V = \gamma - 1$

$$W = \frac{C_V}{R}(p_1V_1 - p_2V_2) = \frac{1}{\gamma-1}(p_1V_1 - p_2V_2)$$

$$W = \frac{1}{1.40-1}\left[3.0\times10^5\,\text{N/m}^2(0.080\,\text{m}^3) - 2.1\times10^6\,\text{N/m}^2(0.020\,\text{m}^3)\right]$$

$$W = \frac{1}{0.40}\left[2.4\times10^4\,\text{N}\cdot\text{m} - 4.2\times10^4\,\text{N}\cdot\text{m}\right] = -4.5\times10^4\,\text{J}$$

Work is performed on the gas in the amount of $4.5\times10^4\,\text{J}$.

19-12 An ideal monatomic gas at $p_a = 3.0\times10^5\,\text{N/m}^2$, $V_a = 0.060\,\text{m}^3$, and $T = 27^\circ\text{C}$ expands adiabatically to $p_b = 2.0\times10^5\,\text{N/m}^2$, $V_b = 0.085\,\text{m}^3$ and then isothermally to $V_c = 0.100\,\text{m}^3$. What is the final temperature, pressure, and work performed by the gas? Show these paths on a p-V diagram.

Solution: First calculate the temperature at the end of the adiabatic expansion using the ideal gas law $\frac{p_a V_a}{T_a} = \frac{p_b V_b}{T_b}$ or $T_b = T_a \frac{p_b}{p_a}\frac{V_b}{V_a} = 300\,\text{K}\frac{2.0\times10^5\,\text{N/m}^2}{3.0\times10^5\,\text{N/m}^2}\frac{0.060\,\text{m}^3}{0.085\,\text{m}^3} = 140\,\text{K}$

The gas has moved along an adiabat from the 300K isotherm to the 140K isotherm (point a to point b on Fig. 19-5).

The pressure at the end of the isothermal expansion can be calculated from the ideal gas law with the temperature a constant, $p_b V_b = p_c V_c$, or

$$p_c = p_b (V_b / V_c) = 2.0\times10^5\,\text{N/m}^2 \left(0.085\,\text{m}^3 / 0.100\,\text{m}^3\right) = 1.7\times10^5\,\text{Pa}$$

The temperature of the gas at point c is 140K, and the pressure is 1.7×10^5 Pa.

The work performed is positive in both parts of the expansion. During the adiabatic portion of the expansion, the work is the area under the a-b portion of the curve (see problem 19-11).

$$W_{a-b} = \frac{1}{\gamma - 1}\left[p_a V_a - p_b V_b\right]$$

$$W_{a-b} = \frac{1}{0.40}\left[3.0\times10^5\,\text{N/m}^2(0.060\,\text{m}^3) - 2.0\times10^5\,\text{N/m}^2(0.085\,\text{m}^3)\right]$$

$$W_{a-b} = \frac{1}{0.40}\left[1.8\times10^4\,\text{J} - 1.7\times10^4\,\text{J}\right] = 2500\,\text{J}$$

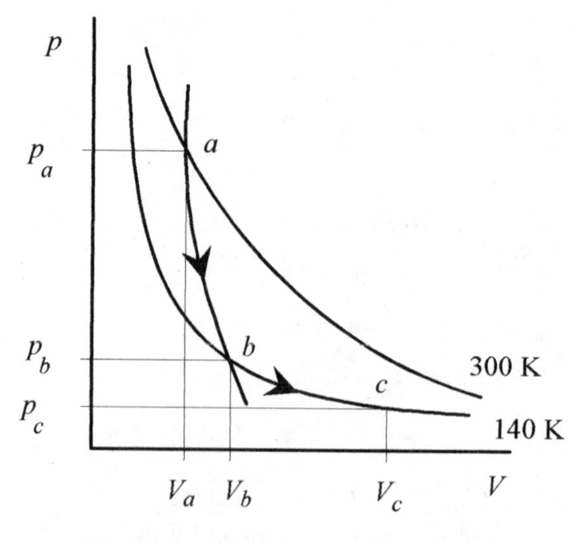

Fig. 19-5

In the isothermal portion of the expansion the work is the area under the b-c curve (equation 19-3).

$$W = \int_{V_b}^{V_c} p\,dV = nRT \ln \frac{V_c}{V_b}$$

The nRT is equal to $p_b V_b$, so

$$W = p_b V_b \ln(V_c / V_b)$$

This is the work performed in the isothermal expansion. Along the isotherm $p_b V_b / T_b = p_c V_c / T_c$ reduces to $p_b V_b = p_c V_c$ because $T_b = T_c$.

$$W_{b-c} = 2.0 \times 10^5 \frac{\text{N}}{\text{m}^2} (0.085\text{m}^3) \ln \frac{0.100\text{m}^3}{0.085\text{m}^3} = 2760\text{J}$$

The total work performed by the gas is 5260J.

Look over this problem again. It is probably the most difficult one you will encounter, and if you understand all the steps and can do them on your own, you understand the first law of thermodynamics very well.

CHAPTER 20

SECOND LAW OF THERMODYNAMICS

The first law of thermodynamics deals with heat, work, and internal energy. The second law deals with the direction and efficiencies of thermodynamic processes. Many thermodynamic processes proceed in one direction but not the reverse. Applying the brakes to stop a car is an excellent example of a thermodynamic process (the conversion of mechanical energy to heat) that is not reversible.

It is possible to convert mechanical energy into heat with 100% efficiency. It is not possible to convert heat into mechanical energy with 100% efficiency.

Heat Engines

A **heat engine** is a device that converts thermal energy to other (usually mechanical or electrical) forms of energy. The engine takes a working substance through a cycle where heat is absorbed from a hot source and expelled to a cold source. The cycle may involve taking on or expelling heat, expansion, compression, or a phase change. Figure 20-1 depicts a heat engine extracting heat from a hot source, performing work, and expelling heat to the cold source. In this engine the working substance goes through a cycle so that over one cycle $\Delta U = 0$. In real engines, such as the typical automobile engine, the working substance is the air-fuel mixture continuously supplied to the engine. In this case the working substance is not physically the same but has the same characteristics. According to the first law

$$W = Q_H - Q_C \qquad \textbf{(20-1)}$$

The **thermal efficiency** of an engine is defined as the ratio of the work performed to the heat taken in, or

$$e = \frac{W}{Q_H} = \frac{Q_H - Q_C}{Q_H} = 1 - \frac{Q_C}{Q_H} = 1 - \frac{|Q_C|}{|Q_H|} \qquad \textbf{(20-2)}$$

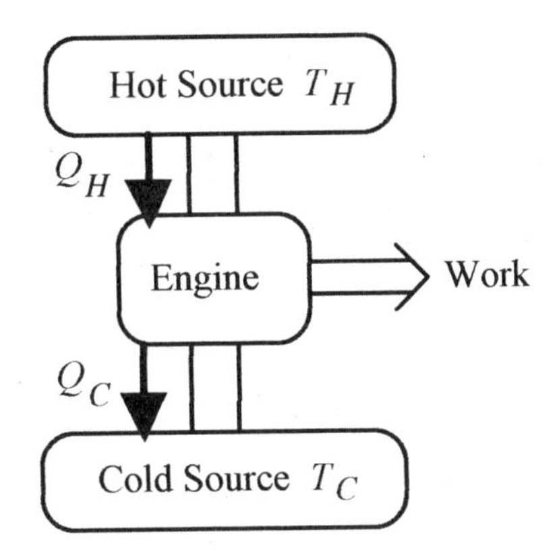

Fig. 20-1

To avoid the confusion of determining the algebraic sign of heat entry or exit of the system, the thermal efficiency, *e*, is defined with absolute value signs. According to this definition an engine that converted all the heat taken in to work would be 100% efficient.

One way of stating the **second law of thermodynamics** is to say that: A heat engine cannot have an efficiency of 100%. In terms of the engine, all the heat taken in cannot be converted to work. The first law of thermodynamics says that we cannot get more energy out of a thermodynamic system than we put in, and the second law says we cannot hope to break even in extracting energy (mechanical or electrical) from heat.

20-1 What is the efficiency of an engine that takes in 300 J and expels 100 J over each cycle?

Solution: $e = 1 - |Q_C|/|Q_H| = 1 - (100/300) = 0.67 \Rightarrow 67\%$

20-2 If an (gasoline) automobile engine performs 10 kW of work and is 28% efficient, what is the rate at which heat is taken in and expelled?

Solution: It doesn't matter whether we take the heat in and the work performed for one cycle or the rates for heat in and work performed. The efficiency is $e = W / Q_H$ so the rate of heat in is

$$Q_H = \frac{W}{e} = \frac{10\,\text{kW}}{0.28} = 36\,\text{kW}$$

From the efficiency equation 20-2, $\frac{Q_C}{Q_H} = 1 - e$, so the heat expelled is

$$Q_C = Q_H(1 - e) = 36\,\text{kW}(1 - 0.28) = 26\,\text{kW}$$

Carnot Cycle

The Carnot cycle or engine operates in an ideal reversible cycle between two temperatures. Figure 20-2 shows the two isothermal and two adiabatic paths for the working substance. For the cycle to be reversible the paths have to be reversible. A reversible path or process is one that can be made to go backwards or forward. The working substance following these isotherms and adiabats can be taken in either direction by changing the conditions. For example, along the isotherm from A to B changing the volume slowly (in either direction) will make the system move up or down the isotherm.
A similar argument holds for the adiabats.

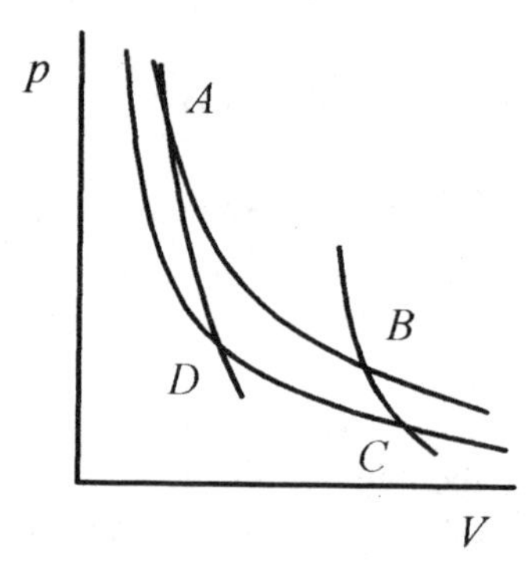

Fig. 20-2

Take an ideal gas as the working substance through a complete cycle.
Path $A \to B$ is an isothermal expansion at T_H. The gas is placed in contact with a heat reservoir that provides Q_H while the gas does work in expanding from V_A to V_B.

This work according to equation 19-3 is $Q_H = W_{AB} = nRT_H \ln \frac{V_B}{V_A}$

Path $B \to C$ is an adiabatic expansion.
Path $C \to D$ is an isothermal compression at T_C. The gas is placed in contact with a heat reservoir that removes Q_C while work is done on the gas compressing it from V_C to V_D.

This work according to equation 19-3 is $Q_C = W_{CD} = -nRT_C \ln \frac{V_C}{V_D}$

Path $D \to A$ is an adiabatic compression.

In this (cyclic) process the working substance has been taken from T_H to T_C and back to T_H. The area bounded by these four curves is a visual measure of the inefficiency of the Carnot engine.

The efficiency of the Carnot engine is $1 - |Q_C|/|Q_H|$ so form $\frac{|Q_C|}{|Q_H|} = \frac{T_C}{T_H} \frac{\ln(V_C/V_D)}{\ln(V_B/V_A)}$

The logarithmic part of this expression can be simplified by looking at the adiabatic relations. The adiabatic processes follow

$$T_H V_B^{\gamma-1} = T_C V_C^{\gamma-1} \qquad \text{and} \qquad T_H V_A^{\gamma-1} = T_D V_D^{\gamma-1}$$

Dividing, $\left(\frac{V_B}{V_A}\right)^{\gamma-1} = \left(\frac{V_C}{V_D}\right)^{\gamma-1}$ or $\frac{V_B}{V_A} = \frac{V_C}{V_D}$ making $\frac{\ln(V_C/V_D)}{\ln(V_B/V_A)} = 1$

and $|Q_C / Q_H|$ reduces to T_C / T_H and the efficiency is $e = 1 - \frac{T_C}{T_H}$

20-3 A steam engine is operated with steam at atmospheric pressure and 100^oC. Steam leaves the engine at slightly above outside temperature, 35^oC. What is the efficiency of the engine?

Solution: $e = 1 - 308/373 = 0.17 \Rightarrow 17\%$

20-4 For the situation of problem 20-3 increase the pressure in the boiler so that the temperature of the steam is $450\,\text{K}$. What is the efficiency of the engine?

Solution: $e = 1 - (308/450) = 0.32 \Rightarrow 32\%$

20-5 A heat engine takes in $2000\,\text{J}$ of heat and performs $400\,\text{J}$ of work each cycle. What is the efficiency, amount of heat rejected, and the power output for a cycle rate of $100\,\text{cycles/s}$?

Solution: The amount of heat rejected in each cycle is $2000\,\text{J} - 400\,\text{J} = 1600\,\text{J}$.

The efficiency is $e = 1 - \frac{|Q_C|}{|Q_H|} = 1 - \frac{1600\,\text{J}}{2000\,\text{J}} = 1 - 0.8 = 0.20 \Rightarrow 20\%$.

The power output for the engine is $P = \frac{100\,\text{cycle}}{\text{s}} \frac{400\,\text{J}}{\text{cycle}} = 40\,\text{kW}$.

20-6 An electric generating plant has an output of $50\,\text{MW}$. Steam enters the turbines at 550^oC and exits at 110^oC. The efficiency of the plant is 80% of the theoretical Carnot efficiency. What is the rate of heat consumption and heat rejection?

Solution: The Carnot efficiency is $e = 1 - T_C/T_H = 1 - 383\text{K}/823\text{K} = 0.53$

Eighty percent of this is 43% efficiency. This means that 43% of the input is equal to the $50\,\text{MW}$ output, giving an input requirement of $116\,\text{MW}$. The rate of heat rejection is 57% of the $116\,\text{MW}$ input or $66\,\text{MW}$ (rejected).

Refrigerators and Heat Pumps

Refrigerators and heat pumps perform the same function, they supply mechanical energy (work) to an engine that moves heat from a cold place to a hot place.

A heat pump has as its purpose to heat. The measure of efficiency is called the coefficient of performance and is the ratio of heat transferred to mechanical work input. Heat engines take in heat in order to perform mechanical work. Refrigerators and heat pumps take in mechanical energy in order to move heat from a cold to a hot place (opposite to the normal direction of the process).

$$COP(\text{heat pump}) = \frac{Q_H}{W}$$

Remember that for a Carnot engine $W = Q_H - Q_C$ and that $Q_H/Q_C = T_H/T_C$. For a Carnot engine run as a heat pump the coefficient of performance is

$$COP_C(\text{heat pump}) = \frac{T_H}{T_H - T_C}$$

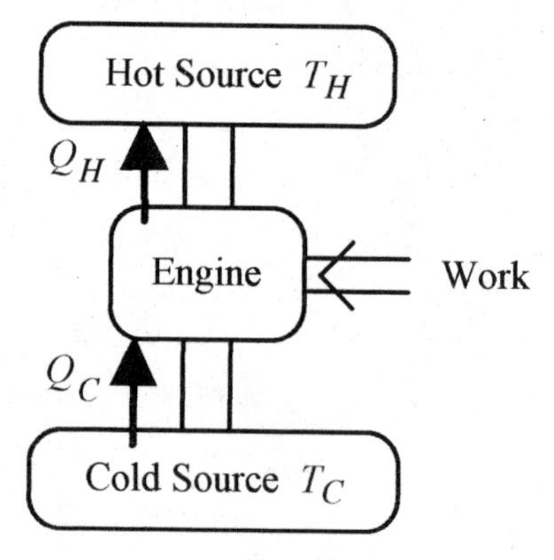

Fig. 20-3

A refrigerator has as its purpose to cool. The coefficient of performance is

$$COP(\text{refrigerator}) = \frac{Q_C}{W}$$

For the Carnot engine run as a refrigerator

$$COP_C(\text{refrigerator}) = \frac{T_C}{T_H - T_C}$$

20-7 What is the coefficient of performance of a refrigerator operating as a Carnot engine between $5^o\,\text{C}$ and $25^o\,\text{C}$?

Solution: $COP_C(\text{refrigerator}) = 278/20 = 14$

Most refrigerators run at 1/2 or less of their Carnot efficiencies.

20-8 Assuming Carnot efficiency calculate the coefficient of performance of a heat pump operating between $20^o\,\text{C}$ and $12^o\,\text{C}$. Find the minimum power requirements to move thermal energy from the $12^o\,\text{C}$ area to the $20^o\,\text{C}$ area at the rate of $20\,\text{kW}$.

Solution: The coefficient of performance for a heat pump is

$$COP_C(\text{heat pump}) = \frac{T_H}{T_H - T_C} = \frac{293}{8} = 37$$

The coefficient of performance is the ratio Q_H/W so $Q_H/W = 37$, or

$$W = \frac{Q_H}{37} = \frac{20\,\text{kW}}{37} = 0.54\,\text{kW}$$

The theoretical minimum power to move the $20\,\text{kW}$ (from cold to hot) is $0.54\,\text{kW}$. Real heat pumps require about twice this theoretical minimum power. Notice, however, that it takes much less power to "move" heat than it does to heat directly. In this case about one kW moves $20\,\text{kW}$. Heat pumps are most efficient in mild climates. As the temperature difference increases the coefficient of performance decreases.

Entropy

Entropy is a measure of disorder. Since perfect order or perfect disorder is hard to define, the definition of entropy is given in terms of the change in entropy. If a system can be taken over a reversible path at constant temperature then the **change in entropy** is the heat necessary to traverse this path divided by the temperature, regardless of the path actually taken

$$\Delta S = \frac{\Delta Q_r}{T} \qquad \textbf{(20-3)}$$

The subscript r is a reminder that the heat transfer is to be measured over a reversible path even though the system may have been taken over an irreversible path.

20-9 What is the change in entropy for $50\,\mathrm{kg}$ of water going from liquid to vapor at $100^o\mathrm{C}$?

Solution: The conversion of water from liquid to vapor is reversible, and the heat required is the heat of vaporization of water ($3.3\times10^5\,\mathrm{J/kg}$), so

$$\Delta S=\frac{3.3\times10^5\,\mathrm{J/kg}(50\,\mathrm{kg})}{373\mathrm{K}}=4.4\times10^4\,\mathrm{J/K}$$

Water vapor has more disorder than water (liquid).

(C) **20-10** Calculate the change in entropy for $200\,\mathrm{kg}$ of water slowly heated from $20^o\mathrm{C}$ to $80^o\mathrm{C}$.

Solution: The basic definition of entropy is $\Delta S=\Delta Q_r/T$. The heating process is surely reversible. Switching to a calculus definition

$$\Delta S=\frac{dQ_r}{T}=\frac{mcdT}{T}$$

and integrating over the temperature range

$$\Delta S=\int_{T}^{T_f}mc\frac{dT}{T}=mc\ln T\Big|_{T_i}^{T_f}=200\,\mathrm{kg}\frac{4200\mathrm{J}}{\mathrm{kg}\cdot{}^o\mathrm{C}}\ln 4=1.16\times10^6\,\mathrm{J/K}$$

Consider the change in entropy during one Carnot cycle. Refer to Fig. 20-2 and the associated discussion of the cycle. Along the isotherm $A\rightarrow B$ an amount of heat Q_H is absorbed at T_H. Path $B\rightarrow C$ is adiabatic as is path $D\rightarrow A$, so no heat is absorbed or rejected in these paths. Along path $C\rightarrow D$ an amount of heat Q_C is expelled at T_C so

$$\Delta S=\frac{Q_H}{T_H}-\frac{Q_C}{T_C}$$

Recalling that $Q_C/Q_H=T_C/T_H$

$$\Delta S=\frac{Q_H}{T_H}-\frac{1}{T_C}\frac{Q_H T_C}{T_H}=0$$

The total entropy change for one Carnot cycle is zero. This is also true for any reversible cycle.

CHAPTER 21

MECHANICAL WAVES

Mechanical waves travel in a medium. The medium and the mechanism for producing the wave determine its properties. There are two types of mechanical waves.

Transverse waves are those where the mechanical motion is perpendicular to the direction of the wave. A wave snapped on a rope in tension is a transverse wave: pieces of the rope go up and down, and the wave travels along the rope. Waves on water are transverse: the water goes up and down, and the wave travels along the surface of the water.

Longitudinal waves are waves in the direction of the motion. An audio loudspeaker creates a varying pressure wave by physically moving the (column of) air between the speaker and your ear.

Waves move with a speed determined primarily by the medium. Waves also transport energy.

Transverse Waves on a String

First consider the periodic sinusoidal waves that propagate on a stretched string. If we took a "snapshot" (Fig. 21-1) of the wave we would see a sine wave with **amplitude**, the maximum displacement from equilibrium, and **wavelength**, the length for one complete cycle, or from crest to crest or trough to trough. The velocity of the wave is the wavelength divided by the time for the wave to pass a point on the string.

$$v = \frac{\lambda}{T} = \lambda f \qquad \textbf{(21-1)}$$

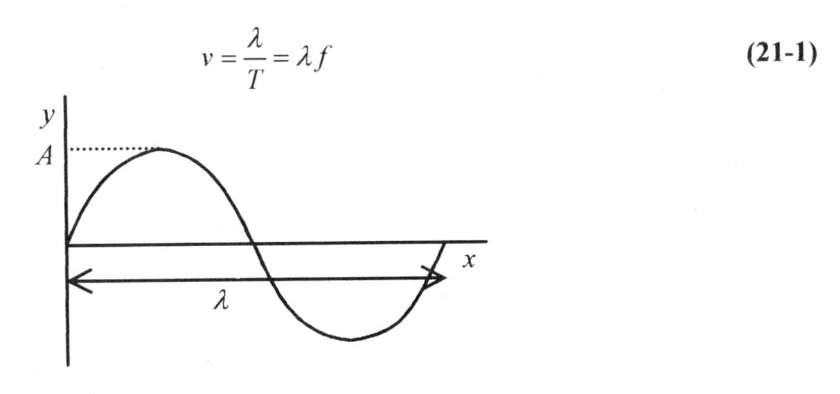

Fig. 21-1

This time for the wave to pass a point on the string, or for a piece of the string to execute one sine wave, is called the **period**. The reciprocal of the period is called the **frequency**.

21-1 The end of a stretched rope is being vibrated up and down at $120\,\text{Hz}$. A "snapshot" of the wave indicates a wavelength of $0.50\,\text{m}$. What is the speed of the wave?

Solution: The wave has velocity given by $v = \lambda f = 0.50\,\text{m}(120\,\text{Hz}) = 60\,\text{m/s}$.

Take a stretched string being vibrated with an amplitude A at a frequency of $10\,\text{Hz}$, or period $0.10\,\text{s}$. In this discussion do not be concerned with how the stretched string is terminated. The displacement of the end of the string is

$$y = A\sin(2\pi t/T) = A\sin(2\pi f t) = A\sin(\omega t)$$

The displacement is a sine function, and sine functions are periodic in 2π, so when t has gone from zero to $0.10\,\text{s}$, the time for one period, the oscillation has gone through one (2π) cycle. One 2π cycle is shown in Fig. 21-1. Writing the argument of the sine function as $2\pi f t$ is more compact, and replacing $2\pi f$ by ω is even more convenient.

From the basic definition of $v = x/t$, the time for a wave to move a distance is x/v.

The displacement of a piece of the rope at any distance along the rope and at any time is

$$y = A\sin 2\pi \frac{1}{T}\left(t - \frac{x}{v}\right) = A\sin 2\pi f\left(t - \frac{x}{v}\right)$$

The velocity v is also equal to λf so another often convenient form is

$$y = A\sin 2\pi\left(\frac{t}{T} - \frac{x}{\lambda}\right) \qquad \textbf{(21-2)}$$

In many problems the wave number, $k = 2\pi/\lambda$, is often used and combining with $\omega = 2\pi f$

$$y = A\sin(\omega t - kx) \qquad \textbf{(21-3)}$$

This equation describes the wave on the string at any particular time. It also describes the motion of a piece of the string at any position on the string. The minus sign is for a wave moving in the positive direction. A plus sign would be for a wave moving in the negative direction.

21-2 A sinusoidal wave is propagated on a string with amplitude $0.050\,\text{m}$, wavelength $0.20\,\text{m}$, and frequency of $40\,\text{Hz}$. Write an expression for the displacement of the string as a function of the time and position on the string.

Solution: The period of the wave is $1/40\,\text{s} = 0.025\,\text{s}$. The displacement is according to equation 21-2

$$y = (0.050\,\text{m})\sin 2\pi\left(\frac{t}{0.025\,\text{s}} - \frac{x}{0.20\,\text{m}}\right)$$

21-3 For the situation described in problem 21-2, what is the expression for the time variation of a point on the string at $x = 0.010\,\text{m}$?

Solution: The expression for y with $x = 0.10\,\text{m}$ is

$$y = (0.050\,\text{m})\sin 2\pi\left(\frac{t}{0.025\,\text{s}} - 0.50\right)$$

If at $t = 0$, a wave is initiated from the point $x = 0$ in the positive direction, then according to this expression, it takes $t = (0.025\,\text{s})(0.50) = 0.0125\,\text{s}$ for the wave to reach $0.010\,\text{m}$. After this $0.0125\,\text{s}$ time interval, the point $x = 0.010\,\text{m}$ oscillates in a sinusoidal manner. Look back at the numbers for wavelength and period in problem 21-2 and notice that it takes $1/2$ of a period for the wave to travel $1/2$ of a wavelength.

21-4 For the situation of problem 21-2, what is the expression for the space variation of the wave at $t = 0.25\,\text{s}$?

Solution:

$$y = (0.050\,\text{m})\sin 2\pi\left(10 - \frac{x}{0.20\,\text{m}}\right)$$

There are two speeds associated with this description of a wave on a string. The speed that the wave moves along the string is called the **phase velocity**. This is the velocity of a crest of the wave as it moves in the positive direction.

C For a wave traveling at constant speed, the displacement must be a constant as a function of space and time, so $\sin(\omega t - kx)$ must equal a constant, and therefore $\omega t - kx$ must equal a constant. Taking the total derivative of $\omega t - kx$ yields the **phase velocity** dx/dt.

$$\omega dt = k dx \qquad \text{or} \qquad \frac{dx}{dt} = \frac{\omega}{k} = \frac{2\pi f}{2\pi/\lambda} = \lambda f = \frac{\lambda}{T}$$

The speed of the wave along the string is the length of the wave divided by the time for one oscillation.

(C) The **transverse velocity** is the up and down velocity of a piece of the string. This is $\partial y/\partial t$. We use the ∂ notation rather than the d because it is possible to take derivatives of y with respect to both x and t. For a sinusoidal wave described by equation 21-3

$$\frac{\partial y}{\partial t} = A\,\omega \cos(\omega t - kx)$$

This derivative shows that the up and down motion of a piece of the string is sinusoidal. The maximum velocity (of a piece of the string) occurs at its equilibrium position, where $\cos(\omega t - kx) = 0$, and is equal to $A\,\omega$.

Wave Equation

(C) If the equation for the wave $y = A\sin(\omega t - kx)$ is differentiated twice with respect to t and twice with respect to x a very interesting result is obtained:

$$\frac{\partial^2 y}{\partial t^2} = -A\,\omega^2 \sin(kx - \omega t)$$

$$\frac{\partial y}{\partial x} = Ak\cos(kx - \omega t) \qquad \text{and} \qquad \frac{\partial^2 y}{\partial x^2} = -Ak^2 \sin(kx - \omega t)$$

Now

$$\frac{\partial^2 y}{\partial t^2} \Big/ \frac{\partial^2 y}{\partial x^2} = \frac{-A\,\omega^2 \sin(kx - \omega t)}{-Ak^2 \sin(kx - \omega t)} = \frac{\omega^2}{k^2}$$

But ω/k is v, the velocity, so

$$\frac{\partial^2 y}{\partial x^2} = \frac{1}{v^2}\frac{\partial^2 y}{\partial t^2} \qquad \textbf{(21-4)}$$

which is called the **wave equation**, a most important equation in physics. Any function of x and t that satisfies this relation propagates in the positive x direction with velocity v. This analysis is equally valid with longitudinal waves (sound) with the added complication that y describes the displacement of the particle from its equilibrium but in the direction of the wave.

Transverse Wave Speed

The speed of a transverse wave on a string depends on the tension and the mass per unit length. Common experience is that a wave "snapped" on a string moves faster when the string is under greater tension and slower when a heavier (mass per unit length) string is used.

Look at the crest of the wave as forming part of a circle and follow this crest as it propagates along the string. Take the arc of the circle as length $\Delta\ell$ and the string of mass per unit length μ. Thus the mass of the length $\Delta\ell$ is $\mu\Delta\ell$. Next look at the forces on this piece of string. The tension in the string, F, has the components as shown in Fig. 21-2.

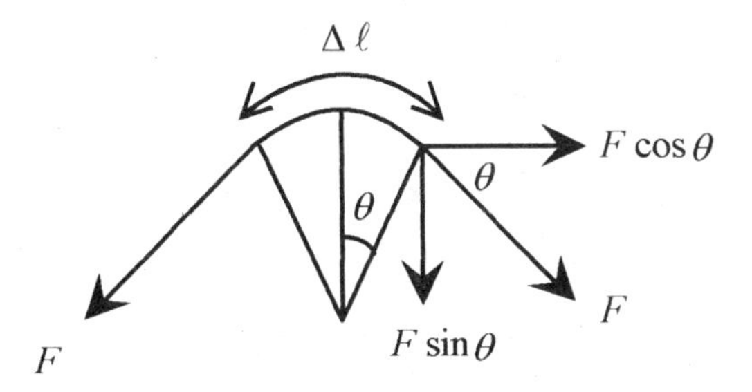

Fig. 21-2

The force, on the piece of string $\Delta\ell$, directed toward the center of the circle is $2F\sin\theta$.

At small angles $\sin\theta \approx \theta \approx \dfrac{\Delta\ell/2}{R} = \dfrac{\Delta\ell}{2R}$, so $2F\sin\theta \approx \dfrac{F\Delta\ell}{R}$.

This center-directed force is the centripetal force $\dfrac{mv^2}{r} = \dfrac{\mu\Delta\ell v^2}{R}$.

Set this center-directed force equal to $\dfrac{F\Delta\ell}{R}$: $\dfrac{F\Delta\ell}{R} = \dfrac{\mu\Delta\ell v^2}{R}$, or

$$v = \sqrt{F/\mu} \qquad \textbf{(21-5)}$$

21-5 What is the speed and wavelength of a $20\,\text{Hz}$ frequency wave on a string of $0.20\,\text{kg/m}$ under tension of $80\,\text{N}$?

Solution: The velocity of the wave is $v = \sqrt{F/\mu} = \sqrt{80\,\text{N}/0.20\,\text{kg/m}} = 20\,\text{m/s}$.

Use the basic relation $v = \lambda f$ to find $\lambda = \dfrac{v}{f} = \dfrac{20\,\text{m/s}}{20\,\text{Hz}} = 1.0\,\text{m}$

Power

(C) When a stretched string is displaced (plucked), a pulse travels along the string. There is a force associated with this displacement. This force acting on an element of the string to produce the displacement and return to equilibrium is the work performed on the element of string or the energy associated with the pulse. This energy per time is the power transported along the string.

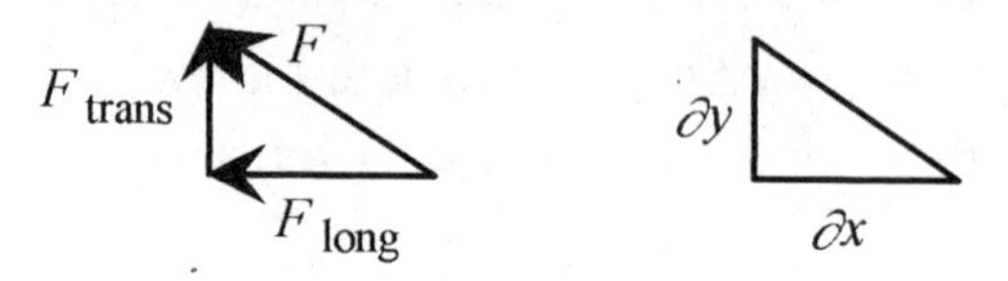

Fig. 21-3

In a stretched string the tension, or force, has longitudinal and transverse components. The components of the force are proportional to the slope as shown in Fig. 21-3. The slope is written in the partial derivative notation. (Remember y is a function of both x and t.)

$$\frac{F_{trans}}{F_{long}} = -\frac{\partial y}{\partial x}$$

For small displacements the longitudinal component of the tension is the static tension in the string ($F_{long} = F$, for small displacements). While the pulse is moving along the string the transverse force is always opposite the slope of the pulse. The transverse force that moves the pulse is opposite (in sign) to the displacement.

$$F_{trans} = -F\frac{\partial y}{\partial x}$$

The power, again from mechanics, is this transverse force times the transverse velocity, or

$$P = -\left(F\frac{\partial y}{\partial x}\right)\left(\frac{\partial y}{\partial t}\right)$$

Take a pulse described by $y = A\sin(\omega t - kx)$ with $\frac{\partial y}{\partial x} = -kA\cos(\omega t - kx)$

and $\frac{\partial y}{\partial t} = \omega A\cos(\omega t - kx)$ so $P = Fk\omega A^2\cos^2(\omega t - kx)$

The average value of the $\cos^2$ function is $1/2$, $\omega = vk$, and $v^2 = F/\mu$, so

$$P_{\text{avg}} = \frac{1}{2}F\frac{\omega}{v}\omega A^2 = \frac{1}{2}\sqrt{F\mu}\,\omega^2 A^2 \qquad \textbf{(21-6)}$$

21-6 A steel wire of length $1.0\,\text{m}$ and mass $4.0\,\text{gm}$ is stretched with a force of $50\,\text{N}$. Waves of amplitude $1.2\times10^{-3}\,\text{m}$ and frequency $1000\,\text{Hz}$ are traveling on the wire. What is the average power of these waves?

Solution: Use the formula

$$P_{\text{avg}} = \frac{1}{2}\sqrt{F\mu}\,\omega^2 A^2 = \frac{1}{2}\sqrt{50\,\text{N}(4.0\times10^{-3}\,\text{kg/m})}(2\pi)^2(10^3\,1/\text{s})^2(1.2\times10^{-3}\,\text{m})^2 = 12.7\,\text{W}$$

21-7 What happens to the power if the frequency is increased by a factor of 10?

Solution: Since the frequency appears in the formula as the square, the power increases by a factor of one hundred.

CHAPTER 22

STANDING WAVES

(STRINGS AND PIPES)

A string stretched between two fixed end points appears to vibrate in a sinusoidal manner when plucked. Blowing across the top of a partially filled bottle of liquid produces a specific frequency sound with the frequency depending on the length of the air column above the liquid. The wave on the string is transverse: the elements of the string move up and down while the wave moves back and forth along the string. The wave in the partially filled bottle is longitudinal: pulses move up and down the column of air in the same direction as the wave. These two very different waves can be analyzed in a similar fashion.

Waves on Strings

When a stretched string is plucked, the pattern appears as a sinusoidal standing wave. The wave pattern is sinusoidal because the only solution to the wave equation (previous chapter) is a sinusoidal function. The wavelength of the waves is determined by the length between the fixed end points of the string. A sine wave on a string reflected at a fixed boundary produces a standing wave where the fixed length of the string is an integral number of half wavelengths. Figure 22-1 depicts the sinusoidal standing waves on strings as observed in laboratory experiments with vibrating strings.

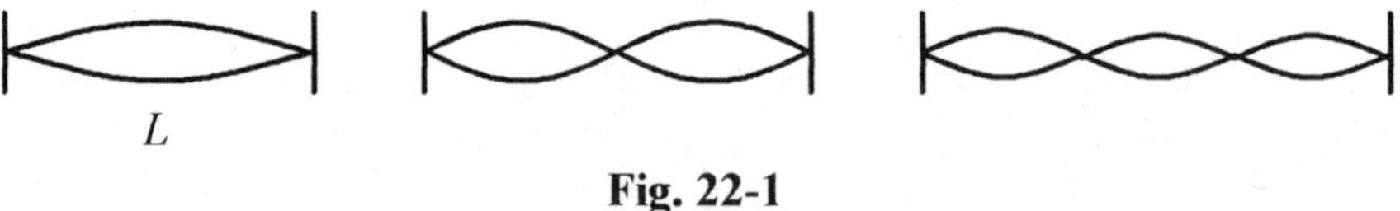

Fig. 22-1

Positions on the string where there is no motion are called **nodes** and positions where there is maximum motion are called **antinodes** or **loops**. The standing wave on the string is produced by two waves, one moving to the right

$$y_1 = A\sin 2\pi(t/T - x/\lambda)$$

and another moving to the left

$$y_2 = -A\sin 2\pi(t/T + x/\lambda)$$

both with the same wavelength and period.

The standing wave is the sum of these two waves

$$y = y_1 + y_2 = A\sin 2\pi(t/T - x/\lambda) - A\sin 2\pi(t/T + x/\lambda)$$

Using two identities from the Introduction, Mathematical Background, the sum is

$$y = \left(2A\cos\frac{2\pi t}{T}\right)\sin\frac{2\pi x}{\lambda} \qquad \textbf{(22-1)}$$

The $\cos 2\pi t/T$ term gives the time dependent motion of a piece of the string at a specific value of x. The $(2A)\sin 2\pi x/\lambda$ term is the amplitude of the up and down motion of the piece of the string at the specific value of x.

Where x is equal to 0 and $\lambda/2$, the $\sin 2\pi x/\lambda$ term is zero, and the string doesn't move. The allowed wavelengths are determined by requiring x/λ to equal $1/2,\ 1,\ 3/2\ .\ .\ .$, that is, where the sine function is equal to zero. This corresponds to $x = \lambda/2,\ \lambda,\ 3\lambda/2\ .\ .\ .\ .$

For a string fixed at end points L distance apart, the allowed wavelengths are then $L = n\lambda/2\,(n = 1,2,3\ .\ .\ .)$. Indexing λ for the allowed wavelengths

$$\lambda_n = \frac{2L}{n}(n = 1,2,3\ .\ .\ .) \qquad \textbf{(22-2)}$$

The allowed frequencies for this stretched string are from $v = \lambda f$ or $v = \lambda_n f_n$, so

$$f_n = \frac{v}{\lambda_n} = n\frac{v}{2L}(n = 1,2,3\ .\ .\ .) \qquad \textbf{(22-3)}$$

Remember that for the stretched string, v is a constant equal to $\sqrt{F/\mu}$. The first frequency (longest wavelength) is called the fundamental and the next frequency the second harmonic or first overtone with subsequent frequencies labeled in a similar manner.

When a stringed musical instrument is plucked, all of these allowed frequencies are prevalent though the overtones are of successively lower amplitude. The difference in sound of stringed instruments plucked at the same fundamental frequency is due to the design of the instrument to enhance or suppress certain of these overtones.

22-1 The third harmonic is excited on a $2.4\,\mathrm{m}$ length of stretched string. What are the positions of the loops and nodes?

Solution: To produce the third harmonic there are two nodes at 0.80 and $1.6\,\text{m}$ so as to divide the string into three segments. The loops are half way between the nodes at 0.40, 1.2, and $2.0\,\text{m}$.

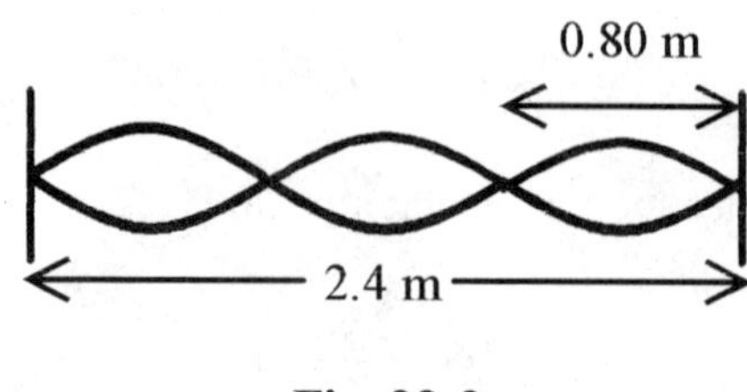

Fig. 22-2

22-2 An $0.040\,\text{kg}$ string $0.80\,\text{m}$ long is stretched and vibrated in a fundamental mode with a frequency of $40\,\text{Hz}$. What is the speed (of propagation) of the wave and the tension in the string?

Solution: The relationship $v = \lambda f$ is used to calculate the speed of the wave. Since the string is vibrating in fundamental mode, the wavelength is $1.6\,\text{m}$ (see Fig. 22-1), so

$$v = \lambda f = (1.6\,\text{m})(40\,\text{Hz}) = 64\,\text{m/s}$$

Having determined the velocity from this basic relation, the tension can be determined from $v = \sqrt{T/\mu}$. The linear density of the string is $\mu = 0.040\,\text{kg}/0.80\,\text{m} = 0.050\,\text{kg/m}$ so

$$T = v^2 \mu = (64\,\text{m/s})^2\, 0.050\,\text{kg/m} = 205\,\text{N}$$

22-3 For the situation described in problem 22-2, add that the amplitude (maximum displacement from the center line of the string) is $2.0\,\text{mm}$. What is the displacement at 0.20, 0.40, and $0.60\,\text{mm}$, and the time for one of these oscillations?

Solution: The oscillation is described in both space and time by equation 22-1 at $x = 0.20\,\text{m}$. The equation is

$$y = (2A\cos 2\pi f t)\sin 2\pi \frac{0.20}{1.6} = (2A\cos 2\pi f t)\sin \frac{\pi}{4}$$

The $\sin \pi/4 = 0.71$ term specifies the fraction of the $2.0\,\text{mm}$ (amplitude) of the oscillation at $x = 0.20\,\text{m}$. This maximum displacement is $1.4\,\text{mm}$. This is the displacement measured from the center line of the string. The height of the envelope or total (up and down) excursion of the piece of string is $2.8\,\text{mm}$.

At $x = 0.40\,\text{mm}$ the sine term becomes $\sin \pi/2 = 1$ and the total excursion is $4.0\,\text{mm}$.

Because of symmetry the maximum excursion at $0.60\,\text{m}$ is the same as at $0.20\,\text{m}$.

The stretched string goes up and down together, that is, all points on the string are vibrating with the same frequency of $40\,\text{Hz}$, so the time for one oscillation is $1/40\,\text{s} = 0.025\,\text{s}$.

22-4 One string on a cello is tightened to produce concert A ($440\,\text{Hz}$) when the string is not touched (shortened by placing a finger on it). The length of the vibrating string is $60\,\text{cm}$ and mass is $2.0\,\text{gm}$. How much must the player shorten the string to play a $660\,\text{Hz}$ note?

Solution: The tension (velocity) is adjusted so that the frequency is $440\,\text{Hz}$ for a wavelength of $1.2\,\text{m}$. The velocity is

$$v = \lambda f = (1.2\,\text{m})(440\,\text{Hz}) = 528\,\text{m/s}$$

If the tension is not changed, the velocity is not changed so the wavelength to produce $660\,\text{Hz}$ is

$$528\,\text{m/s} = \lambda \cdot 660\,\text{Hz} \quad \text{or} \quad \lambda = 0.80\,\text{m}$$

The length of the string to produce this frequency is one-half of a wavelength or $40\,\text{cm}$. The player needs to shorten the string, by pressing the string onto the fingerboard, by $20\,\text{cm}$, from $60\,\text{cm}$ to $40\,\text{cm}$.

Waves in Pipes

If a longitudinal wave is propagated down a gas-filled tube, the wave will be reflected at a closed end in much the same manner as the wave on a string is reflected at a rigid boundary thereby setting up standing waves in the tube. Reflection at this end implies no motion and a displacement node. If the end of the pipe is open, then this end is a displacement antinode. The air is free to move at an open end. Similarly a string with an end free to vibrate has an antinode at the free end.

The simplest example of a standing wave in a tube is an organ pipe. Air is blown across the open end of the organ pipe. The opposite end can be closed or open.

Look first at a closed (at one end) pipe. The open end is a loop (displacement antinode)

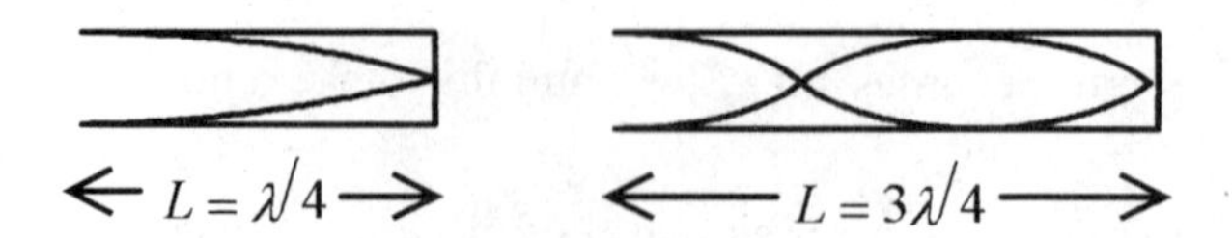

Fig 22-3

and the closed end a displacement node, so the first fundamental frequency has wavelength $4L$ (see Fig. 22-3) The next allowed wave, the second harmonic, as shown in Fig. 22-3 is $4L/3$. The general expression for the wavelengths is

$$\lambda_n = \frac{4L}{n}(n = 1,3,5\ .\ .\ .) \qquad \textbf{(22-4)}$$

and

$$f_n = \frac{nv}{4L}(n = 1,3,5\ .\ .\ .) \qquad \textbf{(22-5)}$$

Next look at an open (both ends) pipe. Both ends have to be loops (displacement

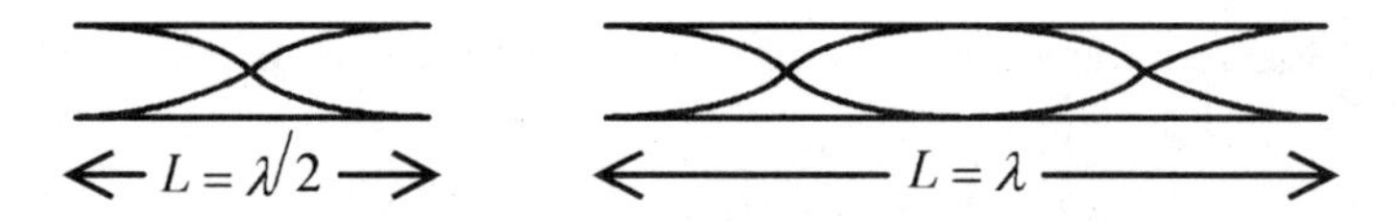

Fig. 22-4

antinodes), and the simplest situation is shown in Fig. 22-4. The second harmonic is the next simplest situation (also Fig. 22-4) with the general expression for the wavelength

$$\lambda_n = \frac{2L}{n}(n = 1,2,3\ .\ .\ .) \qquad \textbf{(22-6)}$$

and frequency

$$f_n = \frac{nv}{2L}(n = 1,2,3\ .\ .\ .) \qquad \textbf{(22-7)}$$

22-5 Standing waves are excited in a $1.0\,\text{m}$ long pipe open at one end, closed at the other. Take the speed of the waves as $340\,\text{m/s}$ and calculate the frequency for the fundamental and the first harmonic.

Solution: The wavelength is determined by the geometry, so look to Fig. 22-3 where the left diagram shows the fundamental wave. If the pipe is $1.0\,\text{m}$ then the wavelength is $4.0\,\text{m}$ and the frequency is

$$f = \frac{v}{\lambda} = \frac{340\,\text{m/s}}{4.0\,\text{m}} = 85\,\text{Hz}$$

The next wavelength is shown in the right diagram and requires a $\lambda = (4/3)\,\text{m}$ and

$$f = \frac{v}{\lambda} = \frac{340\,\text{m/s}}{4/3\,\text{m}} = 255\,\text{Hz}$$

22-6 What is the lowest possible frequency of a $5.0\,\text{m}$ long organ pipe?

Solution: The lowest frequency corresponds to the longest wavelength, and the longest wavelength is for a closed at one end open at the other pipe (see Figs. 22-3 and 22-4). In this situation the wavelength is $20\,\text{m}$ corresponding to a frequency of

$$f = \frac{v}{\lambda} = \frac{340\,\text{m/s}}{20\,\text{m}} = 17\,\text{Hz}$$

This frequency is below what most people can hear.

22-7 An open at both ends organ pipe has two successive harmonics of $567\,\text{Hz}$ and $850\,\text{Hz}$. What is the length of the pipe? Take the speed as $340\,\text{m/s}$.

Solution: For an open at both ends pipe the wavelengths are given by equation 22-6. For successive harmonics (n and $n+1$) two equations can be written in the form $v = \lambda f$.

$$340\,\text{m/s} = (2L/n)567\,\text{Hz} \quad \text{and} \quad 340\,\text{m/s} = (2L/n+1)850\,\text{Hz}$$

or

$$\frac{567}{n} = \frac{850}{n+1} \quad \text{or} \quad n = 2$$

Refer to Fig. 22-4. These frequencies correspond to the second and third harmonics. The lowest frequency ($567\,\text{Hz}$) corresponds to the longest wavelength and using $v = \lambda f$

$$\lambda = \frac{v}{f} = \frac{340\,\text{m/s}}{567\,\text{Hz}} = 0.60\,\text{m}$$

The pipe then is $0.60\,\text{m}$ long.

CHAPTER 23

SOUND

Longitudinal waves in air is called **sound**. The simplest physical picture of sound is a thin membrane vibrating (being driven) sinusoidally with the vibrations being transmitted through the air to another membrane that is set into sinusoidal motion.

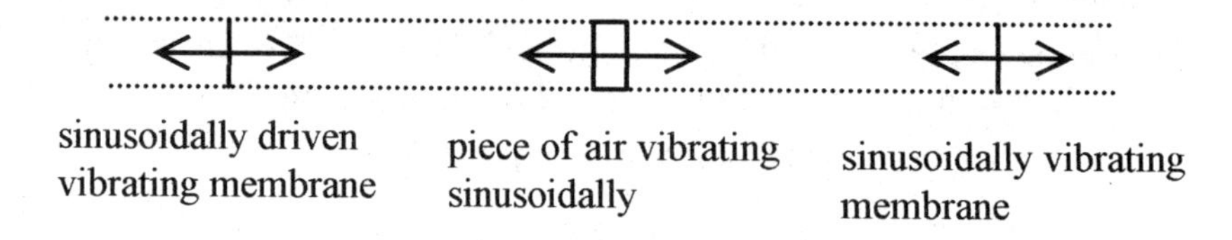

Fig. 23-1

Picture the source as an electrically driven speaker and the detector as your ear that detects the vibrations and transmits their frequency and intensity to your brain.

If the sending membrane executes one sinusoidal oscillation, a region of high pressure followed by a region of low pressure, is propagated through the air to the sensor which executes a similar sinusoidal oscillation as the variations in pressure pass by. (Regions of low pressure are called **rarefactions**.) Remember that the individual air molecules or slices of air between source and detector move only about an equilibrium position while the pressure-rarefaction pulse moves from source to detector.

Take the source as vibrating according to

$$y = A\sin(\omega t - kx) \qquad \textbf{(23-1)}$$

Remember that y is in the same direction as x, and y represents the displacement of a narrow slice of air while x represents the motion of the pulse from source to detector. Go over the picture of sound in your mind as a pressure-rarefaction pulse traveling from source to detector (see Fig. 23-1) until this last sentence becomes clear to you. When you understand this statement you have a good physical image of sound.

The pressure maximum is coincident with the maximum rate of change of y with respect to x. The pressure variation, which is proportional to $(\partial y / \partial x)$, is a cosine function. The slope of y

versus x is a maximum (positive) at $x = 0$, and maximum (negative) at $\pi/2$. Pressure, $(\partial y/\partial x)$ is 90° out of phase with displacement.

$$p = p_{0} \cos(\omega t - kx) \tag{23-2}$$

The ear is very sensitive to pressure variations, capable of detecting pressures of 10^{-5} Pa which is 10^{-10} of an atmosphere!

23-1 At some arbitrary position x graph displacement and pressure as a function of time on the same axis and explain the relationship.

Solution:

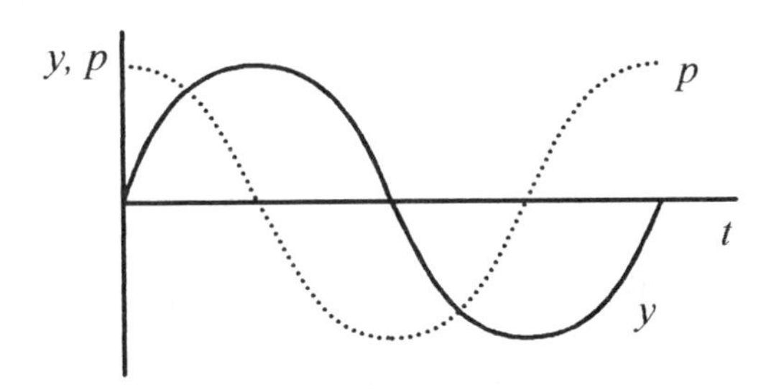

Fig. 23-2

Pick a point x where the displacement versus time is a sine curve (equation 23-1). The pressure curve (equation 23-2) is a cosine curve. Maximum pressure occurs when y is moving to the right with maximum speed. This corresponds to maximum positive slope or the point $t = 0$ on the graph. Minimum pressure (a rarefaction) occurs 180° later where the slope of y versus t has its largest negative value.

Intensity and the Inverse Square Law

Sound waves transport energy. Most sound sources radiate spherically symmetric patterns while most detectors intercept only a certain portion (solid angle) of the radiated energy. Therefore in sound measurements it is convenient to specify **intensity**, the power per unit area, rather than total power.

The power transmitted by a wave on a string is proportional to the square of the amplitude. With techniques similar to those used to determine the power transmitted by a wave on a string, the intensity of a sound wave is

$$I = \frac{p_{0}^{2}}{2\rho v} \tag{23-3}$$

where ρ is the density of air.

For a source of sound, the total power is the intensity times the spherical area corresponding to the point (at a radius out from the source) where the intensity is measured.

$$P_t = I(4\pi r^2) \tag{23-4}$$

23-2 What is the total power output of a source with intensity $0.050\,\mathrm{W/m^2}$ at a distance of $3.0\,\mathrm{m}$ from the source?

Solution: The total power is the intensity times the area of the sphere with radius $3.0\,\mathrm{m}$.

Fig. 23-3

$$P_t = I(4\pi r^2) = (0.050\,\mathrm{W/m^2})4\pi(9.0\,\mathrm{m^2}) = 5.6\,\mathrm{W}$$

23-3 What is the intensity of a $4.0\,\mathrm{W}$ source at $2.0\,\mathrm{m}$?

Solution: The power is $P = 4\pi r^2 I$ so $I = \dfrac{P}{4\pi r^2} = \dfrac{4.0\,\mathrm{W}}{4\pi \cdot 4.0\,\mathrm{m}^2} = 8.0\times 10^{-2}\,\mathrm{W/m^2}$

Consider a spherically symmetric source and intensity measurements on two different concentric spheres. The total power is the same so the power collected over one sphere is the same as the power collected over any other sphere. In equation form this statement is

$$P_1 = P_2 \quad \text{or} \quad 4\pi r_1^2 I_1 = 4\pi r_2^2 I_2 \tag{23-5}$$

leading to

$$\frac{I_1}{I_2} = \left(\frac{r_2}{r_1}\right)^2 \tag{23-6}$$

which is the inverse square law.

23-4 If the intensity from a source is $4.0 \times 10^{-3}\,\mathrm{W/m^2}$ at $12\,\mathrm{m}$, what will the intensity be at $4.0\,\mathrm{m}$?

Solution: Assume a spherically symmetric radiator and equate the powers at the two radii.

$$P_1 = P_2 \quad \text{requires} \quad 4\pi r_1^2 I_1 = 4\pi r_2^2 I_2$$

$$I_1 = \left(\frac{r_2}{r_1}\right)^2 I_2 = \left(\frac{12}{4.0}\right)^2 4.0 \times 10^{-3}\,\mathrm{W/m^2} = 3.6 \times 10^{-2}\,\mathrm{W/m^2}$$

Be careful on problems like this that you don't get the ratio upside down. Develop the habit of checking for the obvious. When you get closer to a source the intensity increases and vice versa.

23-5 How far away can a $50\,\mathrm{W}$ tornado siren be heard? Assume the minimum detectable intensity is $5.0 \times 10^{-6}\,\mathrm{W/m^2}$

Solution: The power is $P = 4\pi r^2 I$ so $r = \sqrt{\dfrac{P}{4\pi I}} = \sqrt{\dfrac{50\,\mathrm{W/m^2}}{4\pi 5.0 \times 10^{-6}\,\mathrm{W}}} = 890\,\mathrm{m}$

The Decibel

The human ear is responsive to sound intensities over a wide dynamic range, 10^{10} or more in intensity levels. A person's perception of "loudness" is that increasing intensity by a factor of 10 approximately doubles the perceived "loudness." This property of the human ear suggests that sound intensity, or level, should be measured with a logarithmic scale. The logarithmic decibel scale for measuring loudness is defined as

$$\beta = (10\,\mathrm{dB})\log(\mathrm{I}/\mathrm{I_o}) \qquad \textbf{(23-7)}$$

where $I_o = 1.0 \times 10^{-12}\,\mathrm{W/m^2}$ is the approximate threshold for human hearing.

23-6 Most people experience pain at sound levels above $100\,\mathrm{dB}$. What is the intensity associated with $100\,\mathrm{dB}$?

Solution: Equation 23-6 needs to be solved for *I*, and this requires switching from a logarithmic to an exponential equation. See the Introduction, Mathematical Background for a review of logarithmic and exponential equations.

First write $\beta/10 = \log(I/I_o)$ and remember that $x = \log y$ has the solution $y = 10^x$. The use of "log" implies base 10.

Switch from a logarithmic equation to an exponential one and write $I/I_o = 10^{\beta/10}$.

For $\beta = 100$, $\quad I = 10^{10} I_o = (10^{10})1.0 \times 10^{-12}\,\mathrm{W/m^2} = 1.0 \times 10^{-2}\,\mathrm{W/m^2}$

Practice switching logarithmic equations to exponential equations and vice versa and solving for the various variables. This is not a common algebraic operation, and you need sufficient practice to develop proficiency with exponents and logarithms.

23-7 What is the dB loss when you double your distance from a spherically symmetric constant-output power source?

Solution: Since the source is constant power output and spherically symmetric the inverse square law applies. Take r_1 and $r_2 = 2r_1$ and apply power equals power.

$$4\pi r_1^2 I_1 = 4\pi r_2^2 I_2 = 16\pi r_1^2 I_2 \quad \text{or} \quad I_1 = 4I_2$$

The dB levels at r_1 and r_2 are

$$\mathrm{dB}_1 = 10\log\frac{I_1}{I_o} \quad \text{and} \quad \mathrm{dB}_2 = 10\log\frac{I_2}{I_o}$$

and the difference is

$$\mathrm{dB}_1 - \mathrm{dB}_2 = 10\log\frac{I_1}{I_o} - 10\log\frac{I_2}{I_o} = 10\left[\log I_1 - \log I_o - \log I_2 + \log I_o\right] = 10\log\frac{I_1}{I_2} = 10\log 4 = 6.0$$

Doubling the distance from the source results in a 6.0 dB loss.

The Doppler Effect

The **Doppler effect** is the apparent increase or decrease in frequency of a source when the source and observer are moving toward or away from one other.

Fig. 23-4

First consider a stationary source and a moving observer. In Fig. 23-4 pressure pulses are depicted as equally spaced and moving radially out from the source. An observer at rest would observe vt/λ waves passing in a time t (vt is the length of the collection of waves or "wave train" traveling from source to observer and λ is the wavelength, so the quotient is the number of waves). The number of waves divided by the time is the frequency $f = vt/\lambda t$ resulting in $v = \lambda f$.

If the observer is moving toward the source with a velocity v_o he will intercept $v_o t/\lambda$ more waves, and the apparent frequency is

$$f' = \frac{vt/\lambda + v_o t/\lambda}{t} = \frac{v + v_o}{\lambda} = f\frac{v + v_o}{v} = f\left(1 + \frac{v}{v_o}\right)$$

If the observer is moving away from the source, the sign of $v_o t/\lambda$ is negative. The apparent frequency then corresponds to the negative sign in equation (23-7).

$$f' = f\left(1 \pm v/v_o\right) \tag{23-8}$$

23-8 What is the apparent frequency of a siren emitting a $300\,\mathrm{Hz}$ sinusoidal signal that you are approaching at $30\,\mathrm{m/s}$?

Solution: Use equation 23-8 with the plus sign because you are approaching the source and intercepting more waves than if you were stationary.

$$f' = 300\,\mathrm{Hz}(1 + 30/340) = 326\,\mathrm{Hz}$$

23-9 What is the apparent frequency of the siren in problem 23-8 after you have passed it and are moving away?

Solution: When you are moving away the frequency is lower because you are intercepting fewer wave fronts than if you were stationary so use the minus sign in equation 23-7 .

$$f' = 300\,\mathrm{Hz}(1 - 30/340) = 274\,\mathrm{Hz}$$

Now consider a moving source and a stationary observer. Figure 23-5 depicts the pressure pulses as packed closer together in the direction the source is moving. This effectively shortens the wavelength to a stationary observer.

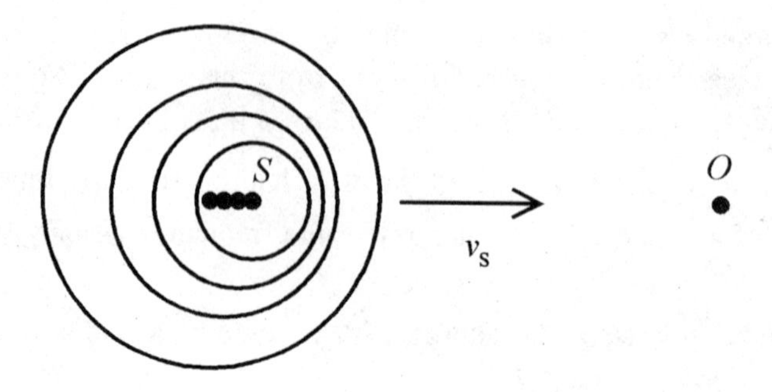

Fig. 23-5

The effective wavelength is $\lambda' = (v - v_S)/f$, and the apparent frequency is

$$f' = \frac{v}{\lambda'} = f\frac{v}{v - v_S} = f\left(\frac{1}{1 - v_S/v}\right)$$

For a source moving away from an observer the effective wavelength is $(v + v_S)/f$, so combining

$$f' = f\frac{v}{v \mp v_S} \tag{23-9}$$

The two expressions for source and observer moving (equations 23-7 and 23-8) can be combined

$$f' = f\frac{v \pm v_O}{v \mp v_S} \tag{23-10}$$

The way to keep the signs straight is to remember that on approach the frequency is increased and on departure the frequency is decreased.

23-10 A railroad train traveling at $30\,\text{m/s}$ emits a note at $400\,\text{Hz}$. What frequency is heard by an observer a) on the train, b) standing beside the track, and c) on another train approaching at $35\,\text{m/s}$?

Solution: a) On the train the observer hears the $400\,\text{Hz}$.

b) Beside the track with the train approaching the observer hears a higher frequency

$$f' = f\frac{v}{v - v_S} = 400\,\text{Hz}\frac{340}{310} = 439\,\text{Hz}$$

And with the train moving away the observer hears a lower frequency

$$f' = f\frac{v}{v+v_S} = 400\,\text{Hz}\frac{340}{370} = 368\,\text{Hz}$$

c) On the other train the observer hears a higher frequency due to the source approaching, $(v - v_S)$ term, and observer approaching, $(v + v_O)$ term.

$$f' = f\frac{v+v_O}{v-v_S} = 400\,\text{Hz}\frac{375}{310} = 484\,\text{Hz}$$

23-11 Submarines use $1000\,\text{Hz}$ signals to avoid collisions. A submarine traveling at $5.0\,\text{m/s}$ observes another submarine on the same course behind it and detects the collision avoidance signal from this submarine at $1005\,\text{Hz}$. What is the speed of the other submarine? The velocity of sound in water is $1500\,\text{m/s}$.

Solution: This is a Doppler effect problem with both source and observer moving. The hard part of the problem is getting the signs correct. First note that the frequency is higher than emitted so the submarines are closing. If the observer were stationary the formula would be $f' = f[v/(v - v_S)]$. If the observer were moving away from a stationary source the formula would be $f' = f[(v - v_O)/v]$.

So combining, the correct formula is $$f' = f\frac{v - v_O}{v - v_S}$$

and putting in the numbers $1005\,\text{Hz} = 1000\,\text{Hz}\left(\dfrac{1500-5}{1500-v_S}\right)$ or $v_S = 12\,\text{m/s}$.

CHAPTER 24

CHARGE AND COULOMB'S LAW

Coulomb's law is the basic force relationship between two charges. This force acts along the line connecting the charges and is repulsive for like charges and attractive for unlike charges.

$$F = \frac{1}{4\pi\varepsilon_o}\frac{q_1 q_2}{r^2} \qquad \textbf{(24-1)}$$

where $\varepsilon_o = 8.9\times10^{-12}\,\mathrm{C^2/Nm^2}$, and $\dfrac{1}{4\pi\varepsilon_o} = 9.0\times10^{9}\,\mathrm{Nm^2/C^2}$

The proportionality constant can, for the moment, be thought of as a constant that makes the force Newtons when the charges are measured in Coulombs and the separation in meters.

24-1 Two charges, $q_1 = 3.0\times10^{-8}\,\mathrm{C}$ and $q_2 = -4.0\times10^{-8}\,\mathrm{C}$, are separated by $6.0\times10^{-3}\,\mathrm{m}$ as shown in Fig 24-1. What is the force of one on the other?

Solution:

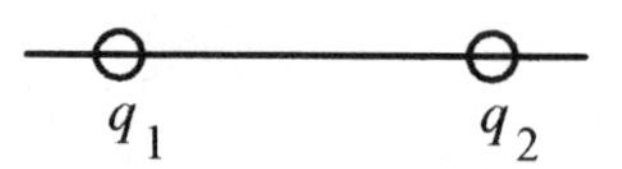

Fig. 24-1

$$F = \frac{1}{4\pi\varepsilon_o}\frac{q_1 q_2}{r^2} = 9.0\times10^{9}\frac{\mathrm{Nm^2}}{\mathrm{C^2}}\frac{12\times10^{-16}\,\mathrm{C^2}}{36\times10^{-6}\,\mathrm{m^2}} = 0.30\,\mathrm{N}$$

This is the force each charge experiences due to the other. The fact that q_1 is positive and q_2 is negative means that the force is directed from one to the other along the line joining their centers.

24-2 Consider a line of charges, $q_1 = 8.0\,\mu\mathrm{C}$ at the origin, $q_2 = -12\,\mu\mathrm{C}$ at $2.0\,\mathrm{cm}$, and $q_3 = 10\,\mu\mathrm{C}$ at $4.0\,\mathrm{cm}$, as shown in Fig. 24-2. What is the force on q_3 due to the other two charges?

Solution: The force on q_3 due to q_1 is repulsive as shown in Fig. 24-2. The magnitude of F_{31} is

$$F_{31} = 9.0 \times 10^9 \frac{\text{Nm}^2}{\text{C}^2} \frac{80 \times 10^{-12}\,\text{C}^2}{16 \times 10^{-4}\,\text{m}^2} = 450\,\text{N}$$

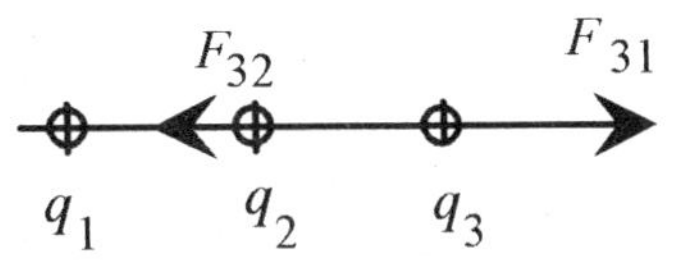

F_{32} F_{31}

q_1 q_2 q_3

Fig. 24-2

It is not necessary to be overly concerned about the sign of the force. The vector makes the direction clear. Drawing a vector diagram is a more sure way of getting the final sign correct than using algebraic signs in the force calculations.

The force on q_3 due to q_2 is attractive, and the magnitude is

$$F_{32} = 9.0 \times 10^9 \frac{\text{Nm}^2}{\text{C}^2} \frac{120 \times 10^{-12}\,\text{C}^2}{4.0 \times 10^{-4}\,\text{m}^2} = 2700\,\text{N}$$

The resultant force is 2250N toward q_1 and q_2.

24-3 Arrange three charges in the form of an equilateral triangle as shown in Fig. 24-3 and find the force on q_3 due to the other two.

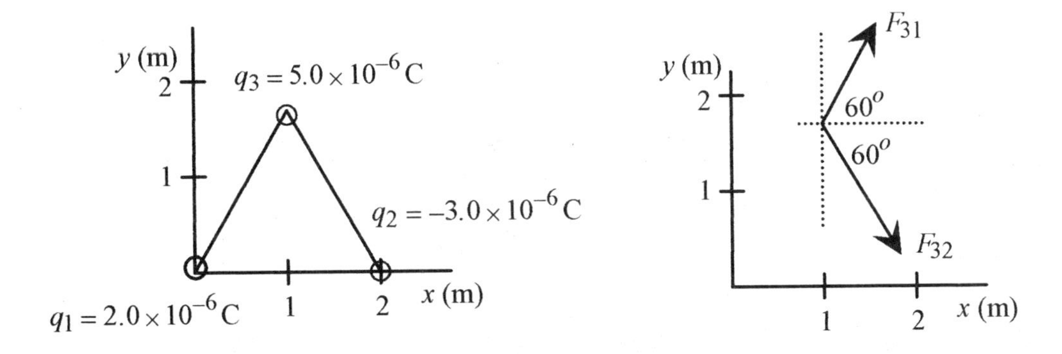

Fig. 24-3

Solution: This is a more complicated vector problem. First calculate the force on q_3 due to q_1 and then due to q_2. Adding these vector forces is called **superposition**. The forces can be added as vectors as long as the force due to one charge does not affect the force due to another charge. In this case the force on q_3 due to q_1 does not influence the force on q_3 due to q_2.

The force on q_3 due to q_1 is

$$F_{31} = 9.0 \times 10^9 \frac{\text{Nm}^2}{\text{C}^2} \frac{2.0 \times 10^{-6}\text{C} \cdot 5.0 \times 10^{-6}\text{C}}{4.0\text{m}^2} = 0.0225\text{N}$$

and is directed as shown in Fig. 24-3. Likewise the force on q_3 due to q_2 is

$$F_{32} = 9.0 \times 10^9 \frac{\text{Nm}^2}{\text{C}^2} \frac{3.0 \times 10^{-6}\text{C} \cdot 5.0 \times 10^{-6}\text{C}}{4.0\text{m}^2} = 0.0338\text{N}$$

and is directed as shown. Referring to Fig. 24-3, the x and y components of the resultant force can be written. The x-component of these two forces is

$$F_x = F_{31}\cos 60^o + F_{32}\cos 60^o = (0.0112 + 0.0169)\,\text{N} = 0.0281\text{N}$$

The y-component of these two forces is

$$F_y = F_{31}\sin 60^o - F_{32}\sin 60^o = (0.0195 - 0.0292)\,\text{N} = 0.0097\text{N}$$

Second Solution: Using unit vector notation the forces are written as

$$\mathbf{F}_{31} = F_{31}\cos 60^o \boldsymbol{i} + F_{31}\sin 60^o \boldsymbol{j} = (0.0112\boldsymbol{i} + 0.0195\boldsymbol{j})\,\text{N}$$

$$\mathbf{F}_{32} = F_{32}\cos 60^o \boldsymbol{i} - F_{32}\sin 60^o \boldsymbol{j} = (0.0169\boldsymbol{i} - 0.0292\boldsymbol{j})\,\text{N}$$

and combine to

$$\mathbf{F}_3 = (0.0281\boldsymbol{i} - 0.0097\boldsymbol{j})\,\text{N}$$

Rather than worry about the algebraic signs of angles other than 0° to 90° draw the picture and keep the angles under 90°. This way it is easy to "see" the sign of the various components. Most mistakes in problems like this are algebraic sign mistakes. The way to avoid these mistakes is to draw and label a vector diagram, more than one if necessary.

As an exercise find the magnitude and direction of the resultant.

A classic charge problem is one where charged balls are hung by strings from a common point with the question being how far apart the balls separate for a given equal charge.

24-4 Consider two conducting balls both of mass m and equal charge q suspended by non-conducting cords of equal length, ℓ as shown in Fig 24-4. How does the separation of the balls depend on charge, mass, and length?

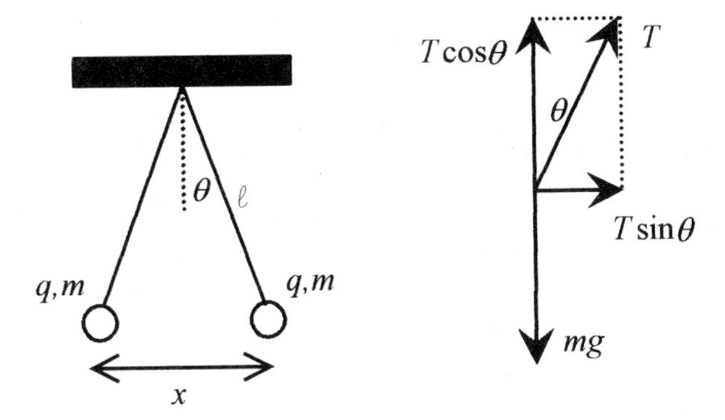

Fig. 24-4

Solution: The arrangement is shown in the accompanying drawing along with the vector diagram. This is a force balance problem. The mechanical force is due to gravity and is directed down. The electrical force is the Coulomb force and is directed along the line of centers of the conducting balls, which for this geometry is horizontal.

On the vector diagram the mg force is balanced by the vertical component of the tension in the cord, or $T\cos\theta = mg$. The horizontal component of the tension is $T\sin\theta$, and replacing T with $mg/\cos\theta$ the horizontal component is

$$T\sin\theta = \frac{mg\sin\theta}{\cos\theta} = mg\tan\theta$$

But $\tan\theta$ can be written in terms of the geometry as $\tan\theta = \frac{x/2}{\ell} = \frac{x}{2\ell}$ and $T\sin\theta = \frac{mgx}{2\ell}$

The horizontal component of the tension in the cord is balanced by the Coulomb force of

repulsion $$F = \frac{1}{4\pi\varepsilon_o}\frac{q^2}{x^2} = T\sin\theta = \frac{mgx}{2\ell}$$

or $$x = \left[\frac{q^2\ell}{2\pi\varepsilon_o mg}\right]^{1/3}$$

24-5 What is the force of attraction between sodium and chlorine ions in salt molecules if each ion carries one electronic charge and the separation is $2.8\times10^{-10}\,\text{m}$?

Solution: $$F = \frac{1}{4\pi\varepsilon_o}\frac{q^2}{r^2} = 9.0\times10^9\,\frac{\text{Nm}^2}{\text{C}^2}\frac{(1.6\times10^{-19}\,\text{C})^2}{(2.8\times10^{-10}\,\text{m})^2} = 2.9\times10^{-9}\,\text{N}$$

CHAPTER 25

THE ELECTRIC FIELD

The **electric field** is defined as the force per unit charge $E = F/q_o$ and as such is a vector. Electric fields can be superposed - added as vectors. The units of the electric field are $\mathrm{N/C}$ or $\mathrm{Volt/meter}$. In terms of charges the electric field has magnitude

$$E = \frac{1}{4\pi\varepsilon_o}\frac{q}{r^2} \tag{25-1}$$

and direction, the direction of a unit positive charge placed at r.

Discrete Charges

A classic problem that illustrates the vector nature of electric fields is the calculation of the E field on the axis of a dipole.

25-1 An electric dipole is depicted as equal and opposite charges separated by a distance $2a$ as shown in Fig. 25-1. Find the electric field at a distance r along the bisector of the center line of the charges.

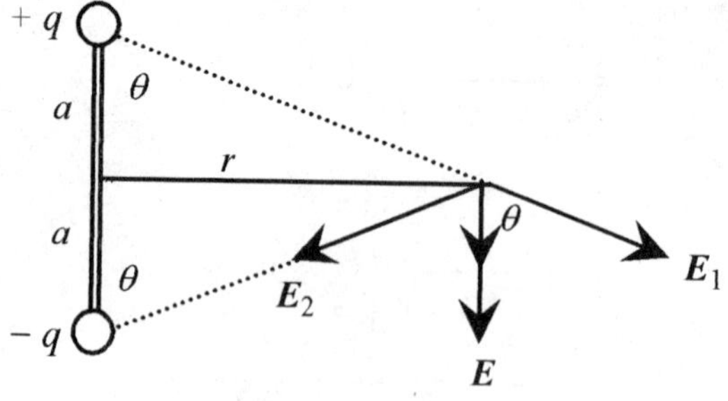

Fig. 25-1

Solution: First draw the diagram and note the direction of the electric field vectors. The horizontal components of E_1 and E_2 add to zero. The vertical component is made up of two vectors of equal magnitude, $E_1 \cos\theta$. The magnitude of the vertical vector is $2E_1 \cos\theta$.

From the geometry, $\cos\theta = \dfrac{a}{\sqrt{a^2 + r^2}}$ and $E_1 = \dfrac{q}{4\pi\varepsilon_o(a^2 + r^2)}$ so

$$E = \frac{1}{4\pi\varepsilon_o}\frac{2aq}{(a^2 + r^2)^{3/2}} \tag{25-2}$$

This is the magnitude of the electric vector. The direction of the vector is down or parallel to a line drawn from $+q$ to $-q$. If r is large compared to a (often the case) then

$$E \approx \frac{1}{4\pi\varepsilon_o}\frac{2aq}{r^3} \tag{25-3}$$

and for large distances the field falls off as $1/r^3$. The product $2aq$ is called the dipole moment $p = 2aq$. It is interesting to note that for a point charge the E field is radial (in or out) and falls off as $1/r^2$ while for the dipole E is tangential to the radius and falls off as $1/r^3$.

25-2 Calculate the electric field due to two charges, $q_1 = 4.0\times10^{-6}\,\text{C}$ and $q_2 = -3.0\times10^{-6}\,\text{C}$, separated by 3.0 m and at a point 3.0 m opposite q_2 as shown in Fig. 25-2.

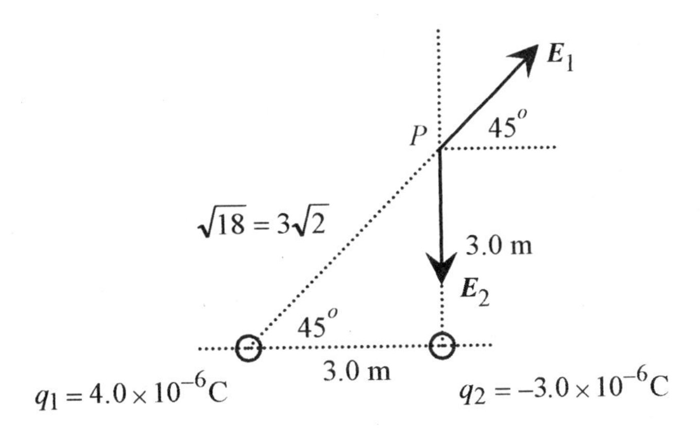

Fig. 25-2

Solution: The electric field due to the first charge is

$$E_1 = \frac{1}{4\pi\varepsilon_o}\frac{q}{r^2} = 9.0\times10^9\,\frac{\text{Nm}^2}{\text{C}^2}\,\frac{4.0\times10^{-6}\,\text{C}}{18\,\text{m}^2} = 2.0\times10^3\,\text{N/C}$$

The electric field due to the second charge is

$$E_2 = \frac{1}{4\pi\varepsilon_o}\frac{q}{r^2} = 9.0\times10^9\,\frac{\text{Nm}^2}{\text{C}^2}\,\frac{3.0\times10^{-6}\,\text{C}}{9.0\,\text{m}^2} = 3.0\times10^3\,\text{N/C}$$

The direction of the fields due to these two charges is shown on the diagram. The vectors are written in component form

$$E_1 = (2.0\times10^3\cos45^o\,\boldsymbol{i} + 2.0\times10^3\sin45^o\,\boldsymbol{j})\,\text{N/C} = (1.4\times10^3\boldsymbol{i} + 1.4\times10^3\,\boldsymbol{j})\,\text{N/C}$$

$$E_2 = (0\boldsymbol{i} + -3.0\times10^3\,\boldsymbol{j})\,\text{N/C}$$

with resultant $\boldsymbol{E_R} = (1.4\times10^3\boldsymbol{i} + -1.6\times10^3\,\boldsymbol{j})\,\text{N/C}$

As an exercise verify that in number plus angle form this vector is $\boldsymbol{E_R} = 2.1\times10^3\angle -49^o\,\text{N/C}$.

25-3 Find an expression for the electric field due to a collection of charges of equal magnitude all placed in a line and separated by a distance d. The charges have alternate signs, and the field to be calculated is the field on a line normal to the line of charges and opposite the central positive charge. Additionally, look for approximate expressions for long distances away from the line of charges.

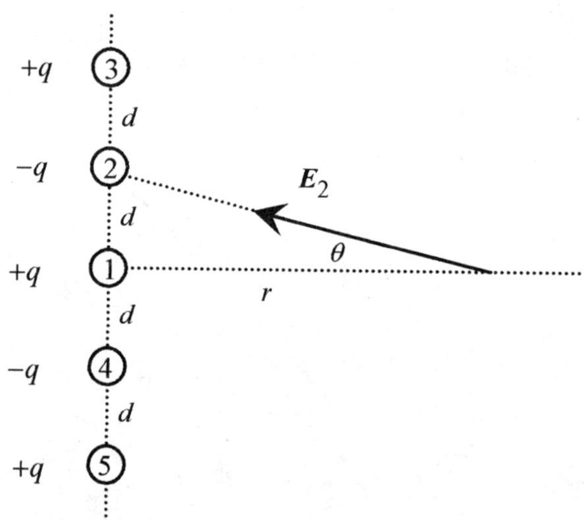

Fig. 25-3

Solution: The electric fields are labeled to correspond with the charges. The fields at a point r distance away from the line of charge are:

$$E_1 = \frac{1}{4\pi\varepsilon_o}\frac{q}{r^2} \qquad E_2 = E_4 = \frac{1}{4\pi\varepsilon_o}\frac{(-q)}{r^2+d^2} \qquad E_3 = E_5 = \frac{1}{4\pi\varepsilon_o}\frac{q}{r^2+4d^2}$$

Look at the geometry associated with $\boldsymbol{E_2}$ and notice that this vector can be written in terms of a horizontal and a vertical component. The vector $\boldsymbol{E_4}$ also has vertical and horizontal components.

The vertical components of E_2 and E_4 add to zero while the horizontal components add along *r*. The same is true of vectors E_3 and E_5. The horizontal component of E_2 is E_2 times the cosine of the angle θ, thus the contribution of E_2 and E_4 is two times the magnitude of E_2 times the cosine of the angle θ.

E_2 and E_4 contribute $2E_2 \cos\theta = \dfrac{1}{4\pi\varepsilon_o}\dfrac{-2q}{r^2+d^2}\dfrac{r}{\sqrt{r^2+d^2}}$ pointed in along *r*.

Similarly E_3 and E_5 contribute $\dfrac{1}{4\pi\varepsilon_o}\dfrac{2qr}{r^2+d^2}\dfrac{r}{\sqrt{r^2+4d^2}}$ also pointed out along *r*

Combining the contributions from all the charges

$$E = \frac{1}{4\pi\varepsilon_o} q\left[\frac{1}{r^2} - \frac{2r}{(r^2+d^2)^{3/2}} + \frac{2r}{(r^2+4d^2)^{3/2}}\right]$$

If *r* is comparable to *d* then this formula is appropriate. However if *r* is large compared to *d*, then the *r* terms predominate in the denominators and the expression is approximately

$$E \approx \frac{1}{4\pi\varepsilon_o} q\left[\frac{1}{r^2} - \frac{2}{r^2} + \frac{2}{r^2}\right] \approx \frac{1}{4\pi\varepsilon_o}\frac{q}{r^2}$$

This formula shows that at long distances the array looks like one positive charge.

Linear Charge

So far we have looked at the electric fields that result from discrete charges. If the charges are spread out, such as over a surface, then the definition of electric field has to be modified to a calculus definition where the contribution of each element of charge to the total field is summed.

$$dE = \frac{1}{4\pi\varepsilon_o}\frac{dq}{r^2} \qquad \textbf{(25-4)}$$

(C) **25-4** A problem that illustrates this is the calculation of the electric field on the axis of a ring of positive charge. Set up the problem as shown in Fig. 25-4.

Solution: Each element of charge dq produces an element of electric field dE at a point on the bisector of a diameter of the ring. This dE has components $dE\cos\theta$ along the direction of the bisector and $dE\sin\theta$ perpendicular to the bisector. An element of dq at one spot on the ring produces $dE\sin\theta$, and across the diameter of the ring a similar element (of dq) produces a

$dE \sin\theta$ component that cancels this one. Because of this symmetry the $dE \sin\theta$ components add to zero.

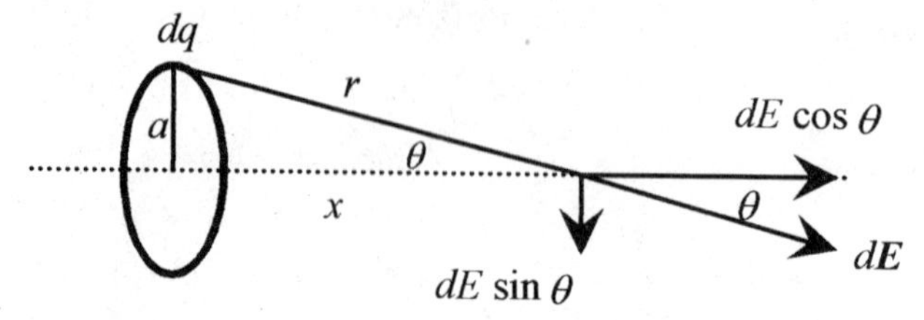

Fig. 25-4

The magnitude of the electric field due to each element of charge can be written as

$$dE = \frac{1}{4\pi\varepsilon_o} \frac{dq}{a^2 + x^2}$$

but it is $dE\cos\theta$ which ultimately produces the field pointing outward from the center of the ring. From the geometry $\cos\theta = x/\sqrt{a^2+x^2}$, so the differential electric field component is

$$dE\cos\theta = \frac{1}{4\pi\varepsilon_o} \frac{dq}{a^2+x^2} \frac{x}{\sqrt{a^2+x^2}}$$

This expression can be summed (integrated) without the need for formal integration. The left side is just the total electric field, the sum of the $dE\cos\theta$ contributions. The right side is the sum of all the elements of dq, which is just q, so

$$E = \frac{1}{4\pi\varepsilon_o} \frac{qx}{(a^2+x^2)^{3/2}}$$

Notice that at $x = 0$, $E = 0$, as it should and that for $x >> a$, $E \approx \frac{1}{4\pi\varepsilon_o}\frac{q}{x^2}$, again as it should. Far away the ring of charge should look like a point.

Surface Charge

(C) **25-5** Calculate the electric field vertically out from a uniformly charged disk of radius R and charge density (charge/area) σ.

Solution: The calculus definition of the electric field is that each element of charge produces an element of electric field according to equation 25-4 where r is the distance from the element to the point where dE is calculated. Because of the symmetry of the problem the dE is produced by a dq in the form of a ring of charge of radius r and width dr. Refer to Fig. 25-5 and note that the

dE produced by this ring of charge is a cone with horizontal components adding to zero. A piece of the ring producing a piece of dE has a corresponding piece across a diameter. The horizontal components of these corresponding pieces add to zero. The vertical component is $dE\cos\theta$ with $\cos\theta$ determined from the geometry as $\cos\theta = x/\sqrt{x^2+r^2}$. Summing the contributions over r produces the electric field pointing away (and perpendicular to) the charged disk. The differential element of charge is $dq = (2\pi r dr)\sigma$ ($2\pi r dr$ is the differential element of area), thus

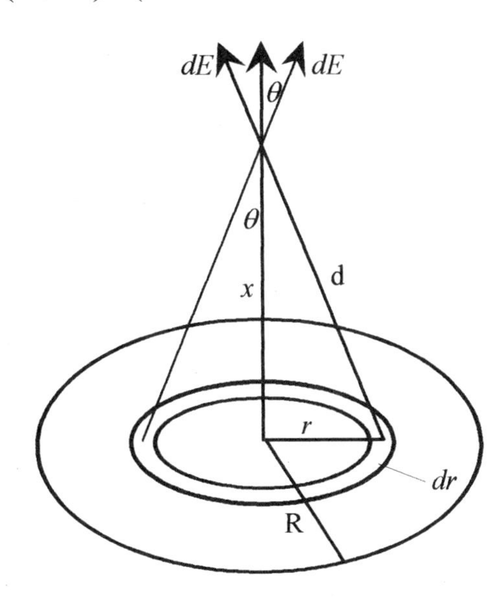

Fig. 25-5

$$E = \frac{1}{4\pi\varepsilon_o}\int_0^R \frac{\sigma 2\pi r dr}{x^2+r^2}\frac{x}{\sqrt{x^2+r^2}} = \frac{\sigma x}{2\varepsilon_o}\int_0^R \frac{r dr}{(x^2+r^2)^{3/2}}$$

Remember the integral is over r and not x. Make a change of variable with $u = x^2 + r^2$ and $du = 2r dr$. The integral transforms according to

$$\int \frac{r dr}{(x^2+r^2)^{3/2}} = \frac{1}{2}\int \frac{du}{u^{3/2}} = -\frac{1}{\sqrt{u}} = -\frac{1}{\sqrt{x^2+r^2}}$$

Changing the sign and switching the limits

$$E = \frac{\sigma x}{2\varepsilon_o}\left[\frac{1}{\sqrt{x^2+r^2}}\right]_R^0 = \frac{\sigma x}{2\varepsilon_o}\left[\frac{1}{x\sqrt{1+r^2/x^2}}\right]_R^0 = \frac{\sigma}{2\varepsilon_o}\left[1 - \frac{1}{\sqrt{1+R^2/x^2}}\right]$$

When R is large compared to x, the bracket approaches one and the electric field is $\frac{\sigma}{2\varepsilon_o}$.

Applications

25-6 Consider an electron ballistics problem where an electron is projected between two plates (as shown in Fig. 25-6) at an initial velocity of $6.0\times10^6\,\mathrm{m/s}$ and angle of 45°. The electric field is $2.0\times10^3\,\mathrm{N/C}$ and directed up. The length of the plates is 10 cm and separation 2.0 cm. Will the electron strike the plates? And if so, where?

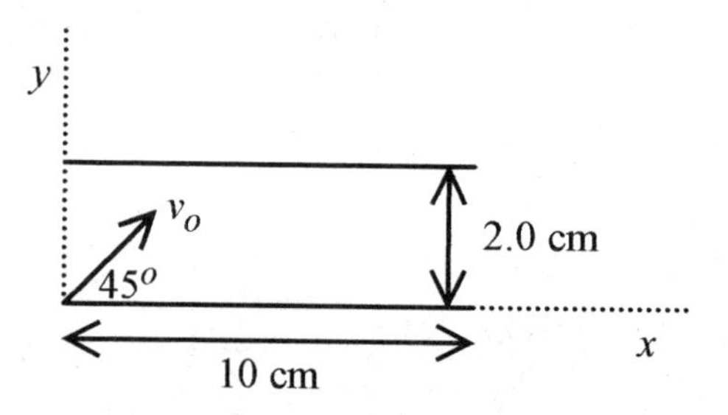

Fig. 25-6

Solution: This is a projectile problem. Look through Chapter 4, Projectile Motion, and refresh your memory on the general approach to projectile motion problems before continuing with this problem. The electronic charge and mass are given on the constants page.

First calculate the acceleration. The force on the electron comes from the electric field, $F = eE$, and this force makes the electron accelerate leading to $a = eE/m$. Using the values for e and m for the electron

$$a = \frac{eE}{m} = \frac{1.6\times10^{-19}\,\mathrm{C}\cdot2.0\times10^3\,\mathrm{N/C}}{9.1\times10^{-31}\,\mathrm{kg}} = 3.5\times10^{14}\,\mathrm{m/s^2}$$

The electric field is directed up, which would be the direction of acceleration of a positive charge. For the negatively charged electron the acceleration is down. Using this logic argument to determine the direction of the acceleration is easier than carrying minus signs throughout the problem.

The most convenient origin for the coordinate system is the edge of the plate where the initial velocity and direction are known with x positive to the right and y positive up. Now write the six equations describing the motion.

$$a_x = 0 \qquad a_y = -3.5\times10^{14}\,\mathrm{m/s^2}$$

$$v_x = v_o\cos45^o = 4.2\times10^6\,\mathrm{m/s} \qquad v_y = 4.2\times10^6\,\mathrm{m/s} - (3.5\times10^{14}\,\mathrm{m/s^2})t$$

$$x = (4.2\times10^6\,\mathrm{m/s})t \qquad y = (4.2\times10^6\,\mathrm{m/s})t - (1.8\times10^{14}\,\mathrm{m/s^2})t^2$$

To determine if the electron hits the upper plate, set $y = 2.0\times10^{-2}$ m and solve for t. Without units

$$1.8\times10^{14}t^2 - 4.2\times10^6 t + 2.0\times10^{-2} = 0$$

and

$$t = \frac{4.2\times10^6 \pm \sqrt{18\times10^{12} - 4(1.8\times10^{14})(2.0\times10^{-2})}}{3.5\times10^{14}} = \frac{4.2\times10^6 \pm 1.9\times10^6}{3.5\times10^{14}} = 0.66\times10^{-8}\,\mathrm{s}$$

The smallest time is taken as this corresponds to the electron striking the top plate. If the solution for t had contained an imaginary number, this would have indicated that the electron would never reach the top plate.

Now with this time calculate the position in x where the electron strikes the plate.

$$x = 4.2\times10^6\,\mathrm{m/s}\cdot0.66\times10^{-8}\,\mathrm{s} = 2.8\times10^{-2}\,\mathrm{m}$$

Another way to solve the problem is to go back to the six equations describing the motion and solve for y as a function of x. First solve the x equation for t and substitute this expression for t into the y equation to obtain

$$y = x - 1.8\times10^{14}\,\frac{\mathrm{m}}{\mathrm{s}^2}\,\frac{x^2}{(4.2\times10^6\,\mathrm{m/s})^2} = x - 10x^2$$

When $y = 0.020$ this equation becomes $10x^2 - x + 0.020 = 0$ with solution

$$x = \frac{1\pm\sqrt{1-4(10)(0.020)}}{2\cdot10} = \frac{1\pm0.45}{20} = 2.8\times10^{-2}\,\mathrm{m}$$

Here again be sure to take the value of x corresponding to the smallest positive time.

Dipoles

A dipole placed in an electric field experiences a torque. The **dipole moment**, defined earlier as $p = 2aq$, can be expanded to define the dipole moment vector of magnitude $2aq$ pointing from $-q$ to $+q$. The force on each charge is qE, and they add to zero so the dipole does not translate.

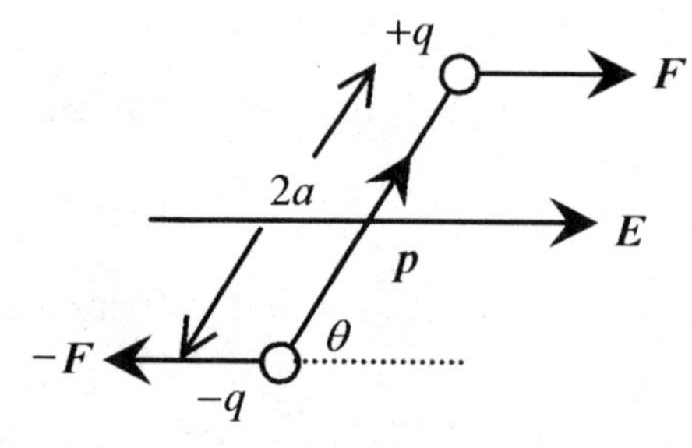

Fig. 25-7

It does, however, rotate due to the torque of magnitude $2aF\sin\theta$. When $\theta = 0$ and the dipole is lined up with the field, the torque is zero. But when $\theta = 90^o$ the torque is $2aF$. Multiply the torque by q/q to obtain

$$\tau = \frac{2aqF}{q}\sin\theta$$

which is a cross product, or

$$\tau = p \times E \qquad \textbf{(25-5)}$$

25-7 A dipole with charge $6.0\times10^{-6}\,\mathrm{C}$ and separation $4.0\times10^{-3}\,\mathrm{m}$ is immersed in an E-field of $3.0\times10^{6}\,\mathrm{N/C}$. Calculate the maximum torque.

Solution: $\tau_{\max} = 2aqE = 2\cdot4.0\times10^{-3}\,\mathrm{m}\cdot6.0\times10^{-6}\,\mathrm{C}\cdot3.0\times10^{6}\,\mathrm{N/C} = 144\times10^{-3}\,\mathrm{Nm}$

If a dipole is subjected to an E-field and the dipole is rotated by that electric field, then work is done on the dipole. The work done by the field must be

$$W = \int_{\theta_0}^{\theta_1} T\,d\theta = \int_{\theta_0}^{\theta_1} pE\sin\theta\,d\theta = -T\cos\theta\Big|_{\theta_0}^{\theta_1}$$

Be careful with the signs and angles.

25-8 Take the original position $\theta_0 = 70^o$, and the final position $\theta_1 = 20^o$ and calculate the work performed on the dipole and the energy stored in the system.

Solution: $$W = -pE\cos\theta\Big|_{70^o}^{20^o} = 0.144\,\mathrm{Nm}(\cos70^o - \cos20^o) = -0.086\,\mathrm{J}$$

The electric field has done $0.086\,\mathrm{J}$ of work on the dipole, and this is the amount of energy stored in the dipole-field system.

CHAPTER 26

GAUSS' LAW

Gauss' law can be tricky. The concepts expressed in mathematical terms often imply considerable mathematical sophistication to work the problems. This is almost always not the case. In general; if you are involved in excessive mathematical manipulations you are doing the problem wrong or do not see an easy way to apply Gauss' law. Remember that Gauss' law represents the **density of flux** lines through an area. Remember also that most of the difficult looking integrals never occur.

Start with a definition of **flux** as

$$\Phi_E = \int \boldsymbol{E} \cdot d\boldsymbol{s} \qquad \textbf{(26-1)}$$

This states that the flux is the sum of the vector (dot) product of electric field over a surface times the area. $\boldsymbol{E}$ represents the electric field, and $d\boldsymbol{s}$ is a vector normal to the surface representing a differential element of surface. The dot product is the component of $\boldsymbol{E}$ parallel to $d\boldsymbol{s}$, that is, normal to the surface, times ds. If $\boldsymbol{E}$ is a constant (most often the case) then the sum of (integral of) $\boldsymbol{E} \cdot d\boldsymbol{s}$ over the surface is the component of $\boldsymbol{E}$ normal to the surface area, times the total surface area.

(C) **26-1** Calculate the flux through a square plate 1.0 m on a side inclined at 120° with respect to a constant electric field of $100 \mathrm{N/C}$. You may at this point want to review the definition of dot product in Chapter 1, Vectors, and the concept of the integral in the Introduction, Mathematical Background.

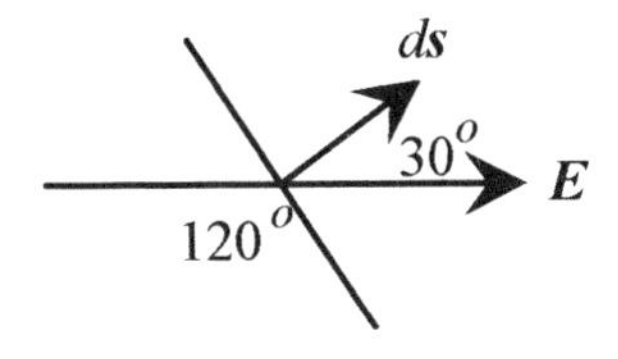

Fig. 26-1

Solution: The surface area *ds* is represented by a vector normal to the surface. The vector product $\boldsymbol{E} \cdot d\boldsymbol{s}$ is $Eds(\cos 30^o)$. Integrating this product

$$\int \mathbf{E}\cdot d\mathbf{s} = EA(\cos\theta) = 100\,\mathrm{N/C}(1.0\,\mathrm{m}^2)\cos 30^o = 87\,\mathrm{Nm}^2/\mathrm{C}$$

This represents the flux through the surface.

The construct (concept) "flux" can be thought of as a collection of lines intercepting this square plate. If the electric field is increased, then there are more "flux lines" per square meter and, if the angle is changed, more or less flux lines are intercepted as the angle is changed.

This integral of a vector product that sounds like a very difficult problem is performed rather easily with an understanding of dot products and integrals.

Electric field lines are a convenient concept or construct to help us visualize the electric field. The electric field lines are envisioned as vectors going from positive charges to negative charges, the direction a unit positive charge would move. The electric field lines between two oppositely charged plates are just lines going from one plate to another. Higher fields would be envisioned as more lines or more lines per cross-sectional area.

In the case of a sphere surrounding a unit positive charge the field lines would be pointing radially out, the direction a unit positive charge would move. On spheres with different radii the electric field would be different; the field line density (number per square meter) would be different. However if we could sum all the electric field lines over the different spheres we would find the same number of lines. Their number density would decrease as the radii were increased but the total number over a sphere would remain the same.

(C) The formal **Gauss' law** connects flux to the charge contained again via an integral

$$\varepsilon_o \Phi_E = q_{\text{enclosed}} \quad \text{or} \quad \varepsilon_o \oint \mathbf{E}\cdot d\mathbf{s} = q_{\text{enclosed}} \tag{26-2}$$

The charge q is the net charge enclosed by the integral. The ε_o can, for the moment, be thought of as a constant that makes the units come out right.

Place a charge q at the center of a sphere and apply Gauss' law $\varepsilon_o \oint \mathbf{E}\cdot d\mathbf{s} = q$.

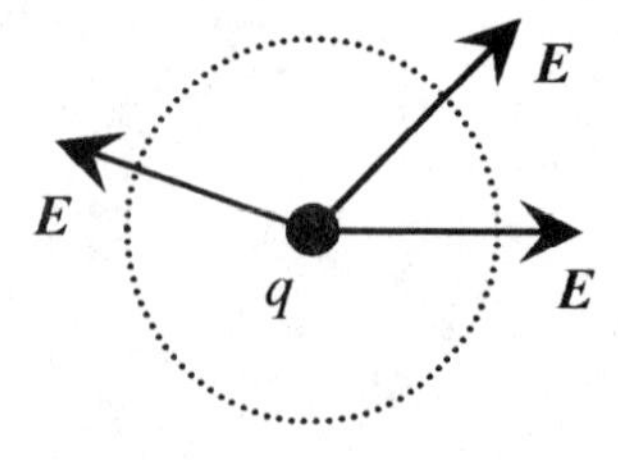

Fig. 26-2

The little circle on the integral sign serves as a reminder to integrate over an enclosed surface. The sphere we are integrating over in this instance, not a physical sphere, is known as a Gaussian surface - a surface where the integral is taken when applying Gauss' law. Now $\boldsymbol{E}$ is a constant over the surface (symmetry) so the integral becomes

$$\varepsilon_o E \oint ds = q \text{ and } \oint ds \text{ is just the surface area of the sphere or}$$

$$\varepsilon_o E(4\pi r^2) = q \qquad \text{and} \qquad E = \frac{q}{4\pi\varepsilon_o r^2} \tag{26-3}$$

which is just the statement we have from Coulomb's law and the definition of the electric field.

26-2 The electric field at the surface of the earth is approximately $150\,\mathrm{N/C}$ directed down. Calculate the sign and magnitude of the charge on the earth.

Solution: Gauss' law states that $\varepsilon_o \oint \boldsymbol{E} \cdot d\boldsymbol{s} = q_{\text{enclosed}}$

Construct a spherical Gaussian surface just outside the physical surface of the earth where E is everywhere normal and of value $150\,\mathrm{N/C}$. The integral of $\boldsymbol{E} \cdot d\boldsymbol{s}$ is the sum of this field over the Gaussian surface. $\boldsymbol{E}$ and $\boldsymbol{ds}$ are 180° so their dot product is negative.

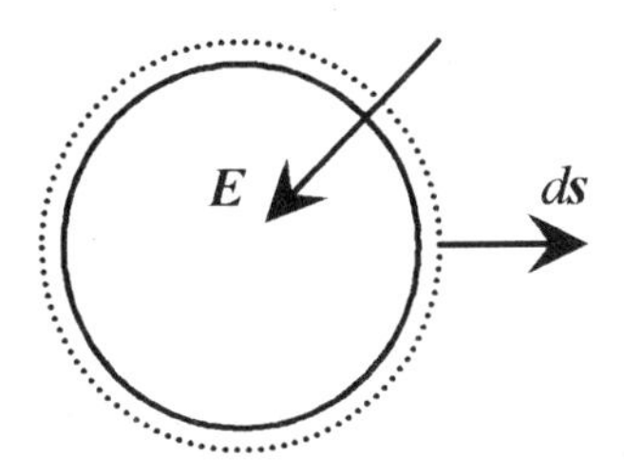

Fig. 26-3

$$q = -\varepsilon_o E(4\pi R_E^2) = -8.9\times10^{-12}\,\mathrm{C^2/Nm^2}\,(150\,\mathrm{N/C})4\pi(6.4\times10^6\,\mathrm{m})^2 = -6.8\times10^5\,\mathrm{C}$$

The field pointing down indicates that the charge is negative as does the dot product being negative. Again notice that this integral over the Gaussian surface is quite easy. These past two problems are very illustrative of the "integration by inspection" approach to problems involving Gauss' law.

26-3 Calculate the sign and magnitude of the charge contained in a cube 10 cm on a side oriented as shown where the E-field is given by

$$E_x = (800\,\mathrm{N/Cm^{1/2}})x^{1/2} \qquad E_y = 0 \qquad E_z = 0$$

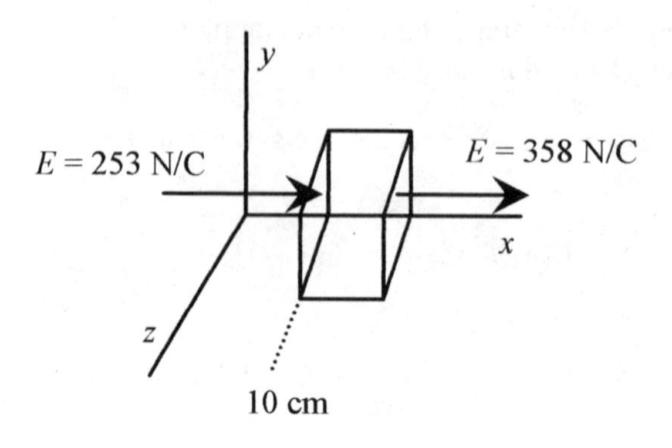

FIG. 26-4

Solution: First calculate the E-fields entering and exiting the cube. Watch the algebraic signs carefully.

$$E_x\big|_{10\text{cm}} = 800(0.10\text{m})^{1/2}\ \text{N/C} = 253\,\text{N/C}$$

$$E_x\big|_{20\text{cm}} = 800(0.20\text{m})^{1/2}\ \text{N/C} = 358\,\text{N/C}$$

There is more flux leaving the cube than there is entering it, so the net charge inside the cube must be positive. On the left side E is to the right and the vector representing ds is to the left, so

$$\int_l \boldsymbol{E}\cdot d\boldsymbol{s} = -(253\text{N/C})(0.010\text{m}^2) = -2.53\text{N/C}$$

Remember that the direction of $d\boldsymbol{s}$ is always outward from the enclosed volume. On the right side E is to the right and the vector representing ds is to the right so

$$\int_r \boldsymbol{E}\cdot d\boldsymbol{s} = 358\text{N/C}(0.010\text{m}^2) = 3.58\text{Nm}^2/\text{C}$$

The integrals over y and z are zero and the integrals over x add to $1.05\text{Nm}^2/\text{C}$. Now q can be calculated

$$q = \varepsilon_0 \oint \boldsymbol{E}\cdot d\boldsymbol{s} = 8.9\times10^{-12}\,\text{C}^2/\text{Nm}^2 \cdot 1.05\text{Nm}^2/\text{C} = 9.3\times10^{-12}\,\text{C}$$

26-4 The total charge (charge density) contained in an electron stream can be calculated with Gauss' law. If the field at the edges of a stream of square cross section $1.0\,\text{cm}$ on a side is $1.0\times10^3\ \text{N/C}$, calculate the charge density in the stream. Assume that the E-field in the direction of the stream is constant.

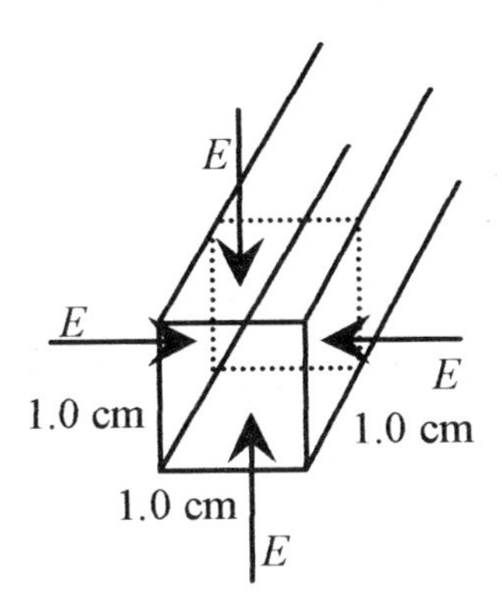

Fig. 26-5

Solution: The field is directed in, so the charge is negative and $\varepsilon_o \oint \mathbf{E}\cdot d\mathbf{s} = (-)q$ reduces to calculating four identical integrals. $\int E\cdot ds = 1.0\times10^3\,\mathrm{N/C}\cdot 1.0\times10^{-4}\,\mathrm{m}^2 = -0.10\,\mathrm{Nm^2/C}$.

The charge in a cube 1.0 cm on a side is now

$$q = \varepsilon_o \int \mathbf{E}\cdot d\mathbf{s} = 8.9\times10^{-12}\,\mathrm{C^2/Nm^2}\,(4)0.10\,\mathrm{Nm^2/C} = 35.6\times10^{-13}\,\mathrm{C}$$

The charge density is

$$\frac{q}{V} = \frac{35\times10^{-13}\,\mathrm{C}}{1.0\times10^{-6}\,\mathrm{m}^3} = 3.56\times10^{-6}\,\mathrm{C/m^3}$$

For the electron charge density we need to introduce the charge of the electron.

$$\frac{e^-}{V} = \frac{3.56\times10^{-6}\,\mathrm{C}}{\mathrm{m}^3}\cdot\frac{e^-}{1.6\times10^{-19}\,\mathrm{C}} = 2.2\times10^{13}\,e^-/\mathrm{m}^3$$

26-5 A Mars probe is traveling radially inward toward the center of the planet. At 600 m above the surface the field is directed up and has value $180\,\mathrm{N/C}$. At 500 m it is still directed up but has value $125\,\mathrm{N/C}$. Find the sign and density of the charge (density) in this region.

Solution: The radius of Mars is sufficiently large so that we can take the region as a cube 100 m on a side. The electric fields are as shown in Fig. 26-6. Gauss' law can be written as

$$\varepsilon_o \oint \mathbf{E}\cdot d\mathbf{s} = \varepsilon_o\left[\int_{\text{sides}} E\cdot ds + \int_{\text{top}} E\cdot ds + \int_{\text{bottom}} E\cdot ds\right] = q_{\text{enclosed}}$$

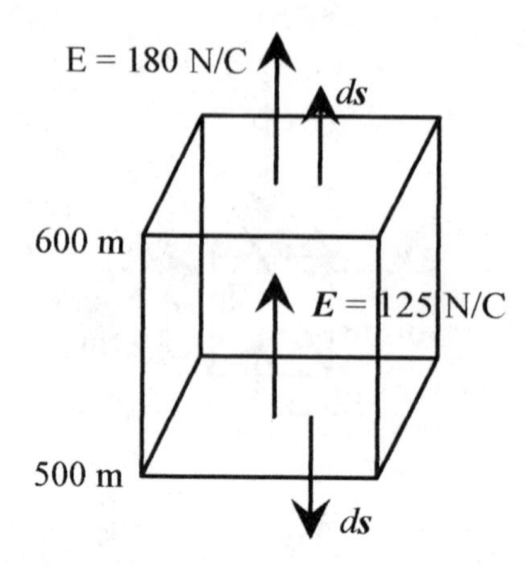

Fig. 26-6

The electric field at the sides is zero. At the top, the electric field and the vector representing *ds* point in the same direction so the dot product has a plus sign. On the bottom, the electric field vector points up while the vector representing *ds* points down giving the dot product a minus sign. Putting in the values for the integrals.

$$\varepsilon_o\left[\text{zero} + 1.8\times10^6\,\text{Nm}^2/\text{C} - 1.25\times10^6\,\text{Nm}^2/\text{C}\right] = q$$

the total charge enclosed within the volume is

$$q = 4.9\times10^{-6}\,\text{C}$$

The charge density then is $4.9\times10^{-12}\,\text{C/m}^3$.

26-6 Calculate the electric field between cylinders carrying charges as shown. The $+q$ is on the inner conductor and $-q$ on the inside of the outer conductor with another $-q$ on the outside of the outer conductor.

Solution: On a surface between the conductors, $\varepsilon_o\oint \boldsymbol{E}\cdot d\boldsymbol{s} = q$ applies to the charge contained and $\varepsilon_o\oint \boldsymbol{E}\cdot d\boldsymbol{s} = E\cdot 2\pi r\ell$, with $2\pi r\ell$ the surface area of the Gaussian cylinder, corresponding to the point where the field is required, so in this region E points inward and has value $E = q/2\pi\varepsilon_o r\ell$.

Note that the electric field at a point between the cylinders is <u>only</u> due to the charge enclosed by the Gaussian surface. Because of the symmetry of the problem, the charge on the outer cylinder produces no net electric field at any point <u>inside</u> the outer cylinder.

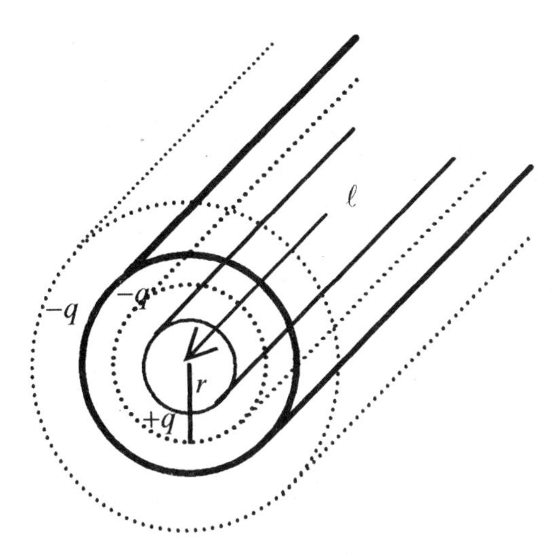

Fig. 26-7

Outside the outer cylinder the E field points inward because the net charge contained is negative, and at any point r (outside the outer cylinder) measured from the center line of the two cylinders $E = q/2\pi\varepsilon_0 r\ell$.

With this particular arrangement of charge, the field has the same algebraic form between the cylinders and outside the outer cylinder but points outward between the cylinders and inward outside the outer cylinder.

CHAPTER 27

ELECTRIC POTENTIAL

In mechanics the use of force to analyze problems works well when the force is a constant. When the force depends on position, then the concepts of work and energy are introduced with work defined as the summation, or integral, of force over distance.

In electricity, potential is introduced for similar reasons. While the electric field is a constant in many problems, it is not a constant, say, in the vicinity of a point charge except on a sphere surrounding that (central) charge. Any motion of charges due to electric force that is not on the surface of the sphere surrounding a central charge has to take account of this non-constancy of the field, or force (field is force per charge).

The **electric potential** or **potential difference** between two points A and B is defined as the work per unit charge necessary to move the unit charge from A to B

$$V_B - V_A = \frac{W_{AB}}{q_o} \tag{27-1}$$

The units of potential are Joule/Coulomb or volt. Usually one point (A) is taken at infinity where $V_A = 0$. This is convenient and correct since the force on the unit charge at infinity (see Coulomb's law) is zero.

Look at the integral (same definition of work as in mechanics) involved in calculating W_{AB} for a charge q_o being brought from infinity to the vicinity of q. This integral is

$$W = \int_A^B \boldsymbol{F}\cdot d\boldsymbol{r} = -\frac{qq_o}{4\pi\varepsilon_o}\int_A^B \frac{dr}{r^2} = \frac{qq_o}{4\pi\varepsilon_o}\frac{1}{r}\bigg|_A^B = \frac{qq_o}{4\pi\varepsilon_o}\left(\frac{1}{r_B}-\frac{1}{r_A}\right)$$

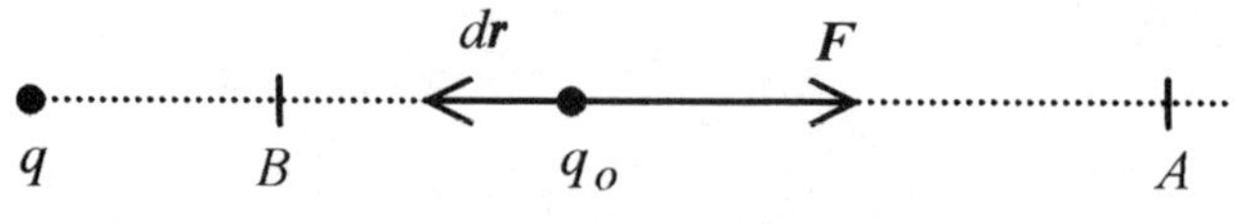

Fig. 27-1

Be careful of the signs here; If q and q_o are positive, the force is repulsive; and if q_o is coming from infinity to r_B, then $\boldsymbol{F} \cdot d\boldsymbol{r}$ is negative. With this definition the potential, or work per unit charge, at any point r radially out from q is

$$V = \frac{q}{4\pi\varepsilon_o r} \tag{27-2}$$

a most convenient definition.

For a constant electric field, such as between two parallel plates, $W_{AB} = Fd = q_o Ed$ and $V_B - V_A = W_{AB}/q_o = Ed$, where d is the distance moved (in the direction of the field).

27-1 In a certain experiment an electric field of $2.0 \times 10^5\,\mathrm{N/C}$ is required for parallel plates separated by $0.0020\,\mathrm{m}$. What voltage will produce this field?

Solution: For a constant field, $V = Ed$ so

$$V = 2.0 \times 10^5\,\mathrm{N/C} \cdot 2.0 \times 10^{-3}\,\mathrm{m} = 4.0 \times 10^2\,\mathrm{J/C} = 400\,\mathrm{V}$$

27-2 An electron placed between two charged parallel plates separated by $2.0 \times 10^{-2}\,\mathrm{m}$ is observed to accelerate at $5.0 \times 10^{14}\,\mathrm{m/s^2}$. What is the voltage on the plates?

Solution: First calculate the force on the electron as

$$F = ma = 9.1 \times 10^{-31}\,\mathrm{kg} \cdot 5.0 \times 10^{14}\,\mathrm{m/s^2} = 4.6 \times 10^{-16}\,\mathrm{N}$$

The electric field required to move this electron is

$$E = F/e^- = 4.6 \times 10^{-16}\,\mathrm{N} / 1.6 \times 10^{-19}\,\mathrm{C} = 2.8 \times 10^3\,\mathrm{N/C}$$

The voltage for this separation is

$$V = Ed = 2.8 \times 10^3\,\mathrm{N/C} \cdot 2.0 \times 10^{-2}\,\mathrm{m} = 57\,\mathrm{V}$$

27-3 What is the electric potential $0.20\,\mathrm{m}$ away from a charge of $1.0 \times 10^{-6}\,\mathrm{C}$?

Solution:

$$V = \frac{q}{4\pi\varepsilon_o}\frac{1}{r} = \frac{9.0 \times 10^9\,\mathrm{N \cdot m^2}}{\mathrm{C^2}}\frac{1.0 \times 10^{-6}\,\mathrm{C}}{0.20\,\mathrm{m}} = 4.5 \times 10^4\,\mathrm{V}$$

27-4 Calculate the electric potential for the same charge but at $0.50\,\mathrm{m}$ away.

Solution: $$V = \frac{q}{4\pi\varepsilon_o}\frac{1}{r} = \frac{9.0\times10^{9}\,\mathrm{N\cdot m^2}}{\mathrm{C^2}}\frac{1.0\times10^{-6}\,\mathrm{C}}{0.50\,\mathrm{m}} = 1.8\times10^{4}\,\mathrm{V}$$

The potential difference between these two positions is 2.7×10^{4} V

27-5 Use the situation (charge and potential calculations) of problems 27-3 and 27-4. If a charge of $5.0\times10^{-6}\,\mathrm{C}$ were moved from a radius of $0.50\,\mathrm{m}$ to a radius of $0.20\,\mathrm{m}$, how much work would be performed?

Solution: The work performed is the potential difference in Joule/Coulomb (or volt) times the charge in Coulomb (observe the units carefully).

$$W = 2.9\times10^{4}\,\mathrm{J/C}\cdot5.0\times10^{-6}\,\mathrm{C} = 0.15\,\mathrm{J}$$

The work is independent of the path between these radii. This is the great value of potential calculations.

27-6 A $5.0\,\mathrm{g}$ conducting sphere with charge of $20\,\mu\mathrm{C}$ hangs by a non-conducting thread in an electric field produced by two plates separated by $8.0\,\mathrm{cm}$. What potential will cause the ball to hang at 25° to the vertical?

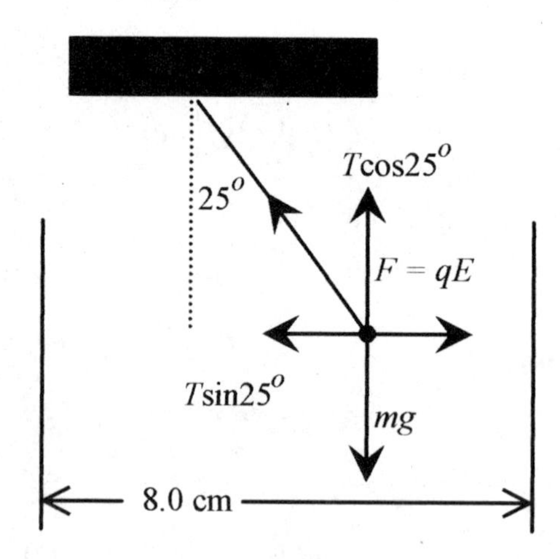

Fig. 27-2

Solution: This is a force-balance problem, the electric force is balanced by the horizontal component of the tension in the thread. Refer to Fig. 27-2 and write the force statements.

$$T\sin25^o = qE \quad \text{and} \quad T\cos25^o = mg$$

Divide one by the other to arrive at

$$\tan 25^o = \frac{Eq}{mg} \quad \text{or} \quad E = \frac{mg}{q}\tan 25^o = \frac{5.0\times10^{-3}\,\text{kg}\cdot 9.8\,\text{m/s}^2}{20\times10^{-6}\,\text{C}}\tan 25^o = 1140\,\text{N/C}$$

The potential is from $E = V/d$ or $V = Ed$

$$V = 1140\,\text{N/C}\cdot 8.0\times10^{-2}\,\text{m} = 91.4\,\text{J/C} = 91.4\,\text{V}$$

27-7 Calculate the electric potential of a spherical (conducting) shell of radius $0.10\,\text{m}$ carrying a charge of $5.0\times10^{-4}\,\text{C}$.

Solution: In calculating the potential on the spherical shell the charge acts as if it were at the center of the sphere.

$$V = \frac{q}{4\pi\varepsilon_o}\frac{1}{r} = \frac{9.0\times10^9\,\text{N}\cdot\text{m}^2}{\text{C}^2}\frac{5.0\times10^{-4}\,\text{C}}{0.10\,\text{m}} = 4.5\times10^7\,\text{V}$$

27-8 Now calculate the electric field just outside this surface.

Solution:

$$E = \frac{q}{4\pi\varepsilon_o}\frac{1}{r^2} = \frac{V}{r} = 4.5\times10^8\,\text{V/m} = 4.5\times10^8\,\text{N/C}$$

As an exercise show that the unit V/m is equal to N/C.

The electric potential due to an array of charges can be calculated by superposing the solutions for each charge. Because the electric potential is not a vector this is usually an easy calculation.

$$V = \sum_n V_n = \frac{1}{4\pi\varepsilon_o}\sum_n \frac{q_n}{r_n} \qquad \textbf{(27-3)}$$

27-9 For the array shown calculate the potential mid-way between q_1 and q_2.

Solution: First calculate the distance from this midpoint to q_3

$$4a^2 - (a/2)^2 = r^2 \qquad a^2(4 - 1/4) = r^2 \qquad a\sqrt{15}/2 = r$$

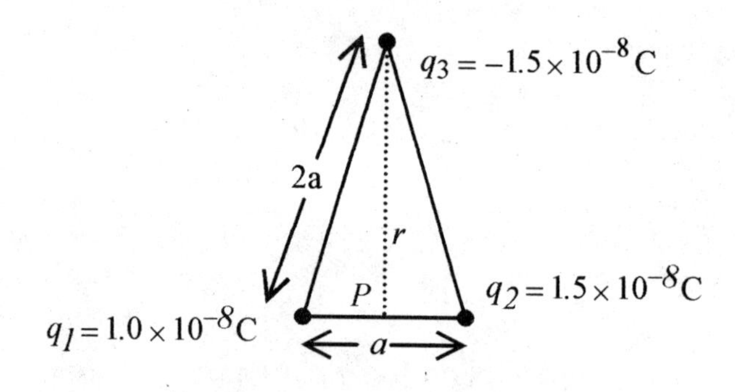

Fig. 27-3

The potential at the point P is the sum of the contributions from each of the charges.

$$V = V_1 + V_2 + V_3 = \frac{1}{4\pi\varepsilon_o}\left[\frac{1.0\times10^{-8}\text{C}}{a/2} + \frac{1.5\times10^{-8}\text{C}}{a/2} - \frac{1.5\times10^{-8}\text{C}}{a\sqrt{15}/2}\right]$$

$$V = \frac{2.0\times10^{-8}C}{4\pi\varepsilon_o a}\left[2.5 - \frac{1.5}{\sqrt{15}}\right] = \frac{9.0\times10^{9}\text{N}\cdot\text{m}^2}{\text{C}^2}\frac{2.0\times10^{-8}\text{C}}{a}[2.1] = \frac{380}{a}\text{V}$$

if a is measured in meters.

27-10 Calculate the electric potential along the x-axis for the triangular array of charges shown.

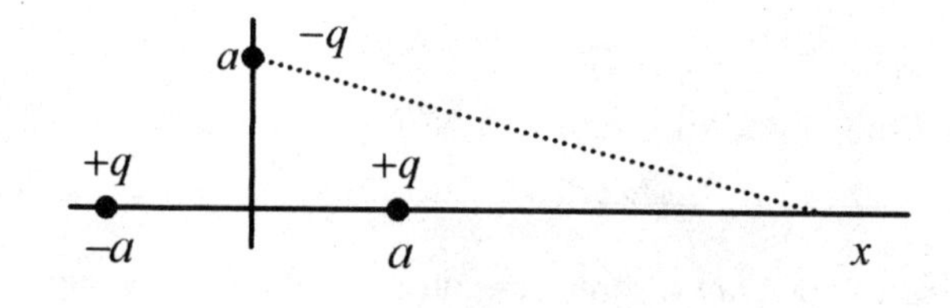

Fig. 27-4

Solution:

$$V = \frac{1}{4\pi\varepsilon_o}\sum_i \frac{q_i}{r_i} = \frac{1}{4\pi\varepsilon_o}\left[\frac{q}{x-a} + \frac{q}{x+a} - \frac{q}{\sqrt{x^2+a^2}}\right]$$

$$V = \frac{1}{4\pi\varepsilon_o}\left[\frac{q(x+a)+q(x-a)}{x^2-a^2} - \frac{q}{\sqrt{x^2+a^2}}\right] = \frac{1}{4\pi\varepsilon_o}\left[\frac{2qx}{x^2-a^2} - \frac{q}{\sqrt{x^2+a^2}}\right]$$

for $x << a$, $V = \frac{1}{4\pi\varepsilon_o}\frac{q}{x}$ which is what we would expect.

A long way away from the collection of charges the collection looks like one positive charge.

A classic problem is the calculation of the potential at a point a distance r and at an angle θ from the center of a dipole. The potential is written by summing the potential from the two charges and using the approximate trigonometric relation between r, r_1, and r_2.

$$V = \frac{1}{4\pi\varepsilon_o}\left[\frac{q}{r_1} - \frac{q}{r_2}\right] = \frac{q}{4\pi\varepsilon_o}\frac{r_2 - r_1}{r_1 r_2}$$

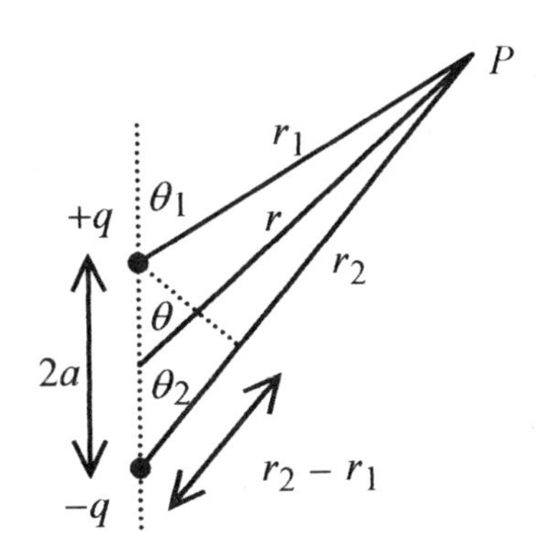

Fig. 27-5

If $r >> a$ then $r_1 r_2 \approx r^2$ and also $\theta_1 \approx \theta_2 \approx \theta$ so $r_2 - r_1 \approx 2a\cos\theta$ and using the dipole definition $p = 2aq$

$$V \approx \frac{1}{4\pi\varepsilon_o}\frac{p\cos\theta}{r^2} \qquad \textbf{(27-4)}$$

27-11 The water molecule H_2O has an asymmetric charge distribution leading to a dipole moment p of $6.2\times10^{-30}\,\mathrm{C\cdot m}$. Calculate the electric potential 1.0 nm a) at right angles to the direction of the dipole moment and b) at 45° to the direction of the dipole moment. (1.0 nm is approximately 10 hydrogen atom diameters)

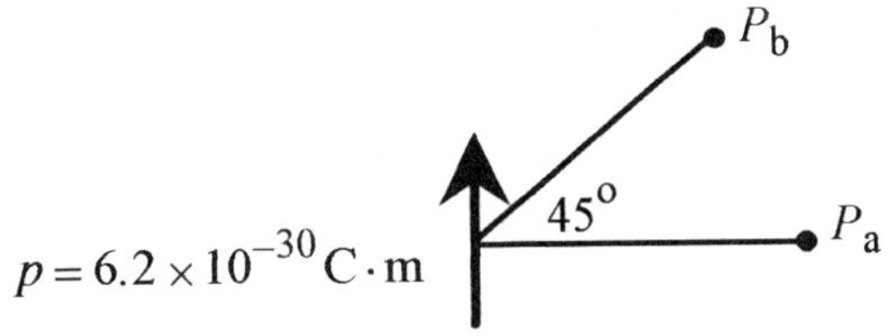

P_b

45° P_a

$p = 6.2\times10^{-30}\,\mathrm{C\cdot m}$

Fig. 27-6

Solution: The approximate expression for the electric potential is $V = \dfrac{1}{4\pi\varepsilon_o}\dfrac{p\cos\theta}{r^2}$, and the electric potentials at P_a and P_b are

$$V_a = \frac{9.0\times 10^9\,\mathrm{N\cdot m^2}}{\mathrm{C^2}}\frac{6.2\times 10^{-30}\,\mathrm{C\cdot m}(\cos 90^o)}{(1.0\times 10^{-9}\,\mathrm{m})^2} = 0\,\mathrm{V} \quad \text{and} \quad V_b = 0.039\,\mathrm{V}$$

(C) **27-12** A positron (positively charged electron) traveling at $3.0\times 10^7\,\mathrm{m/s}$ is projected directly at a stationary proton. What is the distance of closest approach?

Solution: The positron will approach the proton until the energy of the positron is expended in doing work against the Coulomb repulsive force. The positron will stop, then exit the vicinity of the proton having a velocity profile on exit the same as, but in the opposite direction, to the entrance.

$$W = \int_{\infty}^{r_c} \boldsymbol{F}\cdot d\boldsymbol{r} = -\frac{e^2}{4\pi\varepsilon_o}\int_{\infty}^{r_c}\frac{dr}{r^2} = \frac{e^2}{4\pi\varepsilon_o}\frac{1}{r_c}$$

This work is set equal to the kinetic energy $\frac{e^2}{4\pi\varepsilon_o}\frac{1}{r_c} = \frac{mv^2}{2}$ and solved for

$$r_c = \frac{2e^2}{4\pi\varepsilon_o}\frac{1}{mv^2} = 2(1.6\times 10^{-19}\,\mathrm{C})^2\frac{9.0\times 10^9\,\mathrm{N\cdot m^2}}{\mathrm{C^2}}\frac{1}{9.1\times 10^{-31}\,\mathrm{kg}(3.0\times 10^7\,\mathrm{m/s})^2} = 5.6\times 10^{-13}\,\mathrm{m}$$

This is a very close approach being two orders of magnitude smaller than the distance from the nucleus to the first electron orbit radius in the hydrogen atom.

Second Solution: This problem can be done using the concept of the **electron-volt (eV)** unit of energy. An electron accelerated through one volt has an amount of work performed on it equal to one volt (the work per unit charge) times the charge on the electron.

$$1\,\mathrm{eV} = (1\,\mathrm{J/C})1.6\times 10^{-19}\,\mathrm{C} = 1.6\times 10^{-19}\,\mathrm{J}$$

This equation defines the Joule-eV equivalence. An electron accelerated through a potential difference of one volt is said to gain one electron volt in energy. An electron accelerated through 1000 volts would gain 1000 eV in energy. An alpha particle (two positive electronic charges) accelerated through 1000 volts would gain 2000 eV in energy.

In this particular problem the positron with a velocity of $3.0\times 10^7\,\mathrm{m/s}$ has a kinetic energy of

$$KE = \frac{mv^2}{2} = \frac{9.1\times 10^{-31}\,\mathrm{kg}(3.0\times 10^7\,\mathrm{m/s})^2}{2} = 4.1\times 10^{-16}\,\mathrm{J}\frac{1\,\mathrm{eV}}{1.6\times 10^{-19}\,\mathrm{J}} = 2600\,\mathrm{eV}$$

This means that the positron has been accelerated through $2600\,\text{V}$.

A positron with this energy could approach a proton to a radius equivalent to 2600 volt potential.

Set $V = \dfrac{q}{4\pi\varepsilon_o r_c}$ equal to $2600\,\text{V}$ and solve for r_c

$$r_c = \frac{q}{4\pi\varepsilon_o V} = \frac{1.6\times10^{-19}\,\text{C}\cdot 9.0\times10^{9}\,\text{N}\cdot\text{m}^2/\text{C}^2}{2600\,\text{V}} = 5.6\times10^{-13}\,\text{m}$$

Another application of the concept of potential is to calculate the amount of work required to place charges in an array.

27-13 Calculate the work required to arrange the charges as shown in Fig. 27-7. In this problem the subscripts on the q's represent the position of the charges

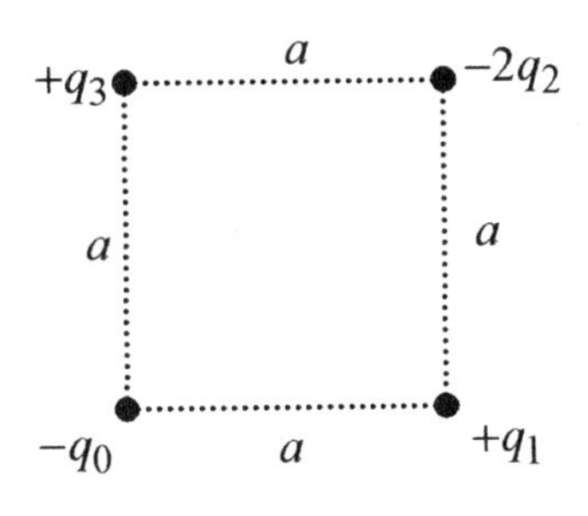

Fig. 27-7

Solution: To make this calculation place q_0 in position (corresponding to the origin of a coordinate system) and calculate the work necessary to place q_1 in position. This is

$$W_{0-1} = -\frac{q^2}{4\pi\varepsilon_o}\left(\frac{1}{r}\right)_{\infty}^{a} = -\frac{q^2}{4\pi\varepsilon_o a}$$

Since the charges are opposite, the electric force is attractive and the work to put this charge in position is negative (this charge has to be restrained as it approaches q_0). Now with these charges in place, add $-2q_2$ with an amount of work

$$W_{0-2} = \frac{2q^2}{4\pi\varepsilon_o}\left(\frac{1}{\sqrt{2}\,a} - \frac{1}{a}\right)$$

There are two components to this work calculation, the work to bring $-2q_2$ to within $\sqrt{2}\,a$ of $-q_0$ and the work to bring $-2q_2$ to within a of $+q_1$. Adding the third charge means taking into account the three charges already in place

$$W_{0-3} = \frac{-q^2}{4\pi\varepsilon_o}\left(\frac{1}{a} - \frac{1}{\sqrt{2}\,a} + \frac{2}{a}\right)$$

The total work is the sum of these, or

$$W = \frac{q^2}{4\pi\varepsilon_o}\left[-\frac{1}{a} + \frac{1}{\sqrt{2}\,a} - \frac{1}{a} - \frac{1}{a} + \frac{1}{\sqrt{2}\,a} - \frac{2}{a}\right] = \frac{q^2}{4\pi\varepsilon_o}\left[-\frac{5}{a} + \frac{\sqrt{2}}{a}\right] = \frac{q^2}{4\pi\varepsilon_o}\left[\sqrt{2} - 5\right]\frac{1}{a}$$

(C) **27-14** Another classic problem has to do with the calculation of the electric potential due to charge on a surface along a line from the center of a disk and perpendicular to the disk.

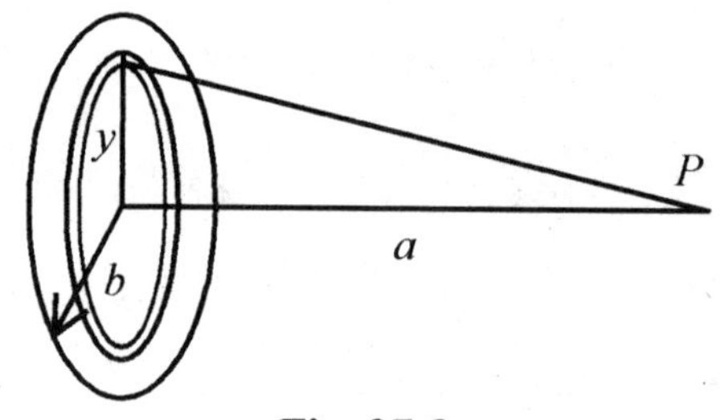

Fig. 27-8

Solution: The potential is

$$V = \frac{1}{4\pi\varepsilon_o}\int \frac{dr}{r} \qquad \textbf{(27-5)}$$

In this problem dq will have to be written in terms of charge per unit area times (differential) area. Start by writing the differential voltage in terms of the differential element of charge, dq, or

$$dV = \frac{1}{4\pi\varepsilon_o}\frac{dq}{r}$$

The charge on the surface is assumed uniform with surface charge density σ (charge per unit area). Take the charge contained in the circular strip, $\sigma(2\pi y)dy$ ($2\pi y$ is the length of the strip and dy is the width). The r is $\sqrt{a^2 + y^2}$, so the contribution of this strip to the electric potential is

$$dV = \frac{\sigma(2\pi y)dy}{4\pi\varepsilon_o\sqrt{a^2 + y^2}}$$

The integration is to be performed in y from zero to b, the radius of the disk

$$V = \frac{\sigma}{4\varepsilon_o}\int_0^b \frac{2y\,dy}{\sqrt{a^2+y^2}}$$

Make a change of variable. Let $u = a^2 + y^2$ and $du = 2y\,dy$ so the integral becomes (and we look at the integral only without the limits)

$$\int \frac{2y\,dy}{\sqrt{a^2+y^2}} = \int u^{-1/2}du = \frac{u^{1/2}}{1/2} = 2\sqrt{a^2+y^2}$$

This integral is in the Introduction, Mathematical Background.

Therefore $$V = \frac{\sigma}{4\varepsilon_o}\left[2\sqrt{a^2+y^2}\right]_0^b = \frac{\sigma}{2\varepsilon_o}\left[\sqrt{a^2+b^2}-a\right]$$

Now look at the limiting case where $b/a << 1$, which is often the case. This is most conveniently done by first rewriting the expression in brackets as

$$V = \frac{\sigma}{2\varepsilon_o}\left[a\sqrt{1+b^2/a^2}-a\right]$$

and then writing the square root as a binomial expansion remembering that $b/a << 1$. The binomial expansion (see the Introduction, Mathematical Background for the binomial expansion) is

$$a\left(1+b^2/a^2\right)^{1/2} = a\left(1+\frac{1}{2}\frac{b^2}{a^2}+\frac{1}{2}\left(-\frac{1}{2}\right)\frac{b^4}{8a^3}+\ \ldots\right)$$ so that

$$V = \frac{\sigma}{2\varepsilon_o}\left[\frac{b^2}{2a}-\frac{b^4}{8a^3}+\ \ldots\right]$$ and finally for $b/a << 1$

$$V = \frac{\sigma\pi b^2}{4\pi\varepsilon_o a} \tag{27-6}$$

This is expected since $\sigma\pi b^2$ is the total charge and a is the distance from that charge. A long way away a disk of charge "looks like" a point charge. The use of the binomial expansion is helpful in two ways. First, it allows a convenient calculation for the situation where $b/a << 1$; and second, the limiting case reduces to the simple expression for the potential due to a point charge, in effect verifying the calculation.

CHAPTER 28

CAPACITANCE

A **capacitor** is a geometric arrangement of conducting plates where charge can be stored. The **capacitance** is a measure of the charge that can be stored per volt. The unit of capacitance, the **farad,** is defined as a Coulomb/Volt. The defining equation for capacitance is $C = q/V$.

The simplest capacitor is two parallel plates of area A and separation d.

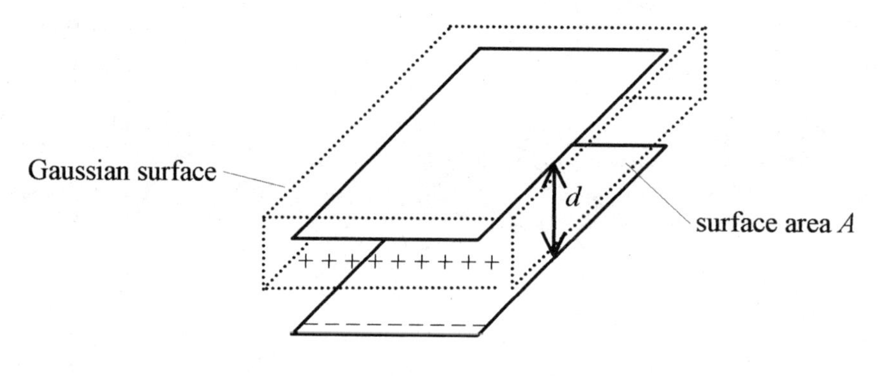

Fig. 28-1

The capacitance of a parallel plate capacitor determines the amount of charge that can be stored on the plates. A battery connected to the plates will place charge on the plates. For this discussion the battery can be viewed as a source of charge (or charge pump) with the amount of charge delivered to the plates determined by the voltage of the battery. Remember that the voltage of the battery is a measure of the amount of work that can be performed on a unit of charge.

When the battery is connected to the plates, we say that positive charge begins to flow onto the positive plate and negative charge onto the negative plate. The first "piece of charge" distributes uniformly over the plate. The positive and negative charges arrange themselves on the plates opposite one another. As more charge is pumped onto the plates the work necessary to place each new "piece of charge" on the plate increases because of the mutual repulsion of the charges already on the plate. Thus the capacity of the device to store charge is directly proportional to the plate area.

The limitation on capacity imposed by the mutual repulsion of like charges on one plate is offset by opposite charges on the other plate. As the distance between the plates is reduced, the attraction between unlike charges makes it easier, in terms of work performed by the battery, to place charge on the plates and the capacity of the device is increased. Thus capacitance is inversely proportional to the separation of the plates.

The capacitance of a parallel plate capacitor is directly proportional to the plate area and inversely proportional to the separation, with a constant added to make the units work out correctly.

$$C = \varepsilon_0 A/d$$

This very simple view of a parallel plate capacitor produces the correct functional relationship as seen in the formal derivation in the following paragraph.

(C) The formal procedure for determining capacitance is to:

1) calculate E using Gauss' law, then

2) calculate V from $V = -\int \mathbf{E} \cdot d\mathbf{l}$, and finally

3) calculate C from the definition

$$C = q/V \qquad \textbf{(28-1)}$$

For the parallel plate capacitor, charge the plates with a battery so that one plate contains positive charge and the other negative charge. The charges will arrange themselves so they are on the insides of the plates opposite one other (unlike charges attract). Construct a Gaussian surface surrounding one plate. The E-field exists only between the plates and is constant over the Gaussian surface. Applying Gauss' law

$$\varepsilon_0 \oint \mathbf{E} \cdot d\mathbf{s} = q \quad \text{or} \quad \varepsilon_0 E A = q \qquad \textbf{(28-2)}$$

E is a constant over the Gaussian surface so the integral of E over this surface is just E times A, the surface area. This is another case of an integral that can be evaluated by looking at it and writing down the answer. In this geometry the relationship between voltage and field is simple, $V = Ed$, so $\varepsilon_0 VA/d = q$ and the capacitance, which is defined as

$$C = q/V = \varepsilon_0 A/d \qquad \textbf{(28-3)}$$

The capacitance is entirely geometry dependent. One thing to notice here is ε_0, which can be expressed in the new units of Farad/meter or F/m.

28-1 A capacitor with circular plates of 0.20m radius separated by 0.0010m is connected to a 100V battery. After a long time the battery is disconnected. What is the charge stored on the plates?

Solution: The capacitance of the arrangement is

$$C = \frac{\varepsilon_0 A}{d} = \frac{8.9 \times 10^{-12}\,\text{F}}{\text{m}} \frac{\pi 0.040\text{m}^2}{0.0010\text{m}^2} = 1.1 \times 10^{-9}\,\text{F}$$

The charge is obtained from the definition of capacitance, equation 28-1

$$q = CV = 1.1 \times 10^{-9}\,\text{F} \cdot 100\text{V} = 1.1 \times 10^{-7}\,\text{C}$$

(C) **28-2** Calculate the capacitance of a cylindrical capacitor of inner radius a, outer radius b, and length ℓ.

Solution: Set up a cylindrical Gaussian surface symmetrical about the inner cylinder and of length ℓ.

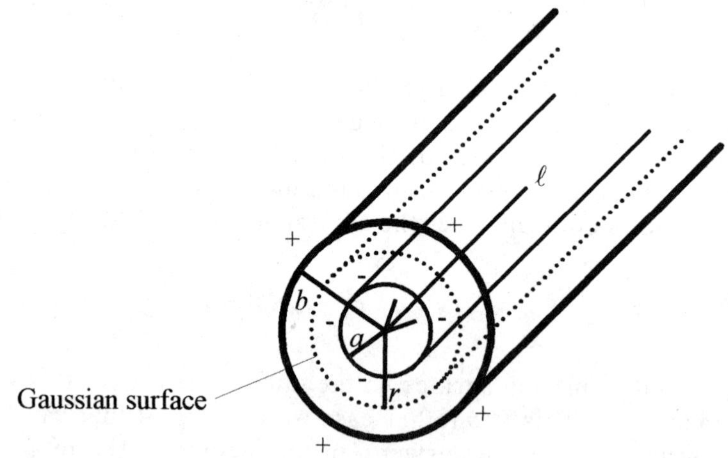

Fig. 28-2

E is constant over this surface and $\varepsilon_0 \oint \mathbf{E} \cdot d\mathbf{s} = q$ or $\varepsilon_0 E(2\pi r)\ell = q$ so that $E = \dfrac{q}{2\pi\varepsilon_0 r\ell}$.

Continuing with the procedure, calculate V.

$$V = -\int_a^b \mathbf{E} \cdot d\mathbf{r} = \frac{q}{2\pi\varepsilon_0 \ell} \int_a^b \frac{1}{r} dr = \frac{q}{2\pi\varepsilon_0 \ell} \ln r \Big|_a^b = \frac{q}{2\pi\varepsilon_0 \ell} \ln \frac{b}{a}$$

and by definition $C = \frac{q}{V}$, so $C = \frac{2\pi\varepsilon_o \ell}{\ln(b/a)}$.

(C) **28-3** Calculate the capacitance of a spherical capacitor with inner radius a and outer radius b.

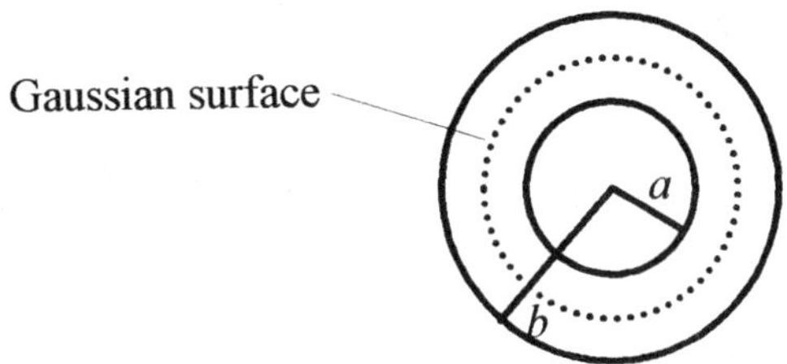

Fig. 28-3

Solution: Set up the Gaussian surface symmetric about the inner sphere. E is a constant over this surface and $\varepsilon_o \oint \mathbf{E} \cdot d\mathbf{s} = q$ or $\varepsilon_o E(4\pi r^2) = q$, so $E = \frac{q}{4\pi\varepsilon_o r^2}$.

Now calculate V from

$$V = -\int_a^b \mathbf{E} \cdot d\mathbf{r} = -\frac{q}{4\pi\varepsilon_o} \int_a^b \frac{dr}{r^2} = \frac{q}{4\pi\varepsilon_o} \frac{1}{r}\bigg|_a^b = \frac{q}{4\pi\varepsilon_o}\left(\frac{1}{a} - \frac{1}{b}\right) = \frac{q}{4\pi\varepsilon_o} \frac{b-a}{ab}$$

and by definition $C = \frac{q}{V}$, so $C = \frac{4\pi\varepsilon_o ab}{b-a}$.

Capacitors as Lumped Circuit Elements

Capacitors are used in electric circuits, so it is necessary to know how capacitors add when used in series and parallel. First look at three capacitors placed in series.

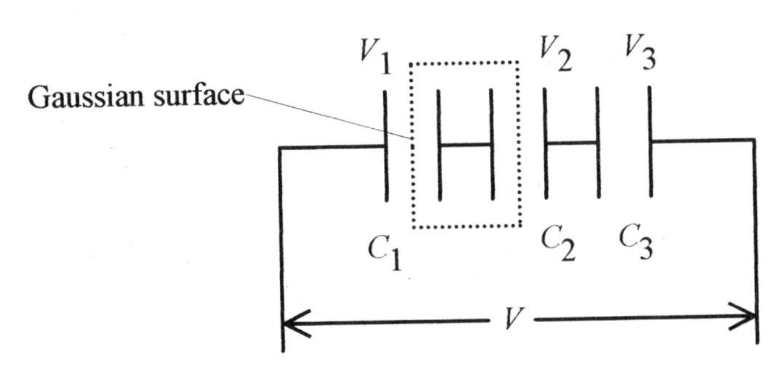

Fig. 28-4

The applied voltage is distributed between the three capacitors so $V = V_1 + V_2 + V_3$. Since there is no way to add charge except to the end plates, those connected directly to the battery, then the charge contained within the Gaussian surface must total zero. The right plate of C_1 has equal but opposite charge to the left plate of C_2. Another Gaussian surface surrounding C_1 also must have total charge zero, so the charge on the plates of C_1 are equal and opposite. Proceeding in this manner the magnitude of the charge on all the plates is the same. Rewriting the equation for the voltage using the definition of capacitance

$$V = V_1 + V_2 + V_3 = \frac{q}{C_1} + \frac{q}{C_2} + \frac{q}{C_3} = q\left(\frac{1}{C_1} + \frac{1}{C_2} + \frac{1}{C_3}\right)$$

Therefore $\frac{1}{C_1} + \frac{1}{C_2} + \frac{1}{C_3}$ can be replaced by an equivalent capacitance $\frac{1}{C}$.

Capacitors placed in series add in a reciprocal manner.

$$\frac{1}{C} = \frac{1}{C_1} + \frac{1}{C_2} + \frac{1}{C_3} + \ \ldots \quad \text{Cs in series} \qquad \textbf{(28-4)}$$

Capacitors placed in parallel add linearly. Look at the expression for C for a parallel plate capacitor: $C = \varepsilon_o A/d$. Adding another parallel capacitor just adds more $\varepsilon_o A/d$ to the capacitor. Therefore, capacitors placed in parallel add linearly.

$$C = C_1 + C_2 + C_3 + \ \ldots \quad \text{Cs in parallel} \qquad \textbf{(28-5)}$$

28-4 A 6.0 μF capacitor and an 8.0 μF capacitor are connected in series. A potential of 200 V is placed across them. Find the charge and potential on each.

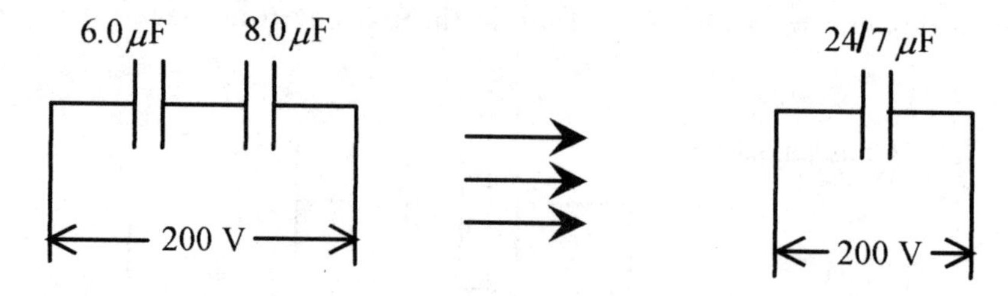

Fig. 28-5

Solution: The circuit is as shown. Since the capacitors are in series the charge on each capacitor is the same. The equivalent capacitance is found from

$$\frac{1}{C}=\frac{1}{C_1}+\frac{1}{C_2}=\frac{1}{6.0\ \mu\text{F}}+\frac{1}{8.0\ \mu\text{F}}=\frac{4}{24\ \mu\text{F}}+\frac{3}{24\ \mu\text{F}}=\frac{7}{24\ \mu\text{F}}$$

so the equivalent capacitance is $(24/7)\ \mu\text{F}$, and the equivalent circuit is as shown.

The charge is easily calculated from $q=CV=(24/7)\ \mu\text{F}\cdot 200\ \text{V}=686\ \mu\text{C}$.
The voltage on each capacitor is

$$V_1=\frac{q}{C_1}=\frac{686\ \mu\text{C}}{6.0\ \mu\text{F}}=114\ \text{V} \qquad V_2=\frac{q}{C_2}=\frac{686\ \mu\text{C}}{8.0\ \mu\text{F}}=86\ \text{V}$$

These voltages add to the 200V applied to the combination.

NOTE: This general approach of reducing circuits to simpler equivalents by applying the rules for lumped circuit elements is at the heart of circuit analysis.

28-5 A $12\mu\text{F}$ capacitor and a $10\mu\text{F}$ capacitor are connected in parallel across a 100V battery. Find the charge and potential on each capacitor.

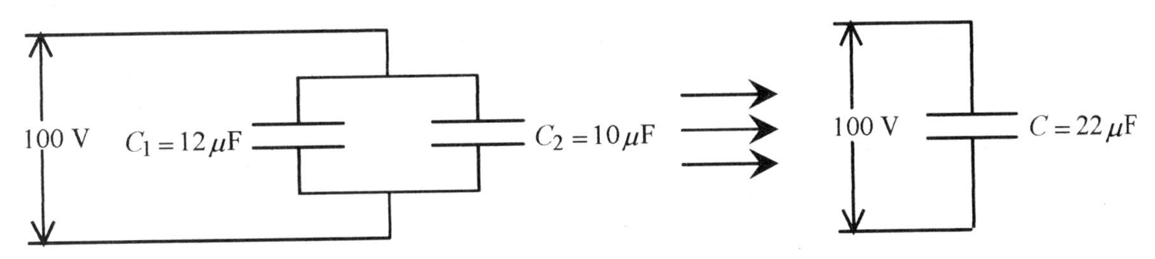

Fig. 28-6

Solution: The circuit is as shown. The voltage across each capacitor is the same, 100V. The charge, however, depends on the capacitance.

$$q_1=C_1V=12\ \mu\text{F}\cdot 100\ \text{V}=12\times10^{-4}\ \text{C} \qquad q_2=C_2V=10\ \mu\text{F}\cdot 100\ \text{V}=10\times10^{-4}\ \text{C}$$

Look at the equivalence. Take this total charge and the applied voltage and find an equivalent capacitance.

$$C=\frac{q}{V}=\frac{22\times10^{-4}\ \text{C}}{100\ \text{V}}=22\ \mu\text{F}$$

This is the sum of the capacitances as expected in a parallel arrangement.

28-6 The analysis of multiple capacitor combinations is done with (sometimes several) equivalent circuits. Consider the circuit shown that has parallel and series combinations.

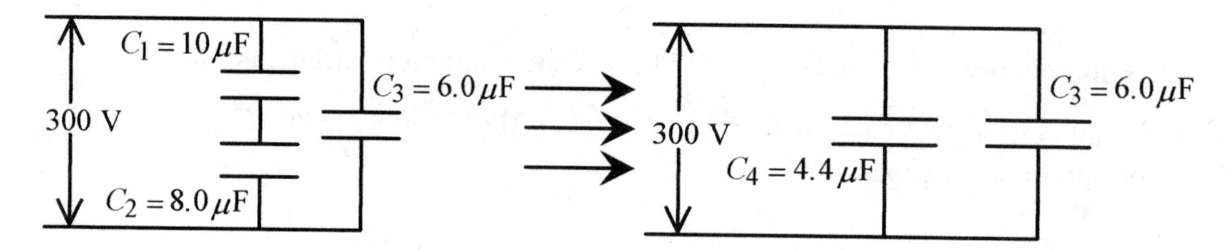

Fig. 28-7

Solution: The first equivalent circuit comes from the combination of C_1 and C_2.

$$\frac{1}{C_4} = \frac{1}{C_1} + \frac{1}{C_2} = \frac{1}{10\,\mu F} + \frac{1}{8.0\,\mu F} = \frac{9.0}{40\,\mu F} \quad \text{or} \quad C_4 = 4.4\,\mu F$$

Now combine C_3 and C_4 to $C_5 = C_3 + C_4 = 4.4\,\mu F + 6.0\,\mu F = 10.4\,\mu F$ allowing construction of the final (simplest) equivalent circuit

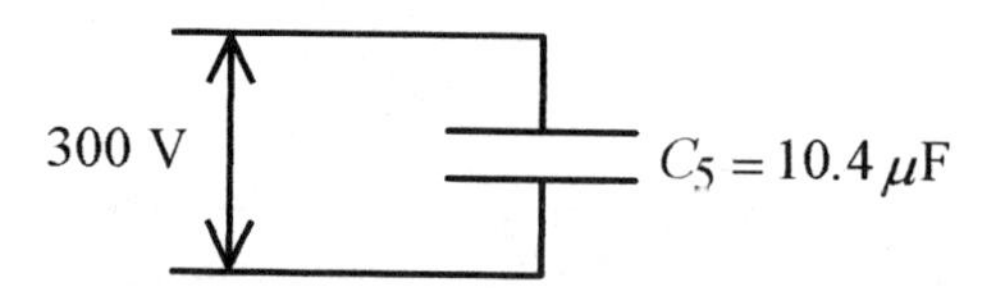

Fig. 28-8

Now calculate the total charge $q_t = C_5 V = 10.4\,\mu F \cdot 300\text{ V} = 3.1 \times 10^{-3}\text{ C}$.

The charge on C_3 is $q_3 = C_3 V = 6.0\,\mu F \cdot 300\text{ V} = 1.8 \times 10^{-3}\text{ C}$.

The remaining charge is 1.3×10^{-3} C, and this is the charge on each of the other capacitors, C_1 and C_2. Remember that the charge on capacitors in series is the same. The voltage on each of these capacitors is

$$V_1 = \frac{q}{C_1} = \frac{1.3 \times 10^{-3}\text{ C}}{10\,\mu F} = 130\text{ V} \qquad V_2 = \frac{q}{C_2} = \frac{1.3 \times 10^{-3}\text{ C}}{8.0\,\mu F} = 163\text{ V}$$

The error here of these voltages not adding up to 300 V is about 2%, which is expected from round off.

28-7 Charge a 12μF capacitor with a 200 V source, then place this capacitor in parallel with an uncharged 7.0μF capacitor and calculate the "new" voltage.

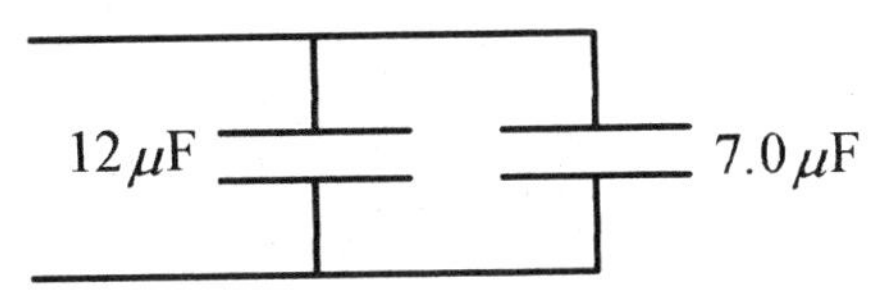

Fig. 28-9

Solution: The 12μF capacitor initially placed on the 200 V source has a charge of

$$q = CV = 12\ \mu\text{F} \cdot 200\ \text{V} = 2.4 \times 10^{-3}\ \text{C}$$

When connected to the 7.0μF capacitor the charge distributes so that the total charge remains the same and the "new" voltages are equal as they must be for capacitors in parallel. Take Q as the amount of charge that moves from the 12μF capacitor to the 7.0μF capacitor. The voltages as expressed by the ratio of q to C must be equal, so write

$$\frac{2.4 \times 10^{-3}\ \text{C} - Q}{12\ \mu\text{F}} = \frac{Q}{7.0\ \mu\text{F}} \quad \text{so} \quad Q = 8.84 \times 10^{-4}\ \text{C}$$

The voltage on each of the capacitors is

$$V_7 = \frac{q}{C} = \frac{8.84 \times 10^{-4}\ \text{C}}{7.0\ \mu\text{F}} = 126\ \text{V} \qquad \text{V}_{12} = \frac{q}{C} = \frac{1.52 \times 10^{-3}\ \text{C}}{12\ \mu\text{F}} = 126\ \text{V}$$

Energy Storage

(C) If the potential of a capacitor is defined as the work required to move an amount of charge from one plate to the other, then in differential form

$$dW = Vdq \quad \text{but} \quad V = \frac{q}{C} \quad \text{so} \quad dW = \frac{q}{C}dq \qquad \textbf{(28-6)}$$

and the integral of this expression to find the total work to move an amount of charge q is

$$W = \int_0^q \frac{q}{C}dq = \frac{q^2}{2C} = \frac{1}{2}CV^2 \qquad (q = CV) \qquad \textbf{(28-7)}$$

This work is equal to the energy stored in the capacitor, U. Another quantity of interest is the energy that can be stored per unit volume. For a parallel plate capacitor, where $C = \varepsilon_o A/d$

$$u = \frac{U}{Ad} = \frac{CV^2}{2Ad} = \frac{\varepsilon_o}{2}\left(\frac{V}{d}\right)^2 = \frac{\varepsilon_o}{2}E^2 \qquad \textbf{(28-8)}$$

28-8 Go back to the previous problem of the $12\,\mu\text{F}$ capacitor charged to $200\,\text{V}$ and then connected to the $7.0\,\mu\text{F}$ capacitor and look at the question of energy.

Solution: The $12\,\mu\text{F}$ capacitor at $200\,\text{V}$ has a stored energy of

$$U = (1/2)CV^2 = (1/2)12\,\mu\text{F}\cdot 40000\,\text{V}^2 = 0.24\,\text{J}$$

NOTE: The units of energy are $\left[\text{FV}^2\right] = \left[(\text{C/V})(\text{J/C})^2\right] = [\text{J}]$.

After the capacitors are connected, the voltage drops to $126\,\text{V}$ and the energy stored in the two capacitors is

$$U = \frac{1}{2}12\,\mu\text{F}\cdot(126\,\text{V})^2 + \frac{1}{2}7.0\,\mu\text{F}(126\,\text{V})^2 = 9.5\,\mu\text{F}(126\,\text{V})^2 = 0.15\,\text{J}$$

Where did the energy go? The difference between the 0.24J and the 0.15J is the kinetic energy needed to transport the charge or the work to move the charge.

Capacitors and Dielectrics

Dielectrics are most conveniently thought of as materials containing dipoles that are "loose" and can line up with an applied electric field. The easiest way to understand dielectrics is to approach them experimentally. Using a parallel plate capacitor as a model observe experimentally that the amount of charge per unit voltage increases with the insertion of various dielectric materials. The easiest way to explain this is to say that $C = \varepsilon_o A/d$ is changed by a factor κ, so $C = \kappa\varepsilon_o A/d$ with this κ called the dielectric constant, a measure of the amount by which the capacity of a capacitor is increased due to a particular dielectric.

The mechanism of increasing the capacitance can be thought of as the dipoles lining up across the separation d with the positive end of the string effectively neutralizing some of the negative charge on the plate and the negative end of the string doing the same to the other plate. The effect of these positive charges in the vicinity of the negative plate allows the negative plate to be negatively charged more easily (with less work) leading to a greater capacity of the device in the presence of the dielectric.

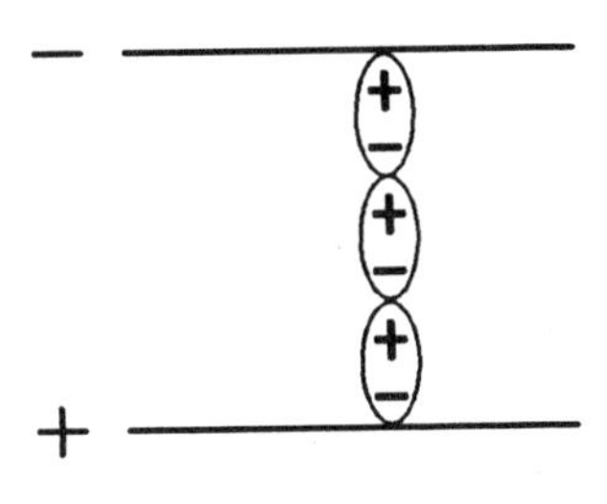

Fig. 28-10

Consider another experiment with dielectrics. Connect a battery in parallel with identical sized parallel plate capacitors except one has a dielectric and the other does not. For the one without the dielectric $C = \varepsilon_0 A/d$. And for the one with the dielectric $C = \kappa\varepsilon_0 A/d$. Since $q = CV$, to carry the same charge the capacitor with the dielectric would have to have V reduced to V_0/κ and likewise E to E_0/κ. Finally looking at the energy density for capacitors with dielectrics

$$u = \frac{CV^2}{2Ad} = \frac{\kappa\varepsilon_0 AV^2}{2Ad^2} = \frac{\kappa\varepsilon_0}{2}E^2 \qquad \textbf{(28-9)}$$

28-9 A parallel plate capacitor of $2.0\times10^{-9}\,\mathrm{F}$ and plate area $0.50\,\mathrm{m}^2$ is connected to $3000\,\mathrm{V}$. Calculate the stored energy.

Solution: $$U = \frac{1}{2}CV^2 = \frac{1}{2}2.0\times10^{-9}\ \mathrm{F}\cdot 9.0\times10^{6}\ \mathrm{V}^2 = 90\times10^{-4}\ \mathrm{J}$$

28-10 The plates (problem 28-8) are physically pulled apart to double the original separation. Calculate the work performed by the outside agent pulling the plates apart, the average force exerted on the plates, and the average power if the separation occurs over 10s.

Solution: First, calculate the original separation from

$$C = \frac{\varepsilon_0 A}{d} \quad \text{or} \quad d = \frac{\varepsilon_0 A}{C} = \frac{8.9\times10^{-12}\ \mathrm{F}}{\mathrm{m}}\frac{0.50\ \mathrm{m}^2}{2.0\times10^{-9}\ \mathrm{F}} = 2.2\times10^{-3}\ \mathrm{m}$$

Substituting the expression for the capacitance of a parallel plate capacitor the total energy is

$$U = \frac{1}{2}CV^2 = \frac{1}{2}\frac{\varepsilon_0 A}{d}V^2$$

If the separation is doubled the energy is halved. The energy density is reduced by one-quarter. (Refer to equation 28-9.) In this case the energy is reduced from 90×10^{-4} J to 45×10^{-4} J. This 45×10^{-4} J energy loss must be the amount of work done in separating the plates. The work is force times distance, so the average force would be

$$F_{avg} = \frac{\text{work}}{\text{distance}} = \frac{45 \times 10^{-4}\ \text{J}}{2.2 \times 10^{-3}\ \text{m}} = 2.0\ \text{N}$$

If this work were done over 10 seconds the power would be

$$P = \frac{\text{work}}{\text{time}} = \frac{45 \times 10^{-4}\ \text{J}}{10\,\text{s}} = 4.5 \times 10^{-4}\ \text{W}$$

CHAPTER 29

CONDUCTIVITY

Electric conductivity is most conveniently thought of as motion by essentially "free" electrons; electrons not bound to atoms. The actual mechanism is more complicated but this simple view is sufficient to understand conductivity in metals. In the absence of an electric field, these free electrons are moving in random patterns with a mean free speed v. They regularly crash into ion cores (metal ions with the conduction electron(s) missing), then move off again with this same mean free speed.

Electric conduction comes about because an electric field is impressed on the metal which slightly modifies this random motion of the electrons giving them a drift velocity in the direction opposite to the field. The accompanying diagram shows the effect of an electric field (dashed lines) on the random motion of a conduction electron in a metal subject to an electric field.

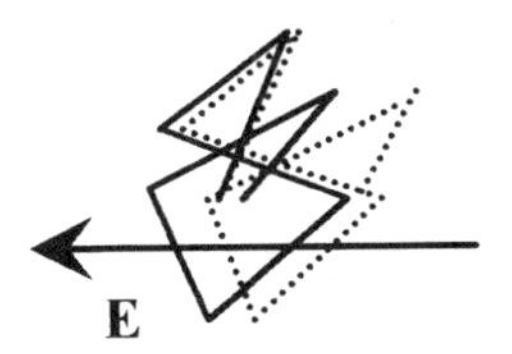

Fig. 29-1

The **mean free speed** is the speed of the electrons as determined by the temperature of the metal, and the **drift velocity** is due to the electric field. This drift velocity is small compared to the mean free speed. The motion of the electron charge carriers is characterized by the drift velocity, the mean free path (the distance between collisions), and the relaxation time (the time between collisions). The drift velocity depends on the applied field. The mean free path and the relaxation time are characteristics of the metal and the physical condition of the metal (temperature, pressure, etc.).

Current

Consider a wire, containing charge, under the influence of an electric field. On a plane through the wire (perpendicular to the charge motion) an amount of charge will pass per unit of time. This is called the **current**

$$i = \frac{q}{t} \quad \text{or} \quad i = \frac{\Delta q}{\Delta t} \quad \text{or} \quad i = \frac{dq}{dt} \qquad \textbf{(29-1)}$$

with unit of ampere = coulomb/second or $A = C/s$.

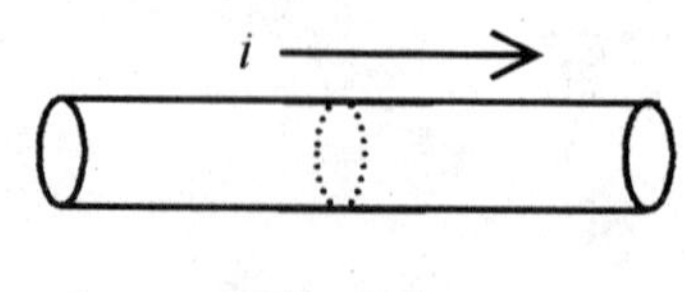

Fig. 29-2

The **current density** is defined as the current per cross-sectional area.

$$j = i/A \qquad \textbf{(29-2)}$$

The current direction is the direction a unit positive charge would move in the field. Despite the charge being carried by electrons, current is viewed as the motion of positive charge. This definition fits with the previous work in potential.

29-1 A current of $5.0\,\text{A}$ exists in a wire of $0.50\,\text{cm}$ diameter. Calculate the current density.

Solution:
$$j = \frac{i}{A} = \frac{5.0\ \text{A}}{\pi(0.25\times 10^{-4}\ \text{m}^2)} = 6.4\times 10^4\ \frac{\text{A}}{\text{m}^2}$$

29-2 Calculate the amount of charge that passes a cross section each second.

Solution: $5.0\ \text{A} = 5.0\,\text{C/s}$, so $5.0\,\text{C}$ passes a cross second each second.

29-3 Calculate the number of electrons per second.

Solution:
$$\frac{5.0\text{C}}{\text{s}}\ \frac{1e^-}{1.6\times 10^{-19}\ \text{C}} = 3.1\times 10^{19}\ e^-\ \text{per second}$$

Notice that a current on the order of 10^{-15} A would amount to 10^4 electrons per second.

29-4 What is the current and current density in a $1.2\,\text{cm}$ diameter gas discharge tube if 3.0×10^{17} electrons and 2.0×10^{16} positive ions pass a cross section of the tube each second?

Solution: The current is due to electrons going one way and positive ions going the other way.

$$i = 3.2\times 10^{17}(1/\text{s})\cdot 1.6\times 10^{-19}\ \text{C} = 0.051\ \text{A}$$

The current density is $j = \dfrac{3.2\times10^{17}(1/\text{s})\cdot1.6\times10^{-19}\,\text{C}}{\pi(0.36\times10^{-4}\,\text{m}^2)} = 453\dfrac{\text{A}}{\text{m}^2}$

The amount of charge in a length of wire is the product of number density of charge carriers (number/volume), cross-sectional area, and length (the volume), and the charge per carrier, or

$$q = n\,A\,\ell\,e \tag{29-3}$$

In the presence of an electric field the time for this charge to traverse the length, ℓ, of this volume is the length divided by the drift velocity, $t = \ell/v_d$. The current is the charge contained in the volume divided by the time for the charges to move the length of the volume. The current is

$$i = \frac{q}{t} = \frac{n\,A\,\ell\,e}{\ell/v_d} = n\,A\,e\,v_d$$

and the current density is

$$j = \frac{i}{A} = n\,e\,v_d \tag{29-4}$$

29-5 A metal conductor with 2.0×10^{25} charge carriers/m^3 and cross-sectional area $1.0\times10^{-4}\,\text{m}^2$ has a current of $4.0\times10^{-2}\,\text{A}$. What is the drift velocity of the charge carriers?

Solution: $v_d = \dfrac{j}{ne} = \dfrac{4.0\times10^{-2}\,\text{A}/1.0\times10^{-4}\,\text{m}^2}{2.0\times10^{25}(1/\text{m}^3)\cdot1.6\times10^{-19}\,\text{C}} = 1.25\times10^{-4}\,\dfrac{\text{m}}{\text{s}}$

Resistance

The **resistance** of a conductor is defined as the voltage divided by the current, $R = V/i$ and has a special unit, the ohm = volt/amp, or in symbol form $\Omega = \text{V}/\text{A}$. The **resistivity** of a material is defined as the electric field divided by the current density

$\rho = \dfrac{E}{j}$, but since $E = \dfrac{V}{\ell}$ and $j = \dfrac{i}{A}$, then $\rho = \dfrac{V/\ell}{i/A} = \dfrac{V\,A}{i\,\ell}$ or $\dfrac{V}{i} = \rho\dfrac{\ell}{A}$ and $R = \rho\dfrac{\ell}{A}$,

which gives resistivity the units of $\Omega\cdot\text{m}$. The resistivity of most metals is the order of $10^{-8}\,\Omega\cdot\text{m}$.

The instantaneous relationship between voltage, current, and resistance is

$$V = iR \tag{29-5}$$

and is known as Ohm's law. Many metals over a wide range of voltage and current obey Ohm's law.

29-6 The resistivity of copper is $1.7 \times 10^{-8}\,\Omega\cdot\text{m}$. What current flows through a 2.0m long copper conductor of $1.0 \times 10^{-4}\,\text{m}^2$ cross section when $20\,\text{V}$ is applied?

Solution: First calculate the resistance: $R = \rho\dfrac{\ell}{A} = \dfrac{1.7 \times 10^{-8}\,\Omega\cdot\text{m}(2.0\text{m})}{1.0 \times 10^{-4}\text{m}^2} = 3.4 \times 10^{-4}\,\Omega$

Now the current: $i = \dfrac{V}{R} = \dfrac{20\,\text{V}}{3.4 \times 10^{-4}\,\Omega} = 5.9 \times 10^4\,\text{A}$

The "free" charge carriers in a conductor experience an electric field when a voltage is applied and this field produces an acceleration $a = eE/m$, the electric force divided by the mass. The charge carriers accelerate until they crash into an ion core, then the process of accelerate and crash begins again. The characteristic time between collisions, which is dictated by the mean free speed, is called the relaxation time. The acceleration times the relaxation time produces the drift velocity.

$$v_d = \frac{eE\tau}{m}, \quad \text{but} \quad v_d = \frac{j}{ne}, \quad \text{so} \quad \frac{j}{ne} = \quad \text{or} \quad \frac{E}{j} = \rho = \frac{m}{ne^2\tau}$$

This expression for resistivity is consistent with Ohm's law. Resistance (and resistivity) is a constant provided the applied field does not change the time between collisions. Remember, the time between collisions is a property of the material and is not appreciably affected by the applied electric field.

29-7 Take a rectangular block of metal $1.0 \times 10^{-4}\,\text{m}^2$ in cross section, 50cm in length with a relaxation time of $3.0 \times 10^{-14}\,\text{s}$. Apply 4.0V for a current of 3.0A. Calculate resistivity, electrons/sec. passing a cross section, current density, the acceleration of the electrons, and the drift velocity.

Solution: The resistivity is $\rho = \dfrac{RA}{\ell} = \dfrac{4.0\text{V}}{3.0\text{A}}\dfrac{1.0 \times 10^{-4}\,\text{m}^2}{50 \times 10^{-2}\,\text{m}} = 2.7 \times 10^{-4}\,\Omega\cdot\text{m}.$

The number of e^-/s passing a cross section is $N = \dfrac{3.0\text{C}}{\text{s}}\dfrac{1e^-}{1.6 \times 10^{-19}\text{C}} = \dfrac{1.9 \times 10^{19}\,e^-}{\text{s}}.$

Calculate the current density. $j = \frac{i}{A} = \frac{3.0\,\text{A}}{1.0 \times 10^{-4}\,\text{m}^2} = 3.0 \times 10^4 \frac{\text{A}}{\text{m}^2}$

Calculate the acceleration of the electrons. $a = \frac{eE}{m} = \frac{1.6 \times 10^{-19}\,\text{C}}{9.1 \times 10^{-31}\,\text{kg}} \frac{4.0\,\text{V}}{50 \times 10^{-2}\,\text{m}} = 1.4 \times 10^{12} \frac{\text{m}}{\text{s}^2}$

Calculate the drift velocity. $v_d = a\,\tau = (1.4 \times 10^{12}\,\text{m/s}^2)(3.0 \times 10^{-14}\,\text{s}) = 4.2 \times 10^{-2}\,\text{m/s}$

NOTE: The mean free speed, the speed of the electrons before the application of the field, is the order of $10^6\,\text{m/s}$.

Power

(C) In electric circuits, energy is transferred when charge moves under the influence of voltage. Since voltage is work/charge, then the energy involved in moving an amount of charge under the influence of voltage is the voltage times the charge. In differential form

$$dU = Vdq = Vidt \qquad \textbf{(29-6)}$$

Power (work/time) dU/dt is $P = Vi$

In a circuit where V, i, and R are related by $V = iR$ this power can be written

$$P = Vi = i^2 R = V^2/R \qquad \textbf{(29-7)}$$

In the case of electrons passing through a metal this energy is manifest as heat in the conductor. The energy transfer is known as Joule's law of heating. The units are $\text{V A} = (\text{J/C})(\text{C/s}) = \text{J/s} = \text{W}$.

29-8 A 1000 W electric heater operates at 115 V. Calculate the current, resistance, and energy generated in one hour.

Solution: The current in the heater is $i = \frac{P}{V} = \frac{1000\,\text{W}}{115\,\text{V}} = 8.7\,\text{A}$.

The resistance is $R = \frac{V^2}{P} = \frac{(115\,\text{V})^2}{1000\,\text{W}} = 13.2\,\Omega$.

The energy generated in one hour is $U = Pt = (1000\,\text{J/s})(3600\,\text{s}) = 3.6 \times 10^6\,\text{J}$.

(C) **29-9** For the situation of problem 29-8, if the voltage is reduced to 110 V (assume no change in resistance) how does the heat output change?

Solution: This is best handled with a variational equation. In this case start with $P = V^2/R$ and ask how does P change with V, while R remains constant. The left side becomes ΔP, and the right side becomes $(1/R)$ times the differential of V^2. Writing in delta format

$$\Delta P = \frac{1}{R}\Delta(V^2) = \frac{1}{R}2V\,\Delta V$$

The $2V\Delta V$ is the difference between $(V+\Delta V)^2$ and V^2.

$$(V+\Delta V)^2 - V^2 = V^2 + 2V\Delta V + \Delta V^2 - V^2 = 2V\Delta V$$

The ΔV^2 term is dropped because it is small compared with $2V\Delta V$. For this case then

$$\Delta P = \frac{1}{R}2V\Delta V = \frac{1}{13.2\,\Omega}2\cdot 115\,\mathrm{V}\cdot 5\,\mathrm{V} = 87\,\mathrm{W}$$

What is the fractional change in power output, $\Delta P/P$?

NOTE: The fractional change is the change divided by the original amount.

$$\frac{\Delta P}{P} = \frac{2V\Delta V}{R}\frac{R}{V^2} = 2\frac{\Delta V}{V} = 2\frac{5}{115} = 0.087$$

There is an 8.7% decline in power for a 4.35% decline in voltage.

Temperature Dependence of Resistivity

The resistance or resistivity of a metal depends on temperature according to

$$R = R_o[1+\alpha(T-T_o)] \quad \text{or} \quad \rho = \rho_o[1+\alpha(T-T_o)] \qquad \textbf{(29-8)}$$

where α is the temperature coefficient of resistance. This increase in resistance is consistent with our view of electric conductivity as charge carriers moving under the influence of an electric field until they crash into much larger ion cores.

As a metal is heated, energy goes into vibration of the ion cores. Charge carriers moving through the metal have a greater chance of intercepting an ion core as the amplitude of the oscillations increases. More collisions means greater resistance.

As an analog, consider the problem of walking blindfolded across a room with all the same size people standing in a random pattern in the room. The number of times you collide with another person depends on the relative cross sections (areas) of you and the people standing in the room. If the people are put into oscillation, perpendicular to your direction of motion, the probability of collision becomes greater. In the time it takes for you to move past each person, they move, increasing the probability of collision. They have a greater effective cross section, with respect to your walking across the room, because of their motion.

29-10 Calculate the resistance of an iron bar when the temperature is raised from $0^{\circ}\,\mathrm{C}$ to $100^{\circ}\,\mathrm{C}$. The initial resistance is $1.43\,\Omega$, and the temperature coefficient of resistivity is $0.00501/\mathrm{C}^{\circ}$.

Solution: $R = R_o[1 + \alpha(T - T_o)] = 1.43\,\Omega\left[1 + 0.00501/\mathrm{C}^{\circ}(100\,\mathrm{C}^{\circ})\right] = 2.15\,\Omega$

Batteries and *emf*

Batteries can be viewed (loosely) as charge pumps. Work is performed on charge as it moves through the battery effectively raising its potential (energy). This potential energy is expended when a path (circuit) is provided for the charge to travel outside the battery. The ability of the battery to do work on the charge passing through it is called the **electromotive force (*emf*)**. The electromotive force is measured in volts. A 10 volt battery can perform 10 Joules of work on each coulomb of charge passing through it and is said to have an emf of 10 V.

A battery connected to a load with resistance, R, causes current to flow in the load according to the $V = iR$ relation. For an amount of charge flowing through the load, the drop in potential (voltage) is the same as the gain in potential (voltage) in the battery. This is a conservation of energy statement: the charge does not go faster and faster as it moves around the circuit; neither does it go slower and slower; it moves at a rate dictated by $V = iR$.

Real batteries have internal resistance. Batteries viewed as charge pumps have resistance associated with moving the charge through the battery. The no load (no current) voltage at the terminals of the battery is the emf of the battery. As current is supplied by the battery, the internal resistance lowers the voltage at the terminals. In circuits, batteries are usually depicted as a voltage source in series with a resistor.

29-11 A battery with an ***emf*** of $10\,\mathrm{V}$ has internal resistance of $0.70\,\Omega$. What is the voltage at the terminals of the battery when $1.0\,\mathrm{A}$ is being drawn by an external load?

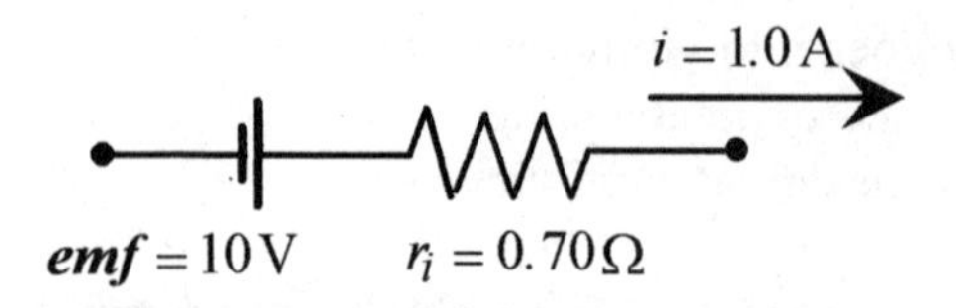

Fig. 29-3

Solution: The voltage drop inside the battery is $i r_i = 1.0\,\text{A} \cdot 0.70\,\Omega = 0.70\,\text{V}$, so the output voltage is reduced to $9.30\,\text{V}$.

When batteries with internal resistance are connected to external loads, the battery plus internal resistance is usually shown enclosed either by a dashed line or shading and the emf specified by the symbol ***emf.***

29-12 A battery of $12\,\text{V}$ ***emf*** has internal resistance of $0.50\,\Omega$ and is connected to a $6.0\,\Omega$ load. What is the terminal voltage of the battery under these conditions and the power dissipated in the external resistor?

Solution: The total current comes from $\boldsymbol{emf} = i(r_i + R)$ or $i = \dfrac{12\,\text{V}}{6.50\,\Omega} = 1.85\,\text{A}$.

At this current the voltage drop due to the internal resistor is $1.85\,\text{A} \cdot 0.50\,\Omega = 0.92\,\text{V}$, so the output voltage or voltage applied to the $6.0\,\Omega$ resistor is $11.08\,\text{V}$.

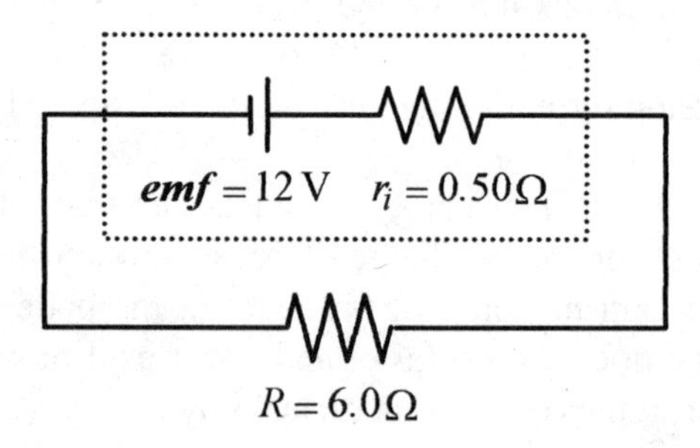

Fig. 29-4

The power dissipated in the external resistor is $P = Vi = 11.08\,\text{V} \cdot 1.85\,\text{A} = 20.5\,\text{W}$.

CHAPTER 30

RESISTORS IN D.C. CIRCUITS

In this chapter we will look first at series and parallel resistors, then at how series and parallel combinations are analyzed by writing successively simpler equivalent circuits.

Series Resistors

Take three resistors in series as shown in the accompanying diagram.

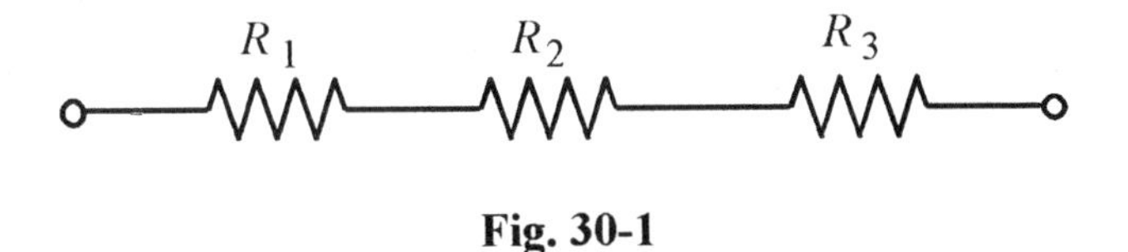

Fig. 30-1

If a voltage, V, is applied to this combination the current through each resistor is the same. The voltage statement for each of the resistors is $V_1 = IR_1, V_2 = IR_2, V_3 = IR_3$. The individual voltages across each resistor must add to the total so $V = V_1 + V_2 + V_3 = I(R_1 + R_2 + R_3)$. The equivalent resistor for this combination is $R = R_1 + R_2 + R_3$. Resistors in series add linearly.

30-1 What is the equivalent resistance of $3.0\,\Omega$ and $4.0\,\Omega$ resistors placed in series? What is the current in each resistor if $10\,\text{V}$ is applied to the combination?

Solution: The equivalent resistance is $R_{eq} = 3.0\,\Omega + 4.0\,\Omega = 7.0\,\Omega$. The current is the same in each resistor and has value $I = V/R_{eq} = 10\,\text{V}/7.0\,\Omega = 1.4\,\text{A}$.

Parallel Resistors

Place three resistors in parallel as shown in Fig. 30-2. If a voltage is applied to this combination, the voltage across each resistor is this applied voltage and the currents are determined by

$V = I_1R_1, V = I_2R_2$, and $V = I_3R_3$. These currents must add up to the total current delivered to the combination.

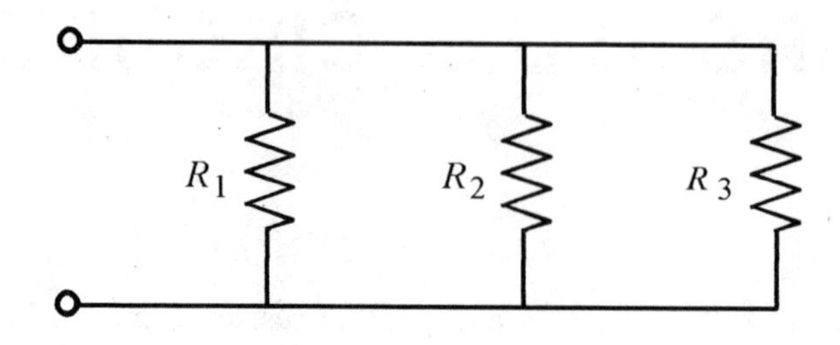

Fig. 30-2

$$I = I_1 + I_2 + I_3 = \frac{V}{R_1} + \frac{V}{R_2} + \frac{V}{R_3} = V\left(\frac{1}{R_1} + \frac{1}{R_2} + \frac{1}{R_3}\right) = \frac{V}{R_{eq}}$$

An equivalent resistor (for this combination) then would have the form

$$\frac{1}{R_{eq}} = \frac{1}{R_1} + \frac{1}{R_2} + \frac{1}{R_3}$$

Resistors in parallel add reciprocally.

30-2 What is the current in each resistor for a parallel combination of $3.0\,\Omega$, $3.5\,\Omega$, and $4.0\,\Omega$ resistors with $8.0\,\text{V}$?

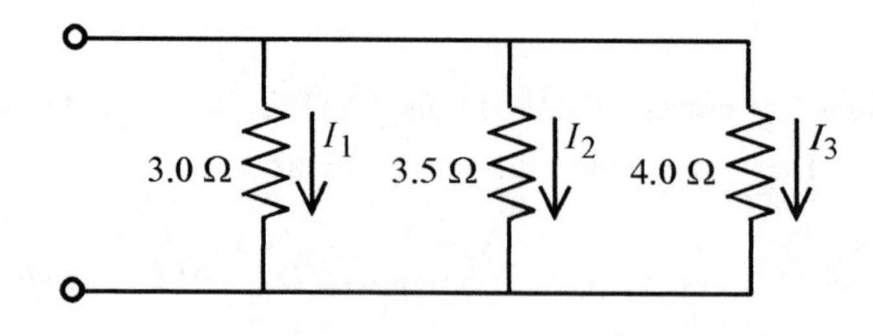

Fig. 30-3

Solution: $I_1 = \frac{8.0\,\text{V}}{3.0\,\Omega} = 2.7\,\text{A}$ $\quad I_2 = \frac{8.0\,\text{V}}{3.5\,\Omega} = 2.3\,\text{A}$ $\quad I_3 = \frac{8.0\,\text{V}}{4.0\,\Omega} = 2.0\,\text{A}$

The total current delivered by the power source is $I = 2.7\text{A} + 2.3\text{A} + 2.0\text{A} = 7.0\text{A}$. Now check the total by using the equivalence

$$\frac{1}{R_{eq}} = \frac{1}{3.0\Omega} + \frac{1}{3.5\Omega} + \frac{1}{4.0\Omega} = \frac{36.5}{42\Omega} \quad \text{or} \quad R_{eq} = 1.15\Omega$$

Using the equivalent resistance (1.15Ω) the total current is $I = V/R_{eq} = 8.0\,\text{V}/1.15\Omega = 7.0\,\text{A}$.

Series-Parallel Combinations

30-3 Calculate the equivalent resistance of the series-parallel combination shown in Fig. 30-4.

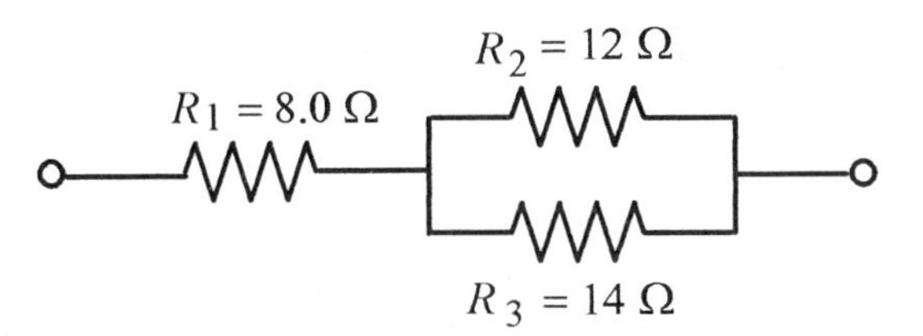

Fig. 30-4

Solution: The first step in this circuit analysis is to find the equivalent of the parallel combination.

$$\frac{1}{R_4} = \frac{1}{R_2} + \frac{1}{R_3} = \frac{1}{12\Omega} + \frac{1}{14\Omega} = \frac{13}{84\Omega} \quad \text{or} \quad R_4 = \frac{84\Omega}{13} = 6.5\Omega$$

The circuit then becomes two series resistors and they add to a single resistor.

NOTE: The equivalent resistance of any two resistors is lower than either of the two alone.

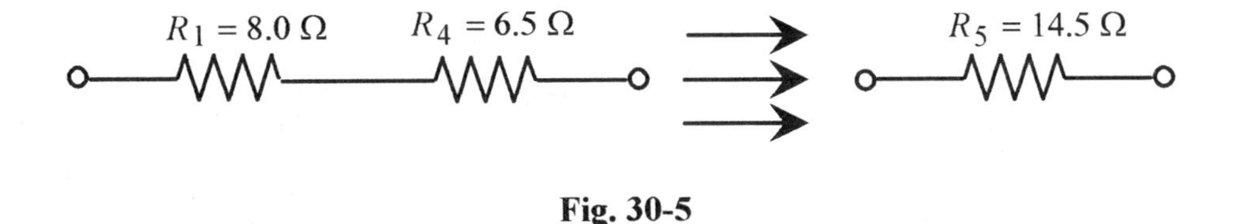

Fig. 30-5

30-4 Apply 20 V across this network and find the voltage across and current through each resistor.

Solution: In order to find the current in a resistor in the original network, it is necessary to work back through the equivalent circuits. Start with the last circuit. Apply 20V to a 14.5Ω resistor for a current of $I = V/R = 20\,\text{V}/14.5\Omega = 1.4\,\text{A}$. Working backwards, this current through R_1

produces a voltage across this resistor of $V_1 = (1.4\,\text{A})R_1 = (1.4\,\text{A})(8.0\Omega) = 11\,\text{V}$. This current passes through the equivalent resistor R_4 producing a voltage across it of $V_4 = (1.4\,\text{A})R_4 = (1.4\,\text{A})(6.5\Omega) = 9.1\,\text{V}$. V_1 and V_4 add to $20\,\text{V}$, the total applied, to within round off error range. The $1.4\,\text{A}$ divides between R_2 and R_3. The voltage across R_2 and R_3 is $V_4 = 9.1$ V so

$$I_2 = 9.1\,\text{V}/12\,\Omega = 0.76\,\text{A} \quad I_3 = 9.1\,\text{V}/14\,\Omega = 0.65\,\text{A}$$

Again this total adds to $1.4\,\text{A}$, within round off error. This analysis produces the voltage across and current through each resistor in the network. The numbers are internally consistent (the currents through R_2 and R_3 add up to the current through their equivalence) giving confidence in the calculation.

30-5 In the circuit of Fig. 30-6 calculate the voltage across, current through, and power requirements of the 8.0Ω resistor when $15\,\text{V}$ is applied to the network.

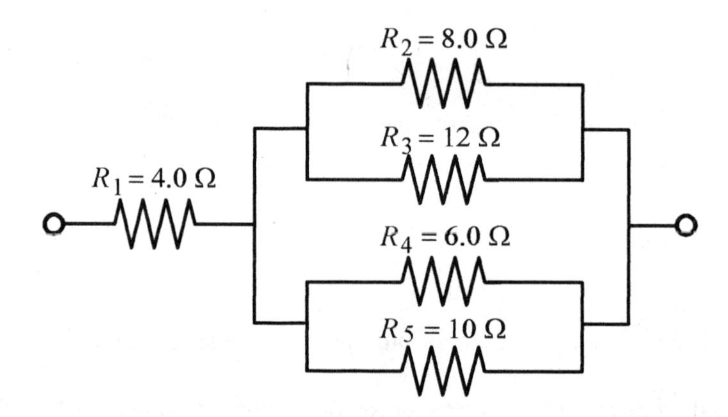

Fig. 30-6

Solution: The first step in the analysis is to find the equivalent of R_2 and R_3.

$$\frac{1}{R_6} = \frac{1}{R_2} + \frac{1}{R_3} = \frac{1}{8.0\Omega} + \frac{1}{12\Omega} = \frac{5}{24\Omega} \quad \text{or} \quad R_6 = \frac{24}{5} = 4.8\Omega$$

A similar analysis on R_4 and R_5 yields

$$\frac{1}{R_7} = \frac{1}{R_4} + \frac{1}{R_5} = \frac{1}{6.0\Omega} + \frac{1}{10\Omega} = \frac{8}{30\Omega} \quad \text{or} \quad R_6 = \frac{30}{8} = 3.8\Omega$$

Now the equivalent can be drawn.

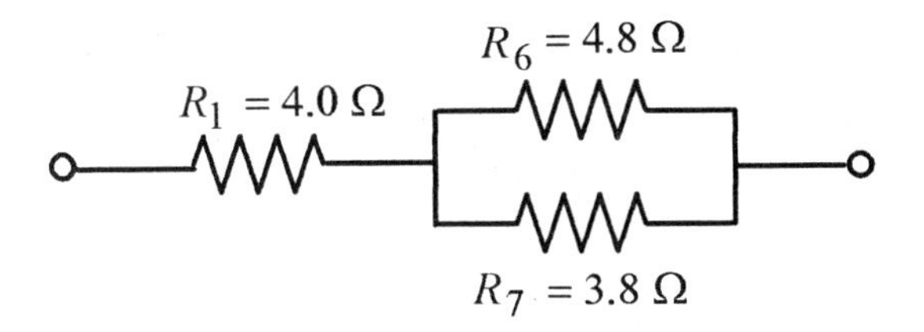

Fig. 30-7

The parallel combination of R_6 and R_7 produces another equivalent.

$$\frac{1}{R_8} = \frac{1}{R_6} + \frac{1}{R_7} = \frac{1}{4.8\Omega} + \frac{1}{3.8\Omega} = \frac{8.5}{18\Omega} \quad \text{or} \quad R_8 = \frac{18}{8.5} = 2.1\Omega$$

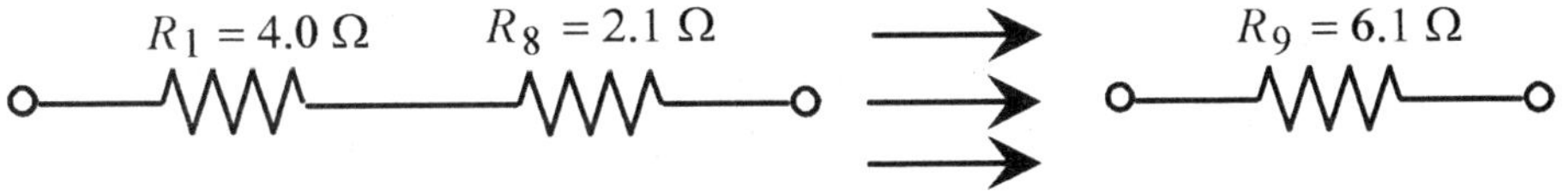

$R_1 = 4.0\ \Omega$ $\quad R_8 = 2.1\ \Omega$ $\quad \Rightarrow \quad R_9 = 6.1\ \Omega$

Fig. 30-8

The total current is $I = V/R_9 = 15\,\text{V}/6.1\Omega = 2.46\,\text{A}$.

This current through the 2.1Ω resistor produces $V_8 = (2.46\,\text{A})R_8 = (2.46\,\text{A})(2.1\Omega) = 5.16\,\text{V}$.

Look back through the equivalent circuits and notice that this voltage is across R_8, R_6, and R_2.

The current through R_2 then is $I_2 = V_8/R_2 = V_2/R_2 = 5.16\,\text{V}/8.0\,\Omega = 0.64\,\text{A}$.

The power dissipated in this resistor is $I^2R = (0.64\,\text{A})^2(8.0\Omega) = 3.3\,\text{W}$.

The general procedure for circuit analysis is to look at the circuit and find parts of it that can be replaced with simpler equivalents and just keep applying the process until the circuit is reduced to one resistor, perform $V = IR$ on that resistor, and move back through the equivalents to obtain the desired information.

CHAPTER 31

KIRCHHOFF'S LAWS

Kirchhoff's laws applied to circuits containing multiple branches, voltage sources, and resistors allow calculation of currents in each branch of the circuit. Start with a very simple circuit consisting of a battery and resistor. The battery polarity is taken as indicated with positive charge

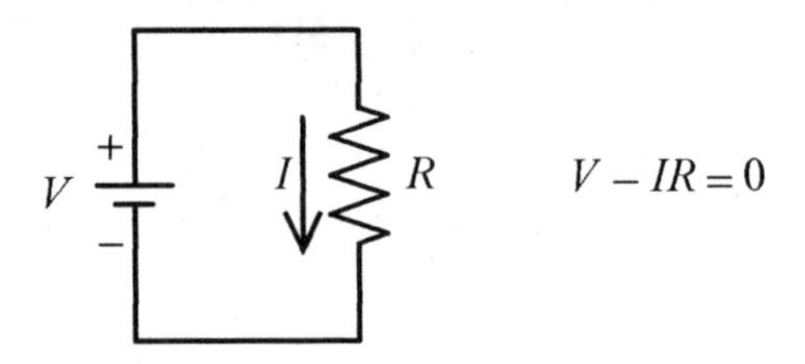

Fig. 31-1

leaving the positive plate. The current is as indicated by the arrow. An amount of charge or current over time starting at the negative plate of the battery is viewed as gaining energy in passing through the battery and losing this same amount of energy in the resistor. The equation for this circuit is shown in Fig. 31-1.

Kirchhoff Voltage Loop Equations

The $V - IR$ equation for this simple circuit illustrates the **first Kirchhoff law**: The algebraic sum of the changes in potential (energy) encountered in a complete traverse of a path is zero. This is analogous to a conservation of energy statement.

The only difficulty in applying Kirchhoff's laws is keeping the algebraic signs correct. When confused, go back to this simple diagram of a battery and resistor and the current flowing (in the external circuit) from the positive side of the battery through the resistor to the negative side of the battery. Write the $V - IR = 0$ statement establishing that when a battery is traversed in a positive direction, there is a gain in energy giving V a plus sign; and when a resistor is traversed in the direction of the current arrow, there is a loss in energy giving IR a negative sign. Every Kirchhoff law problem you do should have this little circuit drawn in a corner of the paper to remind you of the algebraic signs and their importance.

Think of a "piece of charge" traversing the circuit gaining energy as it passes through the battery and losing energy as it passes through the resistor. In this simple circuit the direction of the current is clear. In multiple branch circuits it is not possible to "guess" the correct directions for the currents. Make the best educated guess that you can and let the mathematics tell you whether the currents are positive or negative.

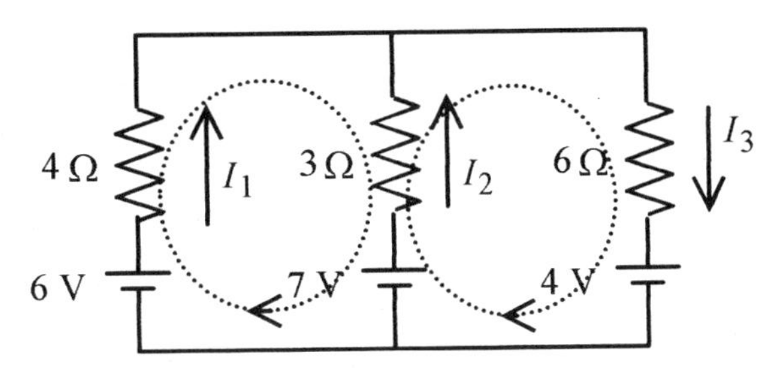

Fig. 31-2

Consider the three branch circuit of Fig. 31-2. Make your best estimate of the current direction for each resistor and draw and label the current arrows. It is not necessary to have the direction correct, it is only necessary to be consistent in applying the sign convention. Now make two loops around this circuit, one around the left and one around the right as indicated by the closed loops. Remember, when traversing a battery from minus to positive write $+V$ and when traversing the resistor in the direction of the current arrow write $-IR$. Around the first loop

$$6-4I_1+3I_2-7=0$$

In traversing this loop start at the low, or negative, side of the 6 V battery and proceed around the loop:

6 is positive

$4I_1$ is negative because the resistor is traversed in the direction of the current arrow.

$3I_2$ is positive because the resistor is traversed in the direction opposite to the current arrow.

7 is negative because the battery is traversed from plus to minus.

Around the second loop

$$7-3I_2-6I_3-4=0$$

Rewrite these two equations

$$\begin{aligned} -4I_1+3I_2 \qquad &= 1 \\ 3I_2+6I_3 &= 3 \end{aligned}$$

Writing the equations this way illustrates very clearly that they cannot be solved. In mathematical language they are two equations in three unknowns. The third equation necessary to solve for the three currents comes from the next Kirchhoff law.

Kirchhoff Current Junction Equations

The **second Kirchhoff law** deals with the currents: the sum of the currents to any junction must add to zero. This law is often forgotten. It is, however, easily remembered via a whimsical law called the "fat wire law." The fat wire law is very simple. The sum of the currents to any junction must equal zero, otherwise the wire will get fat. Applying this second Kirchhoff law to the junction just above the 3Ω resistor

$$I_1 + I_2 - I_3 = 0$$

Now we have three <u>independent</u> equations in three unknowns.

NOTE: For convenience in the discussion, individual equations in a set of equations will be designated by letter symbols. The number of significant figures used in the calculations will be kept to a minimum so as not to obscure the mathematical procedure with numbers. Likewise, units will not be used. Several techniques for solving multiple equations in multiple unknowns will be presented. It is not necessary to know all these techniques. Pick a technique that you feel comfortable with and learn it well.

$$\begin{aligned} (a) \quad & I_1 + I_2 - I_3 = 0 \\ (b) \quad & -4I_1 + 3I_2 = 1 \\ (c) \quad & 3I_2 + 6I_3 = 3 \end{aligned}$$

At this point you may be asking why we did not take another loop to obtain a third equation. Taking another loop in the circuit will only serve to generate another equation which is a linear combination of the first two loop equations. Take a loop around the outside of the circuit and write:

$$6 - 4I_1 - 6I_3 - 4 = 0 \quad \text{or} \quad 4I_1 + 6I_3 = 2$$

This equation is just the difference of the first two loop equations and is not a linearly independent third equation. Therefore it cannot be used to solve for the three currents. As an exercise take a figure eight path around the circuit and prove that this statement is also a linear combination of the first two loop equations.

Solving by Adding and Subtracting

Now to the solution of these three equations. The most popular and direct approach is to add and subtract equations to produce two equations in two unknowns and then further add and subtract to solve for one current and work backwards to find all the currents.

$$
\begin{array}{lr}
\text{multiply } 4\times(\text{a}) & 4I_1+4I_2-4I_3=0 \\
\text{add (b)} & -4I_1+3I_2 \qquad\quad =1 \\
\text{to obtain} \quad (\alpha) & 7I_2-4I_3=1
\end{array}
$$

Combine equation (α) with 2/3 of equation (c) and write (α) and (β)

$$
\begin{array}{l}
(\alpha) \quad 7I_2-4I_3=1 \\
(\beta) \quad 2I_2+4I_3=2 \\
\qquad 9I_2=3 \quad \text{or} \quad I_2=1/3
\end{array}
$$

Place $I_2 = 1/3$ into (c). $\qquad 3(1/3)+6I_3=3 \quad \text{or} \quad I_3=1/3$

Now place $I_2 = 1/3$ and $I_3 = 1/3$ into (a). $\qquad I_1+(1/3)-(1/3)=0 \quad \text{or} \quad I_1=0$

Check these current values in each of the original equations to verify that they are correct.

Solving by Determinants

Another method of solving these equations is with determinants. Using determinants to solve three equations in three unknowns is explained in the Introduction, Mathematical Background. Your calculator may have an algorithm for solving simultaneous equations using determinants (sometimes called Cramer's rule), so this may be your method of choice.

Going back to the equations marked (a), (b), (c) set up the master determinant and expand along the first column. Equation (c) is divided by 3 to reduce the numbers.

$$
D=\begin{vmatrix} 1 & 1 & -1 \\ -4 & 3 & 0 \\ 0 & 1 & 2 \end{vmatrix}=1\begin{vmatrix} 3 & 0 \\ 1 & 2 \end{vmatrix}+4\begin{vmatrix} 1 & -1 \\ 1 & 2 \end{vmatrix}+0\begin{vmatrix} 1 & -1 \\ 3 & 0 \end{vmatrix}=1[6-0]+4[2+1]+0[0+3]=6+12=18
$$

The determinant for I_1 is expanded along the top row

$$
D_{I_1}=\begin{vmatrix} 0 & 1 & -1 \\ 1 & 3 & 0 \\ 1 & 1 & 2 \end{vmatrix}=0\begin{vmatrix} 3 & 0 \\ 1 & 2 \end{vmatrix}-1\begin{vmatrix} 1 & 0 \\ 1 & 2 \end{vmatrix}-1\begin{vmatrix} 1 & 3 \\ 1 & 1 \end{vmatrix}=0[6-0]-1[2-0]-1[1-3]=-2+2=0
$$

The value of $I_1 = \frac{D_{I_1}}{D} = \frac{0}{18} = 0.$

The determinant for I_2 is expanded about the top row

$$D_{I_2} = \begin{vmatrix} 1 & 0 & -1 \\ -4 & 1 & 0 \\ 0 & 1 & 2 \end{vmatrix} = 1\begin{vmatrix} 1 & 0 \\ 1 & 2 \end{vmatrix} - 0\begin{vmatrix} -4 & 0 \\ 0 & 2 \end{vmatrix} - 1\begin{vmatrix} -4 & 1 \\ 0 & 1 \end{vmatrix} = 1[2-0] - 0[-8-0] - 1[-4-0] = 2+4 = 6$$

The value for $I_2 = \frac{D_{I_2}}{D} = \frac{6}{18} = \frac{1}{3}.$

As an exercise set up the determinant for I_3, expand along the top row, and verify the value of I_3 obtained previously.

Solving by Augmented Matrix

A third method of solving three equations in three unknowns is with an augmented matrix. Many calculators that have internal programs for solving simultaneous equations require data to be input in a manner similar to the augmented matrix form. The manipulations of the augmented matrix are similar to adding and subtracting equations to find a solution. Before applying this method to the equations for this problem look at a general set of three equations in three unknowns.

$$\begin{aligned} ax + by + cz &= d \\ ex + fy + gz &= h \\ kx + my + nz &= p \end{aligned}$$

If the first equation were multiplied by $-(k/a)$ and added to the third equation these three equations would read

$$\begin{aligned} ax + by + cz &= d \\ ex + fy + gz &= h \\ (m - kb/a)y + (n - kc/a)z &= p - kd/a \end{aligned}$$

What this accomplishes is to make the coefficient of x zero in the third equation. Now multiply the first equation by $-k/e$ and add to the second equation making the coefficient of x zero in the second equation. After this, the second equation can be multiplied by a constant and added to the third equation with the constant chosen so as to make the coefficient of y in the third equation zero. Once this is accomplished the solutions are easily written down. The procedure is better illustrated with a specific case, the previous problem. These equations already have some coefficients equal to zero making the process easier. This is usually the case with Kirchhoff law analysis

$$\begin{aligned} I_1 + I_2 - I_3 &= 0 \\ -4I_1 + 3I_2 \quad &= 1 \\ I_2 + 2I_3 &= 1 \end{aligned}$$

Multiply the first equation by 4 and add to the second equation generating a new second equation.

$$\begin{aligned} I_1 + I_2 - I_3 &= 0 \\ 7I_2 - 4I_3 &= 1 \\ I_2 + 2I_3 &= 1 \end{aligned}$$

This makes the coefficient of I_1 equal to zero in the second and third equation producing two equations in two unknowns. Now multiply the second equation by $-1/7$ and add to the third equation generating a new third equation.

$$\begin{aligned} I_1 + I_2 - I_3 &= 0 \\ 7I_2 - 4I_3 &= 1 \\ (18/7)I_3 &= 6/7 \end{aligned}$$

Now solve directly with $I_3 = 1/3$. Substituting this into the second equation $I_2 = 1/3$. Putting these two values into the first equation obtains $I_1 = 0$.

The augmented matrix method uses the coefficients from the equations. The adding and subtracting of equations to generate other equations is done by multiplying and adding different rows of the matrix just as if they were equations. As we go through the process imagine that the operations are being performed on equations. Working with the coefficients arranged in an orderly array is easier. Working the matrix so as to produce zeros starting in the lower left hand corner and then moving diagonally across the matrix is easier than working the equations. Start with

$$\left(\begin{array}{ccc|c} 1 & 1 & -1 & 0 \\ -4 & 3 & 0 & 1 \\ 0 & 1 & 2 & 1 \end{array}\right)$$

There is already a zero in the lower left corner so we need to create a zero in the -4 position. Multiply the first row by 4 and add to the second row generating a new second row. This is equivalent to multiplying the first equation by 4 and adding it to the second equation generating a new second equation, which is a linear combination of the first two. This is exactly what is done in solving simultaneous equations by adding and subtracting.

$$\left(\begin{array}{ccc|c} 1 & 1 & -1 & 0 \\ 0 & 7 & -4 & 1 \\ 0 & 1 & 2 & 1 \end{array}\right)$$

Now multiply the third row by −7 and add to the second row generating a new third row.

$$\left(\begin{array}{ccc|c} 1 & 1 & -1 & 0 \\ 0 & 7 & -4 & 1 \\ 0 & 0 & -18 & -6 \end{array}\right)$$

Now go back to the equations where $18I_3 = 6$ or $I_3 = 1/3$.

Put $I_3 = 1/3$ into $7I_2 - 4I_3 = 1$ to obtain $I_2 = 1/3$.

Put $I_2 = 1/3$ and $I_3 = 1/3$ to obtain $I_1 = 0$.

These methods of solving multiple equations in multiple unknowns require considerable number manipulation with the attendant chance for error. The augmented matrix method has the least number manipulation while adding and subtracting may be more familiar to you. All of the techniques work. Pick the one that fits you, and the set of equations you are solving, best and proceed carefully.

Applications

The procedure for solving Kirchhoff's laws problems is the following:

1. Draw the battery and resistor circuit in the corner of your paper to remind you of the sign convention.

2. Place current arrows next to each resistor and label them.

3. Pick a junction and write the current statement.

4. Draw loops and write the loop equations.

5. Solve for the currents and check them in the original equations.

31-1 Solve for the current in each of the resistors in the circuit shown in Fig. 31-3.

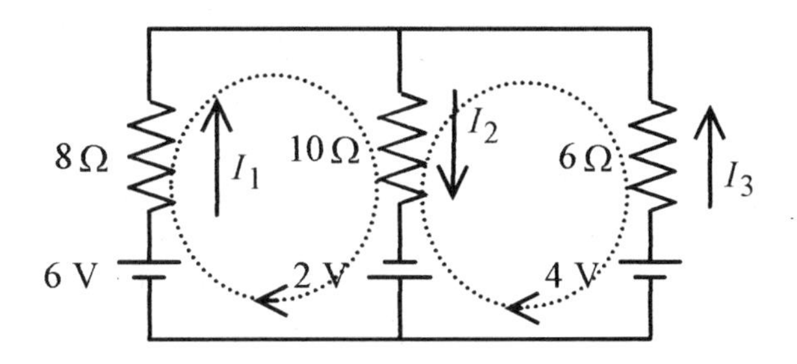

Fig. 31-3

Solution: Follow the procedure for Kirchhoff problems. This is a subject that is conceptually easy but because of the large number of manipulations often hard to actually do. Work this problem as you are following along in the book. Draw the circuit and begin.

Draw the single battery and single resistor with current arrow and $V - IR = 0$ equation.

Place and label the current arrows.

Write the statement of the "fat wire law" for the top center of the circuit.

$$I_1 + I_3 = I_2$$

Write a loop statement for the left side of the circuit.

$$6 - 8I_1 - 10I_2 + 2 = 0$$

Write a loop statement for the right side of the circuit.

$$-2 + 10I_2 + 6I_3 - 4 = 0$$

Now write these three equations in a convenient form for solution.

$$\begin{aligned} I_1 - I_2 + I_3 &= 0 \\ 8I_1 + 10I_2 \quad &= 8 \\ 10I_2 + 6I_3 &= 6 \end{aligned}$$

As an exercise solve these equations and verify that the solutions are $I_1 = 17/47 = 0.36$, $I_2 = 24/47 = 0.51$, and $I_3 = 7/47 = 0.15$. Check these answers in the original equations.

31-2 Consider a more complicated problem as shown in Fig. 31-4 and calculate the currents in the resistors.

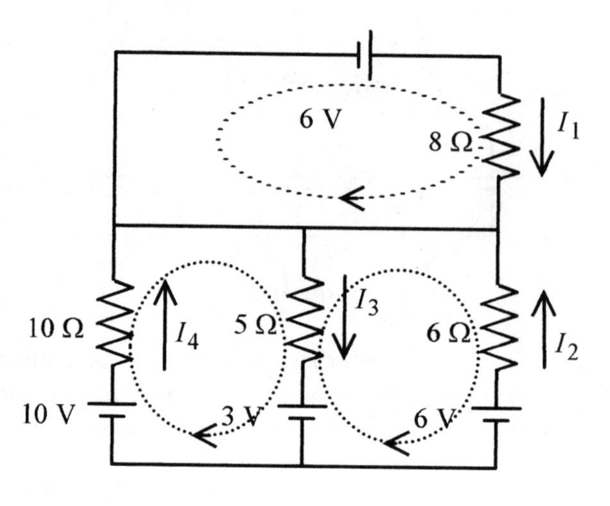

Fig. 31-4

Solution: Draw the battery and resistor circuit and $V - IR$ equation to remind you of the algebraic sign convention.

Note the current arrows. Remember that if one of the currents comes out negative it means only that the current is opposite to the direction of your arrow.

Apply the "fat wire law" to the line across the top of the three resistors.

$$I_1 + I_2 + I_4 = I_3$$

The loop around the top branch of the circuit is $6 - 8I_1 = 0$.

A loop around the lower left branch is $10 - 10I_4 - 5I_3 - 3 = 0$.

A loop around the lower right branch is $3 + 5I_3 + 6I_2 - 6 = 0$.

Rewrite the equations in convenient form for solving.

$$\begin{aligned}
I_1 + I_2 - I_3 + I_4 &= 0 \\
8I_1 &= 6 \\
6I_2 + 5I_3 &= 3 \\
5I_3 + 10I_4 &= 7
\end{aligned}$$

Immediately $I_1 = 3/4 = 0.75$. Work the problem and verify that the other currents are $I_2 = -0.20$, $I_3 = 0.84$, and $I_4 = 0.28$.

31-3 Calculate the currents in the resistors of the circuit shown in Fig. 31-5.

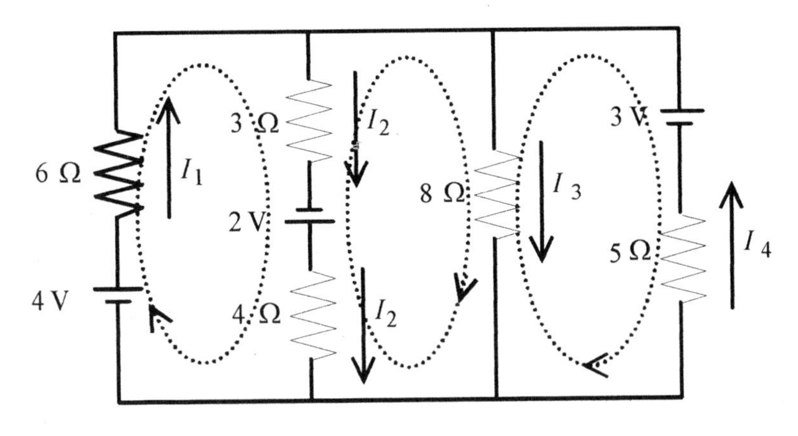

Fig. 31-5

Solution: Draw the battery and resistor circuit and $V - IR$ equation to remind you of the algebraic sign convention. Apply the "fat wire law" to the line across the top of the circuit.

$$I_1 - I_2 - I_3 + I_4 = 0$$

The loop around the left branch is $4 - 6I_1 - 3I_2 + 2 - 4I_2 = 0$.

The middle loop is $4I_2 - 2 + 3I_2 - 8I_3 = 0$.

The loop around the right branch is $-3 + 5I_4 + 8I_3 = 0$.

Rewriting the equations in convenient form for solving

$$\begin{aligned} I_1 - I_2 - I_3 + I_4 &= 0 \\ 6I_1 + 7I_2 &= 6 \\ 7I_2 - 8I_3 &= 2 \\ 8I_3 + 5I_4 &= 3 \end{aligned}$$

Work the problem and verify that the currents are $I_1 = 0.41$, $I_2 = 0.51$, $I_3 = 0.19$, and $I_4 = 0.29$.

Hint: First solve for I_3 using determinants, then the rest of the currents by algebra.

CHAPTER 32

R-C CIRCUITS

Our first look at R-C circuits is with a simple word description of the charging phenomenon. The circuit shown in Fig. 32-1 is appropriate for the study of charging and discharging. After giving a general description of how the circuit works, we will take a closer look at the mathematical description. Throughout the discussion, unless the context and notation indicate otherwise, lower case letters mean time varying quantities and upper case letters constants. For example, q is the symbol for time varying charge on the capacitor while Q is the total charge.

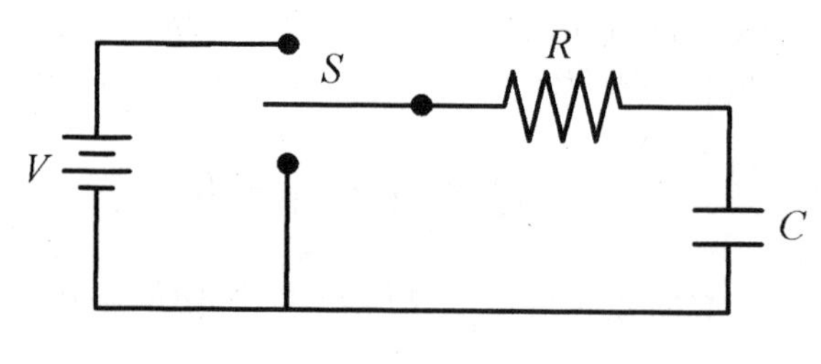

Fig. 32-1

Charging

Assume the capacitor has zero charge and voltage, and place the switch (S) in the charging position (up). When the battery voltage is applied to the R and C in series, current begins to flow. Current through the resistor causes a voltage drop across the resistor. Because of this voltage drop, less than the battery voltage is applied to the capacitor. As current flows, the capacitor charges and less and less current flows until there is no current, no voltage across the resistor and the capacitor is charged to the battery voltage ($Q = CV$).

The charge on the capacitor increases with time starting with zero charge and eventually reaching maximum charge of CV following an exponential function.

$$q = CV\left(1 - e^{-t/RC}\right) \tag{32-1}$$

This function fits our understanding of how the circuit operates since when $t = 0$, q is zero, and when t is very large, $q = CV$, the capacitor is totally charged.

The voltage on the capacitor is

$$q/C = V_C = V\left(1 - e^{-t/RC}\right) \tag{32-2}$$

Here the notation V_C denotes the time varying voltage on the capacitor.

The current in the circuit declines exponentially according to $i = (V/R)e^{-t/RC}$.

With this description of the operation of the charging circuit, we can proceed to a more detailed (mathematical) analysis of the charging situation.

C When the battery voltage is applied to R and C, voltages across these components vary with time. We can, however, write a Kirchhoff-type voltage statement that is correct for the entire time the circuit is charging. This is called an instantaneous voltage statement; it is true at any instant of time. Following Kirchhoff's law

$$V = iR + \frac{q}{C} \quad \text{or} \quad CV - q = RC\frac{dq}{dt} \quad \text{or} \quad \frac{dq}{q - CV} = -\frac{1}{RC}dt$$

This equation can be integrated with a change of variable. Replace $q - CV$ with x so $dq = dx$.

Then $\int \frac{dx}{x} = -\frac{1}{RC}\int dt$. The integral of the left side is $\ln(x) = \ln(q - CV)$, so

$$\ln(q - CV) = -\frac{1}{RC}t + \ln K_1$$

Choosing the constant as $\ln K_1$ is very convenient, allowing further simplification.

$$\ln\frac{q - CV}{K_1} = -\frac{1}{RC}t \quad \text{or} \quad q - CV = K_1 e^{-t/RC}$$

Notice that the logarithm equation goes into an exponential equation. This is not an every day mathematical operation. Review the Introduction, Mathematical Background, and be sure you understand how to go from a logarithmic equation to an exponential one. Rewrite the equation

$$q = CV + K_1 e^{-t/RC}$$

and apply the condition that at $t = 0$, $q = 0$, or $0 = CV + K_1$, or $K_1 = -CV$, so finally

$$q = CV\left(1 - e^{-t/RC}\right)$$

Obtaining this expression for q looks easy, and it is if you remember to choose the constant for convenience in solving the equation, remember how to switch from a logarithmic equation to an exponential equation, and apply the initial conditions correctly.
The voltage across the capacitor is

$$\frac{q}{C} = V_C = V\left(1 - e^{-t/RC}\right)$$

The current in the resistor is

$$i = \frac{dq}{dt} = \frac{V}{R} e^{-t/RC} \qquad \textbf{(32-3)}$$

The voltage across the resistor is

$$V_R = iR = Ve^{-t/RC} \qquad \textbf{(32-4)}$$

Two graphs are particularly helpful in understanding the situation; one of q versus t, and the other of i versus t.

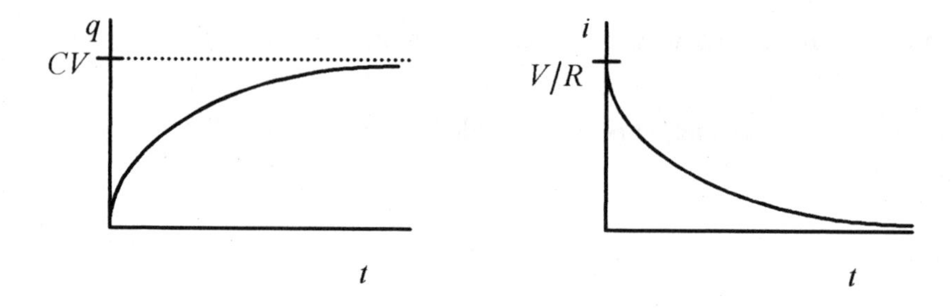

Fig. 32-2

Notice in the curve of q versus t, that at $t = 0$, $q = 0$, and as $t \to \infty$, $q \to CV$.
In the curve of i versus t, at $t = 0$, $i = V/R$, and as $t \to \infty$, $i \to 0$.
At $t = 0$, the charge on the capacitor is zero and the current is maximum; and as t goes to infinity, the current goes to zero and the charge reaches its maximum of CV.

The Time Constant

In biological systems that grow exponentially the systems are often characterized by giving the doubling time, the time for the system to double in number, size, or mass. In electrical systems that grow exponentially the systems are characterized by a time constant, the time to make the exponent of e equal to 1. The time constant for this circuit is RC (see equation 32-1). As an exercise verify that the units of RC are seconds.

Look at equation 32-1 and the q versus t graph. When $t = RC$, $1 - e^{-1} = 0.63$ and the charge on the capacitor has risen to 0.63 of its final value. A similar statement can be made about the voltage on the capacitor. After one time constant, the voltage on the capacitor is 0.63 of the battery voltage.

The current, meanwhile, has in one time constant dropped to $e^{-1} = 0.37$ of its initial value.

32-1 A $10\,\mathrm{k\Omega}$ resistor and a $20\,\mu\mathrm{F}$ capacitor are placed in series with a $12\,\mathrm{V}$ battery. Find the charge on the capacitor, the current, and the voltages on the capacitor and resistor at the instant the switch is closed, $t = 0$.

Solution: At $t = 0$ the charge on the capacitor is zero.

At $t = 0$, the current is $i = V/R = 12\,\mathrm{V}/10 \times 10^3\,\Omega = 1.2 \times 10^{-3}\,\mathrm{A}$.

At $t = 0$, the voltage on the capacitor is zero (it has no charge) and the entire battery voltage of $12\,\mathrm{V}$ is across the resistor.

32-2 For the circuit of problem 32-1, find the time constant, and the charge, current, and V_R, and V_C at a time equal to one time constant.

Solution: The time constant is $RC = 10 \times 10^3\,\Omega \cdot 20 \times 10^{-6}\,\mathrm{F} = 0.20\,\mathrm{s}$.

The charge on the capacitor at $t = 0.20\,\mathrm{s}$ is

$$q|_{t=RC} = CV\left(1 - e^{-1}\right) = 12\,\mathrm{V}(20 \times 10^{-6}\,\mathrm{F})0.63 = 1.5 \times 10^{-4}\,\mathrm{C}.$$

The current at $t = 0.20\,\mathrm{s}$ is

$$i|_{t=RC} = \frac{V}{R}e^{-1} = \frac{12\,\mathrm{V}}{1.0 \times 10^4\,\Omega}0.37 = 4.4 \times 10^{-4}\,\mathrm{A}.$$

The voltage across the capacitor is $V_C = V\left(1 - e^{-1}\right) = 12\,\mathrm{V} \cdot 0.63 = 7.6\,\mathrm{V}$.

The voltage across the resistor is $12\,\mathrm{V} - 7.6\,\mathrm{V} = 4.4\,\mathrm{V}$.

These problems can be deceptively easy. Be sure you know how to manipulate the exponents on your calculator. Don't get a test problem wrong because you did not practice all the steps in the problem and were unfamiliar with manipulating exponents on your calculator.

32-3 For the circuit of problem 32-1, how long does it take for the capacitor to reach 80% of its final charge?

Solution: This problem is similar to radioactive decay problems where we want to know the time for half the material to decay. There is a fair amount of algebraic manipulation that is easier to follow without numbers, so the problem will be worked as far as possible with symbols.

Start with $q = CV\left(1 - e^{-t/RC}\right)$ and note that the final (fully charged) q is equal to CV. Mathematically, when $t \to \infty$, $e^{-t/RC} \to 0$ and $1 - e^{-t/RC} \to 1$ so $q \to CV$.

To find the time for 80% charge set q equal to 80% of the final charge, or $q = 0.80CV$, and solve for t.

$$0.80CV = CV\left(1 - e^{-t/RC}\right) \quad \text{or} \quad 0.80 = 1 - e^{-t/RC} \quad \text{or} \quad e^{-t/RC} = 0.20$$

For convenience switch to positive exponents so $1/e^{t/RC} = 0.20$ or $e^{t/RC} = 1/0.20 = 5$

In order to solve for t, switch the exponential equation to a logarithmic equation. One of the functions of logarithms is to solve for variables in exponents.

$$t/RC = \ln 5 \quad \text{or} \quad t = RC\ln 5$$

Now put in the values for R and C.

$$t = RC\ln 5 = 1.0 \times 10^4\,\Omega \cdot 20 \times 10^{-6}\,\mu\text{F} \cdot \ln 5 = 0.32\,\text{s}.$$

As a check note that $1 - e^{-t/RC} = 1 - e^{-0.32/0.20} = 0.80$

Discharging

After the capacitor is left to charge for a long time (many time constants) the charge is CV. Move the switch to the discharge position (down in Fig. 32-1) where R and C are in series. When the charged capacitor and resistor are placed in series the charged capacitor acts as a battery. The voltage on the capacitor is q/C, and this voltage appears across the resistor as iR. As time goes on, the charge on the capacitor is depleted and the current drops (eventually) to zero. The charge decays according to

$$q = CVe^{-t/RC} \tag{32-5}$$

and the current according to

$$i = \frac{dq}{dt} = \frac{d}{dt}CVe^{-t/RC} = -\frac{V}{R}e^{-t/RC} \tag{32-6}$$

The negative sign indicates that the current in the resistor is in the opposite direction from the charging situation, which must be the case.

Mathematical analysis of the discharge circuit starts with a Kirchhoff's law loop type of statement.

$$iR+\frac{q}{C}=0 \quad \text{or} \quad \frac{dq}{dt}=-\frac{1}{RC}q \quad \text{or} \quad \frac{dq}{q}=-\frac{1}{RC}dt$$

and solving

$$\ln q=(-1/RC)t+\ln K_2$$

Again notice that the choice of constant is very convenient because

$$\ln\frac{q}{K_2}=-\frac{1}{RC}t \quad \text{or} \quad \frac{q}{K_2}=e^{-t/RC} \quad \text{or} \quad q=K_2e^{-t/RC}$$

Now impose the initial condition. At $t=0$, $q=CV$, so $K_2=CV$ and

$$q=CVe^{-t/RC}$$

The charge on the capacitor decays with the same time constant, RC.

32-4 For the circuit described in problem 32-1 placed in the discharge mode, how long does it take for the circuit to discharge to 50% of its original (total) charge?

Solution: Solve equation 32-5 for t when $q=0.50CV$

$$0.50CV=CVe^{-t/RC} \quad \text{or} \quad 0.50=e^{-t/RC}$$

It is more convenient to write 0.50 as 1/2, so when the statement is converted to logarithms,

$$\ln 0.50=-\frac{t}{RC} \quad \text{is} \quad \ln 1-\ln 2=-\frac{t}{RC} \quad \text{and since } \ln 1=0$$

$$t=RC\ln 2=1.0\times10^4\Omega\cdot 20\times10^{-6}\text{F}\cdot\ln 2=(0.20)(0.69\text{s})=0.14\text{s}$$

In terms of time constants this would be $\ln 2$ time constants, or 0.69 of a time constant.

It makes sense that our answer is less than one time constant since it takes less time for the charge to decline to 50% of its initial value than to 37% ($1/e$) of its initial value.

32-5 An RC circuit is observed during discharge to have an initial capacitor potential of 100 V and after 3.0 s to have a potential of 20 V. How long will it take for the capacitor to discharge to 1.0 V.

Solution: The voltage across the capacitor at any time is determined by equation 32-5 rewritten as q/C, or $V_C = V_o e^{-t/RC}$, where V_o is the voltage at $t = 0$.

Take $V_o = 100$, $V_C = 20\,\text{V}$, and $t = 3.0\,\text{s}$ and write

$$20\,\text{V} = 100\,\text{V}\,e^{-3.0/RC} \quad \text{or} \quad 2/10 = e^{-3.0/RC} \quad \text{or} \quad e^{3.0/RC} = 5$$

Switching to logarithms

$$\frac{3.0}{RC} = \ln 5 \quad \text{or} \quad \frac{3.0}{\ln 5} = RC \quad \text{or} \quad RC = 1.9\,\text{s}$$

Knowing this number, the specific decay law for this circuit can be written $V = 100\text{V}\,e^{-t/1.9}$.

Now calculate the time for the voltage to drop to 1.0 V.

$$1.0\,\text{V} = 100\,\text{V}\,e^{-t/1.9} \quad \text{or} \quad (1/100) = e^{-t/1.9}$$

Switching to logarithms

$$-\ln 100 = -(t/1.9) \quad \text{or} \quad \text{t} = 1.9 \cdot \ln 100 = 8.6\text{s}$$

Go back over this problem and note the procedure.

1. After reading the problem, the general law (equation) was written down, $V_C = V_o e^{-t/RC}$.
2. Next the data from the problem (100 V going to 20 V in 3.0 s) was used to find RC.
3. With RC the specific law for this problem was written, $V = 100\text{V}\,e^{-t/1.9}$.
4. Finally, with this specific law the predictive calculation was performed to find the time for the 100 V to decay to 1.0 V.

This analysis procedure is typical of growth and decay problems in general. Be familiar with the steps in this procedure. It will keep you from getting lost and not knowing how to proceed in problems like this.

CHAPTER 33

MAGNETIC FIELDS

The **magnetic field** or ***B* field** is a field that produces a force on a moving charge. This is not the only definition, but it is a good operational definition to start with. The force is given by the vector relation

$$\boldsymbol{F} = q_o \boldsymbol{v} \times \boldsymbol{B} \tag{33-1}$$

Cross products are discussed in detail in the Introduction, Mathematical Background. In a coordinate system the relation is shown in the accompanying figure.

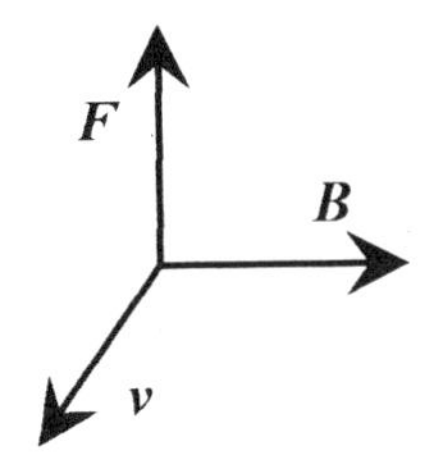

Fig. 33-1

To fix the direction of the force on a positive charge, place the right hand with fingers pointing in the direction of ***v*,** and a line normal to the palm of the hand pointing in the direction of ***B***. Curl the fingers naturally into ***B***. The thumb points in the direction of ***F***. For a negative charge the force is in the opposite direction. The force is a right angles to the plane of ***v*** and ***B***.

Since this is a vector it can be expressed as a cross product in determinant form,

$$\boldsymbol{F} = q_o \boldsymbol{v} \times \boldsymbol{B} = \begin{vmatrix} \boldsymbol{i} & \boldsymbol{j} & \boldsymbol{k} \\ v_x & v_y & v_z \\ B_x & B_y & B_z \end{vmatrix} \tag{33-2}$$

or in magnitude and direction form, $F = vB\sin\phi$, where ϕ is the angle between ***v*** and ***B*** and ***F*** points in a direction normal to the plane of ***v*** and ***B*** as determined by the right-hand rule.

Most problems involve ***v*** and ***B*** at right angles, so the magnitude of the force reduces to the product of q, v, and B with the direction of F determined by the right-hand rule.

The units of B are (from the force equation 33-1) $\mathrm{N \cdot s/C \cdot m}$ or $\mathrm{N/A \cdot m}$, which is called a **Tesla** (T). The Tesla is related to the smaller and widely used unit the **gauss** ($1\ \text{Tesla} = 10^4\ \text{gauss}$).

33-1 Calculate the deflection of the electrons in a typical video display tube due to the earth's magnetic field. Take the electrons with 10 KeV energy. Orient the tube north to south and with the beam at right angles to the earth's B field of 0.40 gauss.

Solution: First calculate the velocity of the electrons using an energy relation $eV = (1/2)mv^2$.

$$10 \times 10^3\,\text{eV}(1.6 \times 10^{-19}\,\text{J/eV}) = (1/2)mv^2 = (1/2)9.1 \times 10^{-31}\,\text{kg} \cdot v^2$$

$$v = \left[\frac{2 \cdot 1.6 \times 10^{-15}\,\text{J}}{9.1 \times 10^{-31}\,\text{kg}}\right]^{1/2} = 5.9 \times 10^7\ \text{m/s}$$

Convert gauss to Tesla: $0.40\,\text{gauss}\left(1\text{T}/10^4\,\text{gauss}\right) = 4.0 \times 10^{-5}\,\text{T}$.

The force on the electron is

$$F = qvB = 1.6 \times 10^{-19}\,\text{C}\left(5.9 \times 10^7\ \text{m/s}\right)4.0 \times 10^{-5}\,\text{T} = 3.8 \times 10^{-16}\,\text{N}$$

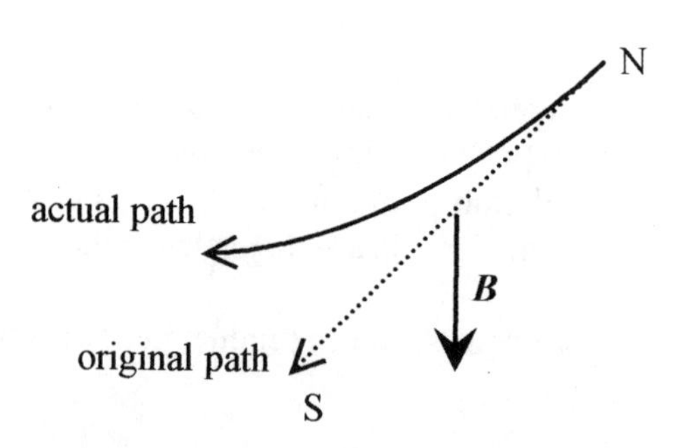

Fig. 33-2

Now for the direction. Place the right hand in the direction of v (out of the paper) and curl the fingers into B (down) so the thumb points to the right (looking at the paper). Because this is a negative charge, the force is to the left.

The electron is deflected in a horizontal plane. B in the vertical direction <u>always</u> produces a force in the horizontal plane and at right angles to the velocity causing the electron to move along a circular path. A mass whirled on the end of a string moves in a circle because the force is at right angles to the velocity (center-directed) throughout the motion.

This last problem is typical of many problems with charged particle trajectories in magnetic fields where the B field is always at right angles to the motion so the vector problem reduces to one of finding the magnitude of the force and then determining the direction via the right-hand rule. This next problem is illustrative of a more difficult problem where the velocity vector has two components. Be sure you have mastered the calculation in problem **33-1**, and above all the direction, before moving on to this more difficult problem.

33-2 Take an electron with velocity (components) $\boldsymbol{v} = 2.0\times10^{6}\boldsymbol{i} - 3.0\times10^{6}\boldsymbol{j}\,\mathrm{m/s}$ and a B field of $\boldsymbol{B} = 4.0\times10^{-3}\boldsymbol{k}\,\mathrm{T}$.

Solution: The force vector in component form is

$$F = 1.6\begin{vmatrix} \boldsymbol{i} & \boldsymbol{j} & \boldsymbol{k} \\ 2.0 & -3.0 & 0 \\ 0 & 0 & 4.0 \end{vmatrix}\times10^{-16}\,\mathrm{N} = 1.6\left[\boldsymbol{i}\begin{vmatrix} -3.0 & 0 \\ 0 & 4.0 \end{vmatrix} - \boldsymbol{j}\begin{vmatrix} 2.0 & 0 \\ 0 & 4.0 \end{vmatrix}\right]\times10^{-16}\,\mathrm{N}$$

$$F = 1.6[-12\boldsymbol{i} - 8.0\boldsymbol{j}]\times10^{-16}\,\mathrm{N} = [-1.9\boldsymbol{i} - 1.3\boldsymbol{j}]\times10^{-15}\,\mathrm{N}$$

It is a bit hard to envision, but a velocity vector in the x-y plane curled into a B-field vector in the z-direction produces a force vector in the x-y plane. Figure 33-3 shows the orientations.

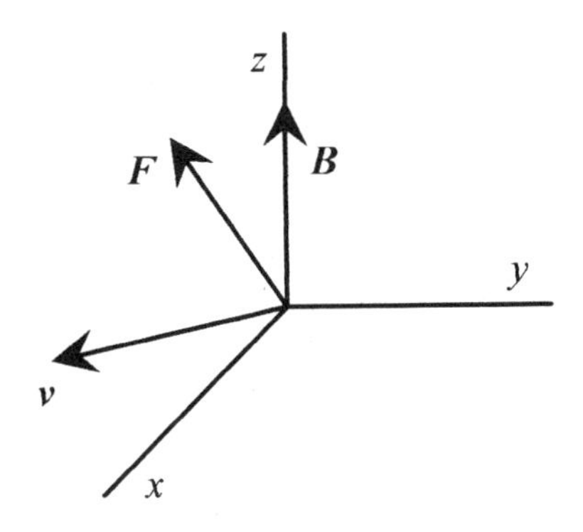

Fig. 33-3

There are several classic phenomena, experiments, and equipment associated with moving charges in B-fields . These are both interesting and instructive. You will probably use one or more of these instruments in your work.

The Cyclotron

If a particle with negative charge, mass, and velocity is injected into a $\boldsymbol{B}$ field in a plane normal to the field the force is at right angles to the velocity and remains at right angles to the velocity throughout the motion causing the charged particle to go in a circle as shown in Fig. 33-4. A positively charged particle would move in a circle in the opposite direction.

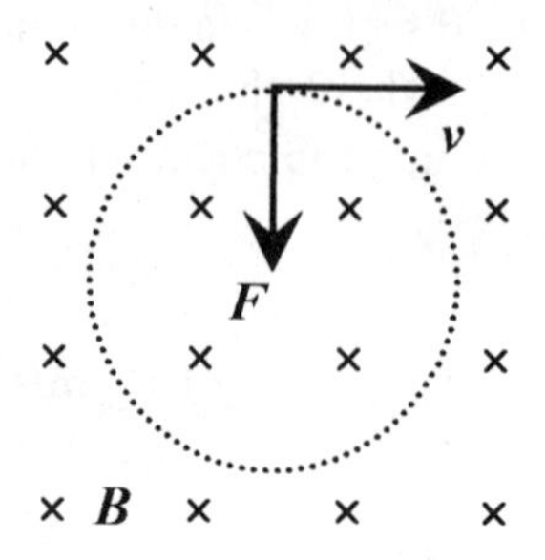

The crosses are the tails of arrows representing the ***B*** field.

Fig. 33-4

The force is at right angles to the velocity. Apply the $v \times \boldsymbol{B}$ law for a negative charge. The magnitude of this force is qvB, and it is always at right angles to the velocity. From mechanics, a motion produced by a force at right angles to the velocity is circular, thus the charged particles move in circular orbits. This qvB force provides the (mechanical force) mv^2/r, so equating these forces

$$qvB = mv^2/r \tag{33-3}$$

There are several relations that follow from this statement, but for the cyclotron the most important is one for the orbit radius

$$r = mv/qB \tag{33-4}$$

and the one for the angular frequency

$$\omega = \frac{v}{r} = \frac{qB}{m} \tag{33-5}$$

33-3 An electron of energy 4.0 KeV is circulating in a plane at right angles to a magnetic field of 3.0T. What is the radius of the orbit?

Solution: The geometry is the same as in Fig. 33-4. First, calculate the velocity from the energy.

$$4.0 \times 10^3\,\mathrm{eV}(1.6 \times 10^{-19}\,\mathrm{J/eV}) = 1/2(9.1 \times 10^{-31}\,\mathrm{kg})v^2 \quad \text{or} \quad v = 3.8 \times 10^7\,\mathrm{m/s}$$

Now calculate *r*

$$r = \frac{mv}{qB} = \frac{9.1 \times 10^{-31}\,\mathrm{kg} \cdot 3.8 \times 10^7\,\mathrm{m/s}}{1.6 \times 10^{-19}\,\mathrm{C} \cdot 3.0\mathrm{T}} = 7.1 \times 10^{-5}\,\mathrm{m}$$

33-4 An α-particle moves in a circle of radius 5.0×10^{-2} m in a magnetic field of 2.0 T. Find the speed of the α-particle and the frequency and period of the motion.

Solution: An α-particle is a helium nucleus, which has two protons and two neutrons. The speed of the α-particle is from the defining equation for the cyclotron.

$$v=\frac{qrB}{m}=\frac{2\cdot1.6\times10^{-19}\,\mathrm{C}\left(5.0\times10^{-2}\,\mathrm{m}\right)2.0T}{4\cdot1.7\times10^{-27}\,\mathrm{kg}}=4.7\times10^{6}\,\mathrm{m/s}$$

The angular velocity is (analogous to mechanics)

$$\omega=\frac{v}{r}=\frac{qB}{m}=\frac{2\cdot1.6\times10^{-19}\,\mathrm{C}\cdot2.0\mathrm{T}}{4\cdot1.7\times10^{-27}\,\mathrm{kg}}=9.4\times10^{7}\,1/\mathrm{s}$$

The frequency, or what is called the cyclotron frequency, is $f=\omega/2\pi=1.5\times10^{7}\,1/\mathrm{s}$.

The period, or time for one transit of the circle is $T=1/f=6.7\times10^{-8}\,\mathrm{s}$.

A **cyclotron** is a device for accelerating charged particles while they are being forced to travel in a circular path by a magnetic field. A source of, say, deuterons is placed at the center of the device. A deuteron is a proton and neutron combination with one electronic charge and 2 amu. An amu (atomic mass unit) is the average mass of a nuclear resident (neutron or proton). The deuterons exit the source with a small velocity to the right. The radius of the circular orbit is given by $r=mv/qB$ and is quite small. The particle enters a semicircular open area between "D" shaped plates and travels in a circular orbit until it exits the "D" area. The voltage between the Dees, as they are called, is alternated sinusoidally so that on leaving the right Dee the particle is accelerated and assumes a new (larger) orbit in the left Dee. When the particle leaves the left Dee the voltage is again switched so that acceleration is accomplished across the Dees. Remember that the angular frequency is qB/m, and is not dependent on velocity!

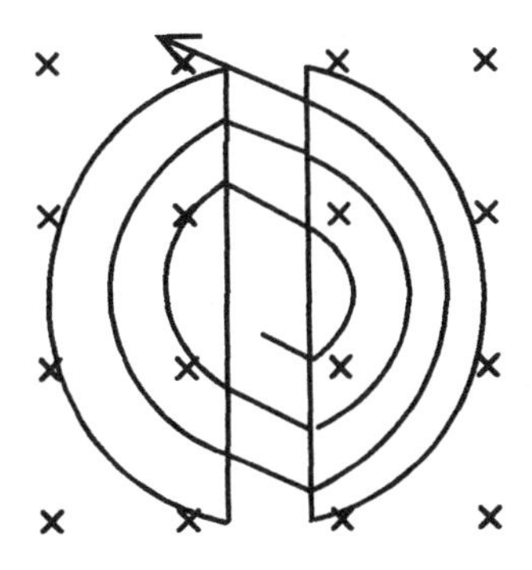

Fig. 33-5

33-5 Calculate the frequency of oscillation of the Dee voltage for a deuteron in a 2.0T magnetic field.

Solution:

$$f = \frac{qB}{2\pi m} = \frac{1.6\times 10^{-19}\,\text{C}\cdot 2.0\text{T}}{2\pi\cdot 2\cdot 1.7\times 10^{-27}\,\text{kg}} = 1.5\times 10^{7}\ 1/\text{s}$$

33-6 What is the maximum energy of a deuteron in this cyclotron with maximum available radius of 1.5m?

Solution: The kinetic energy is $(1/2)mv^2$, but v can be written in terms of the r, q, B, and m according to the basic equation defining the cyclotron orbits.

$$KE = \frac{1}{2}mv^2 = \frac{1}{2}m\left(\frac{qBr}{m}\right)^2 = \frac{(qBr)^2}{2m} = \frac{\left(1.6\times 10^{-19}\,\text{C}\cdot 2.0\text{T}\cdot 1.5\text{m}\right)^2}{2\cdot 2\cdot 1.7\times 10^{-27}\,\text{kg}} = 3.4\times 10^{-11}\,\text{J} = 2.1\times 10^{8}\,\text{eV}$$

The deuteron gained this large amount of energy by making many traverses of the distance between the Dees as the voltage was alternating at the cyclotron frequency.

The Mass Spectrometer

Another application of the circular orbits in a magnetic field is in the mass spectrometer. Ionized atoms are injected into a magnetic field at a known energy (velocity). They take up circular orbits with radii determined by their mass. The ions are detected with a photographic plate or traveling collector.

The ions are accelerated (by a voltage) to a particular energy before they enter the magnetic field region. Their energy is their charge times this accelerating voltage. This is the energy the ions have while orbiting in the magnetic field. (See the previous problem.)

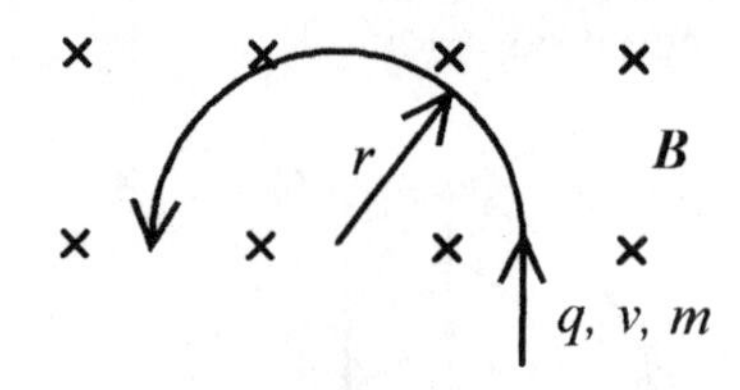

Fig. 33-6

$$qV = \frac{r^2 q^2 B^2}{2m} \quad \text{or} \quad m = \frac{r^2 q B^2}{2V}$$

The ions strike the photographic plate a diameter $x = 2r$ from the injection point, so $m = \frac{qB^2x^2}{8V}$.

33-7 Calculate the mass of a singly charged ion accelerated through $4120\,\text{V}$ and striking the collection area at 7.0×10^{-2} m from the entrance point in a $1.0\,\text{T}$ field.

Solution:

$$m = \frac{qB^2x^2}{8V} = \frac{1.6 \times 10^{-19}\,\text{C}(1.0\,\text{T})^2 49 \times 10^{-4}\,\text{m}^2}{8 \cdot 4120} = 2.4 \times 10^{-26}\,\text{kg} \quad \text{or} \quad 14\,\text{amu}$$

The ion is carbon 14, an isotope of carbon 12.

(C) Another function of this type of mass spectrometer is to separate out isotopes, atoms with the same nuclear charge and number of electrons but with different nuclear mass. The sensitivity of the spectrometer can be found by looking at the variation of x with m. Start with

$m = \frac{qB}{8V}x^2$ and ask how m varies with x: $\Delta m = \frac{qB}{4V}x\Delta x$.

The sensitivity is measured by the ratio of Δm to m.

$$\frac{\Delta m}{m} = \frac{\left(qB^2/4V\right)x\Delta x}{\left(qB^2/8V\right)x^2} = \frac{2}{x}\Delta x \quad \text{or} \quad \Delta m = \frac{2m}{x}\Delta x$$

(C) **33-8** Calculate the resolution required of the photographic plate to differentiate between carbon 12 and carbon 14.

Solution: Use the diameter from the previous problem, 7.0×10^{-2} m, and m as 12 amu with Δm as 2 amu.

$$\Delta x = \frac{x\Delta m}{2m} = \frac{7.0 \times 10^{-2}\,\text{m} \cdot 2\,\text{amu}}{2 \cdot 12\,\text{amu}} = 5.8 \times 10^{-3}\,\text{m}$$

This value is easily obtained on a photographic plate.

The J. J. Thomson Experiment

The J. J. Thomson experiment uses crossed electric and magnetic fields to measure the charge to mass ratio for the electron. This was a classic turn-of-the-century experiment.

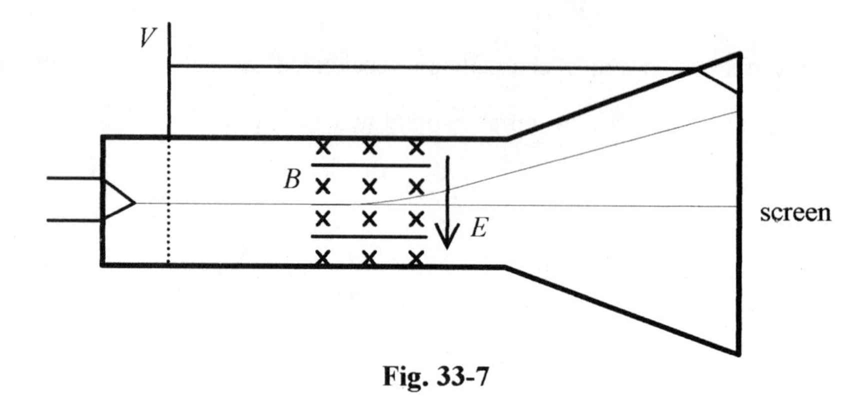

Fig. 33-7

The force equation for moving electrons subject to both electric and magnetic fields is

$$\boldsymbol{F} = q_o \boldsymbol{E} + q_o \boldsymbol{v} \times \boldsymbol{B} \tag{33-6}$$

The electric and magnetic fields are arranged as shown. Verify that a negatively charged particle will experience a "down" force due to the magnetic field and an "up" force due to the electric field. In the tube, the voltage V accelerates the electrons from the hot filament and the electrons enter a field-free region between the accelerating grid, shown as the dotted line, and the video screen.

In the experiment, a deflecting electric field is applied and the spot on the screen is observed to move up on the screen. The B field is added to bring the spot back to zero deflection. In this situation the net force is zero and because of the directions of E, v, and B, $E = vB$.

33-9 An electric field of $1.0 \times 10^3\ \text{V/m}$ and a magnetic field of 0.35T act on a stream of electrons so as to produce zero net force. What is the speed of the electrons?

Solution:

$$v = \frac{E}{B} = \frac{1.0 \times 10^3\ \text{V/m}}{0.35\text{T}} = 2.8 \times 10^3\ \text{m/s}$$

The velocity of the electrons is determined by the accelerating voltage in the tube. The electrons are accelerated before they enter the region of crossed E and B fields and travel through this region and to the screen without further acceleration axially down the tube. The kinetic energy of these electrons is eV, the electronic charge times the accelerating voltage. For zero deflection an energy statement can be made replacing v with E/B.

$$eV = \frac{1}{2}mv^2 = \frac{1}{2}m\frac{E^2}{B^2}$$

from which

$$\frac{e}{m} = \frac{E^2}{2VB^2} \qquad \textbf{(33-7)}$$

The charge to mass ratio for the electron can be determined by the applied voltage and fields in the tube.

33-10 Calculate the charge to mass ratio for charged particles in a J. J. Thomson tube with accelerating voltage of $3000\,\text{V}$ and crossed fields of $E = 6.5 \times 10^4\ \text{V/m}$ and $B = 2.0 \times 10^{-3}\,\text{T}$.

Solution:

$$\frac{e}{m} = \frac{E^2}{2VB^2} = \frac{\left(6.5 \times 10^4\ \text{V/m}\right)^2}{2(3000\,\text{V})\left(2.0 \times 10^{-3}\,\text{T}\right)^2} = 1.76 \times 10^{11}\,\text{C/kg}$$

Verify that this is the correct charge to mass ratio for electrons.

The Hall Effect

The Hall effect is used to determine the algebraic sign and number density, *n*, of charge carriers in a conductor. Set up a slab of unknown material in a magnetic field as shown. In one sample (left

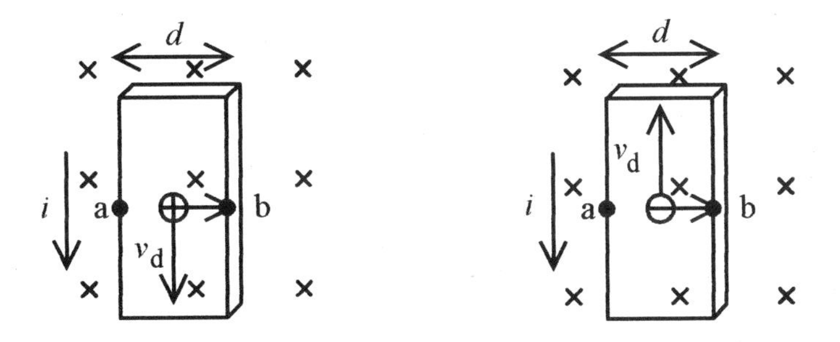

Fig. 33-8

side of Fig. 33-8) assume the current is due to charge carriers that are positive, and, in the other, negative. Apply the $\boldsymbol{v} \times \boldsymbol{B}$ force on each, with the $\boldsymbol{B}$ field directed into the page, and notice that despite the sign of the carriers they will be deflected to the right. This accumulation of carriers will create a "transverse" or Hall voltage, which has a different direction depending on the sign of the charge carriers. In the sample on the left (positive charge carriers) point b is at a higher potential than point a. In the sample on the right (negative charge carriers) point b is at a lower potential than point a. The sign of the voltage produced in the Hall experiment gives the sign of the charge carriers. Charges build up to produce an electric field (the transverse Hall field) that just balances the magnetic force on the charge carriers.

This Hall field is $E_H = V_H/d$.

The Hall field is also the E in the statement for zero force, $0 = q\boldsymbol{E} + q\boldsymbol{v} \times \boldsymbol{B}$.

This produces a Hall field of $E_H = v_d B$.

The current density in the sample is $j = nev_d$.

Combining, $E_H = \dfrac{j}{ne} B$ from which the charge carrier density is

$$n = \frac{jB}{E_H e} \qquad \textbf{(33-9)}$$

33-11 In a Hall experiment the current is $2.0 \times 10^{-3}\,\mathrm{A}$ when $2.0\,\mathrm{V}$ is applied to a semiconductor of length $6.0\,\mathrm{cm}$, width $3.0\,\mathrm{cm}$, and depth $0.20\,\mathrm{cm}$. A magnetic field of $2.0\,\mathrm{T}$ produces a Hall voltage of $2.0 \times 10^{-4}\,\mathrm{V}$. Find the sign of the charge carriers, their number density, and the resistivity of the material.

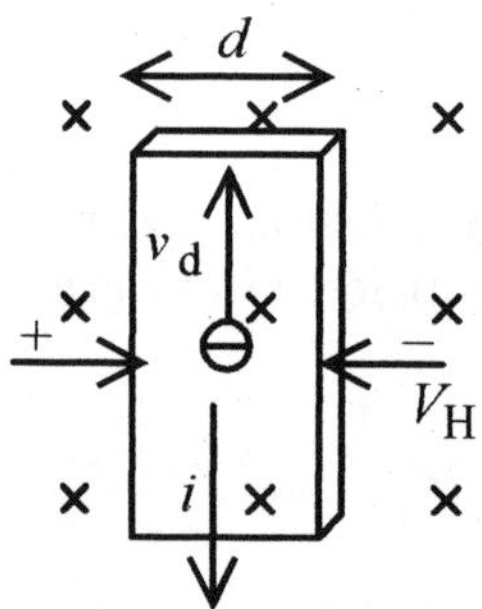

Fig. 33-9

Solution: The sign of the Hall voltage indicates negative charge carriers. Now calculate the current density.

$$j = \frac{i}{A} = \frac{2.0 \times 10^{-3}\,\mathrm{A}}{3.0 \times 10^{-2}\,\mathrm{m} \cdot 2.0 \times 10^{-3}\,\mathrm{m}} = 33\,\mathrm{A/m^2}$$

The Hall field is $E_H = V_H/d = 2.0 \times 10^{-4}\,\mathrm{V}/3.0 \times 10^{-2}\,\mathrm{m} = 6.7 \times 10^{-3}\,\mathrm{V/m}$.

The number density of charge carriers is

$$n = \frac{jB}{E_H e} = \frac{33\,\mathrm{A/m^2} \cdot 2.0\,\mathrm{T}}{6.7 \times 10^{-3}\,\mathrm{V/m} \cdot 1.6 \times 10^{-19}\,\mathrm{C}} = 6.2 \times 10^{22} \frac{1}{\mathrm{m^3}}$$

The resistivity of the sample is easily calculated from the defining statement $E = \rho j$ and remember that this E is not the Hall field but the field driving the current through the sample.

$$2.0 \times 10^{-4}\,\mathrm{V}/6.0 \times 10^{-2}\,\mathrm{m} = \rho(33\,\mathrm{A/m^2}) \quad \text{or} \quad \rho = 1.0 \times 10^{-4}\,\Omega \cdot \mathrm{m}$$

These measurements of the sign of the charge carriers, their number density, and the resistivity are very important in the characterization of semiconductor material.

CHAPTER 34

MAGNETIC FORCES

Magnetic fields exert forces on moving charges. If these charges are confined to wires then the fields can be thought of as exerting forces on current-carrying wires. The vector equation for the force is

$$\boldsymbol{F} = i\boldsymbol{l} \times \boldsymbol{B} \qquad \textbf{(34-1)}$$

where $\boldsymbol{l}$ is the length of the wire taken in the direction of the current. In most instances it is not necessary to do a formal vector cross product because the direction of the current and the magnetic field are perpendicular. The one instance where this becomes important is in considering the torque on a current-carrying wire. This problem is central to several topics and is taken up as a separate exercise.

34-1 A typical problem illustrating the force on a current-carrying wire is the problem of a wire of 0.010kg and 1.0m in length suspended by springs in a $\boldsymbol{B}$ field of strength 30T directed out of the page. What current will cause zero tension in the springs?

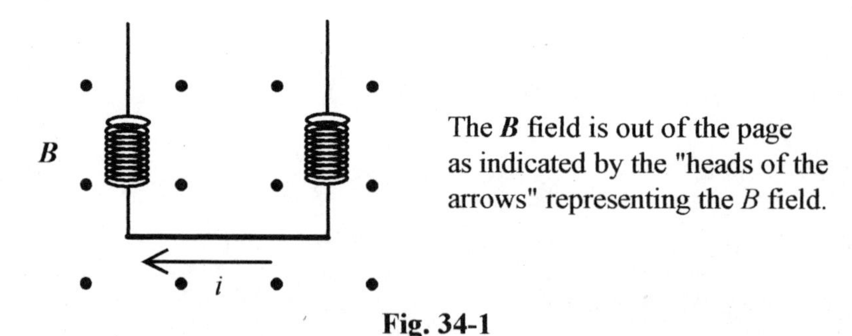

Fig. 34-1

Solution: First consider the direction of the current. The force due to the magnetic field has to be up so the current has to be from right to left. With the right hand, form $\boldsymbol{l} \times \boldsymbol{B}$ with i in both directions to verify this conclusion. The value of the current that makes for zero tension in the springs occurs when the force due to the magnetic field balances gravity. This is a force-balance problem.

$$mg = ilB \quad \text{or} \quad i = \frac{mg}{lB} = \frac{0.010\text{kg} \cdot 9.8\text{m/s}^2}{1.0m \cdot 30\text{T}} = 3.3 \times 10^{-3}\,\text{A}$$

34-2 Another problem illustrating this force is a rod or bar resting on current-carrying rails with a perpendicular magnetic field. Take the field as 0.50 gauss (approximately the earth's field) and the separation as $3.0\,\text{m}$ and calculate the current needed through the bar of mass $0.20\,\text{kg}$ with coefficient of friction 0.070 to make it slide along the rails.

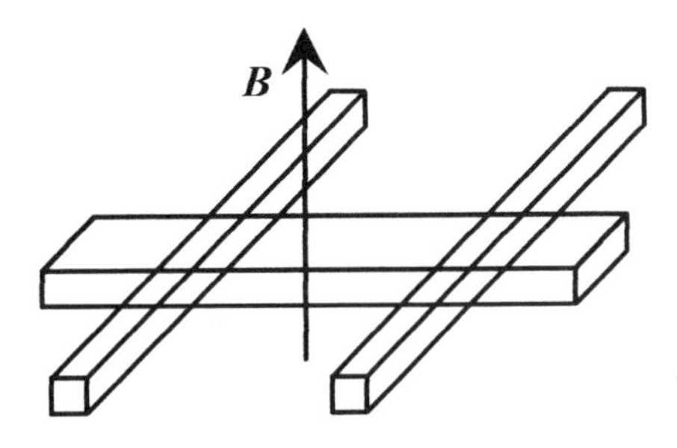

Fig. 34-2

Solution: The orientation of current (either direction in the bar) and magnetic field is such that the force produced by the $il \times \boldsymbol{B}$ will slide the bar along the rails. Verify this with the right-hand rule. The force opposing this sliding is μmg so this is a force-balance problem.

$$ilB = \mu mg \quad \text{or} \quad i = \frac{\mu mg}{lB} = \frac{0.070 \cdot 20\,\text{kg} \cdot 9.8\,\text{m/s}^2}{3.0\,\text{m} \cdot 0.50 \times 10^{-4}\,\text{T}} = 9.1 \times 10^4\,\text{A}$$

A Current-Carrying Loop in a *B* Field

The study of torque on a current-carrying loop leads to the definition of the magnetic moment of a circulating current, the understanding of the meter movement (voltmeter and current meter), and electric motors and generators. A loop of wire placed in a magnetic field and oriented as shown in Fig. 34-3 will have no torque or maximum torque, depending on the orientation (rotation) of the loop with respect to the magnetic field.

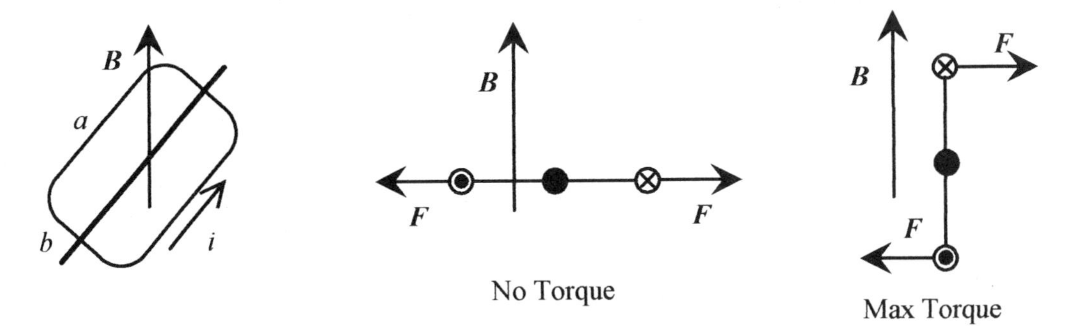

Fig. 34-3

As shown in the middle diagram the piece of wire on the right carrying a current, i, represented by the tail of the current arrow, has a force to the right ($\boldsymbol{l} \times \boldsymbol{B}$) producing no torque. The piece of wire on the left produces a force to the left also producing no torque. In the diagram on the right the forces are in the same direction but for this orientation of the coil the torque is a maximum.

Figure 34-4 shows the torque for angles other than for zero or maximum torque. The torque is the component of this force at right angles to the line connecting the two sides of the coil times $b/2$ (the lever arm) times 2 (two wires).

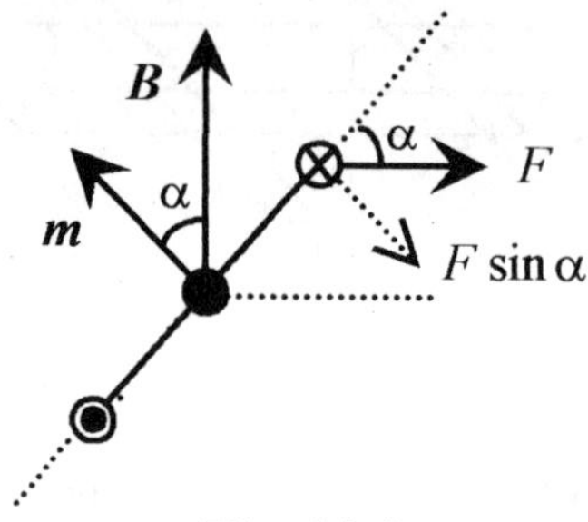

Fig. 34-4

The appropriate component of the force is $F\sin\alpha$. The vector, $\boldsymbol{m}$, the **magnetic moment**, is normal to the plane of the coil and points in the direction of the thumb when the fingers are curled naturally in the direction of the current in the coil. The angle between $\boldsymbol{m}$ and $\boldsymbol{B}$ is the same as the angle between force and lever arm.

$$\tau = 2(b/2)iaB\sin a = iAB\sin a$$

The product ab is the area, A, of the loop. The $\sin\alpha$ term implies a cross product and if iA is taken as the magnitude of $\boldsymbol{m}$, the **magnetic moment,** with the direction as defined in Fig. 34-4 then the vector equation for the torque is

$$\tau = \boldsymbol{m} \times \boldsymbol{B} \quad \text{or} \quad \tau = mB\sin\alpha \quad \text{with direction defined by } \boldsymbol{m} \times \boldsymbol{B} \qquad \textbf{(34-2)}$$

34-3 Find an expression for the torque on a circular coil of radius a carrying a current I placed at an angle to a magnetic field.

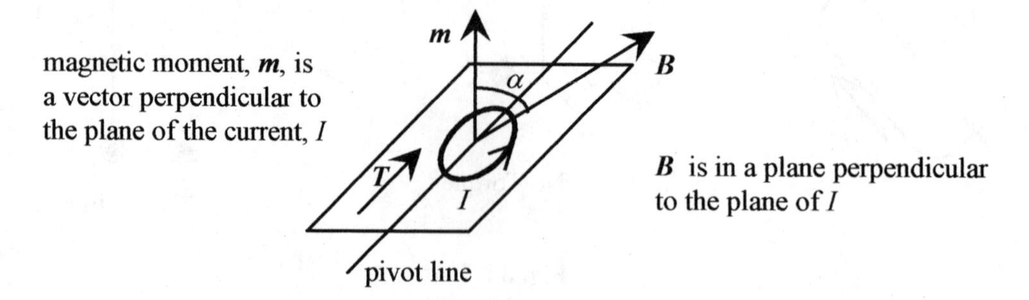

Fig. 34-5

Solution: The definition of the magnetic moment and the expression for the torque based on the magnetic moment is most convenient. The magnetic moment is $m = IA = \pi a^2 I$. The magnitude of the torque is $T = mB\sin\alpha = \pi a^2 I B\sin\alpha$ and the direction is the direction determined by $\boldsymbol{m} \times \boldsymbol{B}$.

34-4 Find the maximum torque on a circular coil of $0.050\,\text{m}$ radius with current of $0.50\,\text{A}$ and in a magnetic field of $0.0050\,\text{T}$.

Solution The maximum torque is

$$\tau_{max} = \pi a^2 IB = \pi(0.050\,\text{m})^2(0.50\,\text{A})(0.0050\,\text{T}) = 2.0 \times 10^{-5}\,\text{N}\cdot\text{m}$$

34-5 The plane of a 10 turn rectangular loop of wire $3.0\,\text{cm}$ by $4.0\,\text{cm}$ carrying $2.0\,\text{A}$ is at an angle of 60° with respect to a magnetic field of $20\,\text{T}$. What is the torque on the loop in this position?

Solution: First calculate the magnetic moment noting that the current is not due to one loop but ten loops. $m = IA = 10 \cdot 2.0\,\text{A} \cdot 12 \times 10^{-4}\,\text{m}^2 = 2.4 \times 10^{-2}\,\text{A}\cdot\text{m}^2$. Refer to Fig. 34-5 and note that if the angle between the plane of the coil and the magnetic field is 60°, then the angle between $\boldsymbol{m}$ and $\boldsymbol{B}$ is 30°, so the torque is

$$\tau = mB\sin\alpha = 2.4 \times 10^{-2}\,\text{A}\cdot\text{m}^2 \cdot 20\,\text{T} \cdot \sin 30^o = 0.24\,\text{N}\cdot\text{m}$$

(C) **34-6** An interesting problem is to ask the question, "What dimensions of rectangle will produce maximum torque for the same length of wire?"

Solution: Take a rectangle with length ℓ, width w, and perimeter equal to $p = 2\ell + 2w$. The perimeter relating ℓ to w is the constraint in the problem. The torque is $T = mB = IAB = I\ell wB$. Before looking for a maximum we need to write ℓ or w in terms of the constraint p.

$$\tau = Iw\frac{p-2w}{2}B = \frac{IB}{2}\left(pw - 2w^2\right)$$

Now take two derivatives of T with respect to w

$$\frac{d\tau}{dw} = \frac{IB}{2}(p-4w) \qquad \frac{d^2T}{dw^2} = -2IB$$

The first derivative is zero when $p = 4w$ and putting this requirement in for p in the constraint equation $(2\ell + 2w = 4w)$ requires that $\ell = w$. The rectangle must be a square. The second derivative is always negative indicating that the curve is everywhere concave down and that this point is a maximum.

d'Arsonval Meter Movement

A schematic of a d'Arsonval galvanometer is shown in Fig. 34-6. The torque on a loop in a magnetic field offset by the torque of a spiral spring as shown here is the basis for mechanical meter movements. Galvanometers are mid-reading meters. Current meters use parallel (with the coil) resistors to change the scale of the meter. Voltmeters use large series resistors to take the small currents necessary to activate the meter to measure voltage.

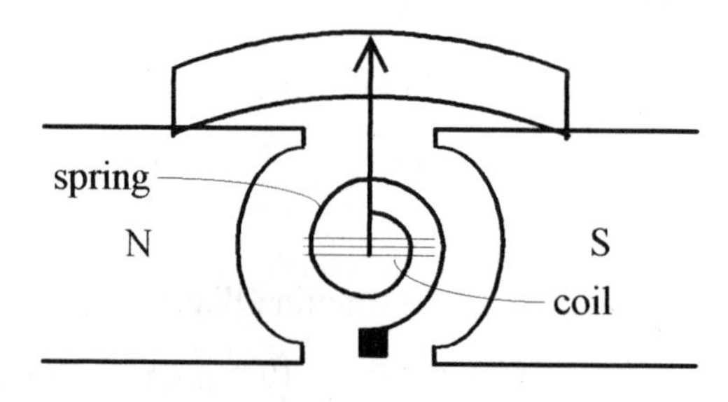

Fig. 34-6

The plane of the coil is parallel to the magnetic field so that a small current in the coil produces a torque on the coil which is offset by the torque of the spiral spring. With this arrangement, the rotation of the coil is proportional to the current. The external magnetic field (supplied by the permanent magnet) is curved (shaped) so the field is parallel to the plane of the coil through an angle of typically sixty degrees. Meter movements are rated as to how much current causes full scale deflection. A movement with a full scale deflection corresponding to $1.0\,\mu\text{A}$ is typical.

Energy Storage in Magnetic Materials

Magnetic dipoles placed in a magnetic field experience a torque. A simple model of a magnetic dipole is a molecule with a circulating current. These dipoles are rotated depending on the restoring torque of the material (structure). Picture a collection of magnetic dipolar molecules in a crystalline structure rotated by an external magnetic field thus stressing the structure. Consider the work done on a magnetic dipole by the application of an external magnetic field. This work is equivalent to the energy stored in the dipole-field combination.

C The energy of the system is taken as zero when $\boldsymbol{m}$ and $\boldsymbol{B}$ are at right angles ($\alpha = 90^o$). The energy at a position α, analogous to both the electrical and mechanical relationship between torque, angle, and energy, is

$$U = \int_{90^o}^{\alpha} \tau d\alpha = \int_{90^o}^{\alpha} IAB\sin\alpha d\alpha = -mB\cos\alpha\Big|_{90^o}^{\alpha} = -mB\cos\alpha$$

In vector notation

$$U = -\boldsymbol{m} \cdot \boldsymbol{B} \qquad \textbf{(34-3)}$$

34-7 A coil of 5.0 x 8.0 cm dimensions and carrying a current of 0.0020A is rotated 45° by a 20T magnetic field from parallel to the plane of the field to an angle of 45° with respect to the field (see Fig. 34-5). Calculate the energy of the system.

Solution: The energy of the system is

$$U = -mB\cos\alpha\Big|_{90^o}^{45^o} = -mB\cos45^o = -40\times10^{-4}\,\text{m}^2\cdot0.0020\,\text{A}\cdot20\text{T}\cdot\cos45^o = -1.1\times10^{-4}\,\text{J}$$

This is the energy stored in the system. An external agent would have to perform this amount of work to move the dipole (coil) back to parallel with the external magnetic field.

CHAPTER 35

AMPERE'S LAW

Ampere's law describes the magnetic field produced by a current-carrying wire. Two statements give a complete description of the magnetic field around a wire. The **strength** of the field is given by

$$B = \frac{\mu_o I}{2\pi r} \qquad \textbf{(35-1)}$$

where $\mu_o = 4\pi \times 10^{-7}\ \mathrm{Wb/A \cdot m}$ is the (magnetic) permeability. The **direction** of the field is (a vector) tangent to a circle about the current-carrying wire normal to the plane of the current. With the thumb of the right hand in the direction of the current the fingers naturally curl in the direction of the ***B*** fields tangent to concentric circles. The drawing on the left side of Fig. 35-1 is a view of a current coming out of the paper showing the field as a vector tangent to the circle and the drop off of the magnetic field with radius. The diagram on the right side is a three dimensional view of the magnetic field around a current-carrying wire.

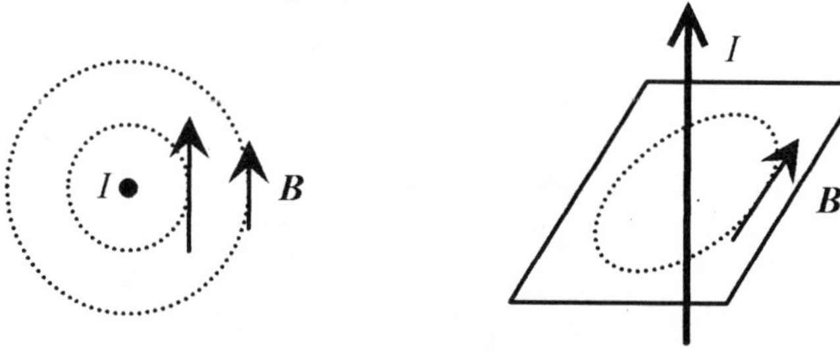

Fig. 35-1

(C) Ampere's law is written in the form of an integral

$$\oint \boldsymbol{B} \cdot dl = \mu_o i \qquad \textbf{(35-2)}$$

where i is the current contained within the path of the integral. This integral sign with a circle means that the integral is to be performed over a closed path. In this case the path is a circle in a plane normal to the current (see Fig. 35-1). The circular path is centered about the wire where the magnetic field is a constant. Over this circular path the magnetic field vector and dl, the vector representing an element of length of the circle are parallel so the sum of (integral of) their dot product, $\boldsymbol{B} \cdot d\boldsymbol{l}$, is the constant B times the circumference of the circle or $B(2\pi r)$. Remember that the dot product is the product of the projection of one vector onto another.

$$2\pi rB = \mu_o i \quad \text{or} \quad B = \frac{\mu_o i}{2\pi r}$$

35-1 Calculate the magnetic field $7.0\,\text{m}$ below a power line carrying $12000\,\text{A}$ of current.

Solution:
$$B = \frac{\mu_o i}{2\pi r} = \frac{4\pi \times 10^{-7}\,\text{Wb/A}\cdot\text{m}(12000\,\text{A})}{2\pi \cdot 7.0\,\text{m}} = 3.4 \times 10^{-4}\,\text{T}$$

This is comparable to the earth's magnetic field. Assuming the power line is parallel to the surface of the earth then the magnetic field produced by this power line is also parallel to the surface of the earth.

35-2 Two long parallel wires separated by $0.60\,\text{m}$ carry antiparallel currents (the currents are physically parallel but in opposite directions) of $7.0\,\text{A}$ each. What is the resulting field along the line between the wires and $0.40\,\text{m}$ from the lower wire?

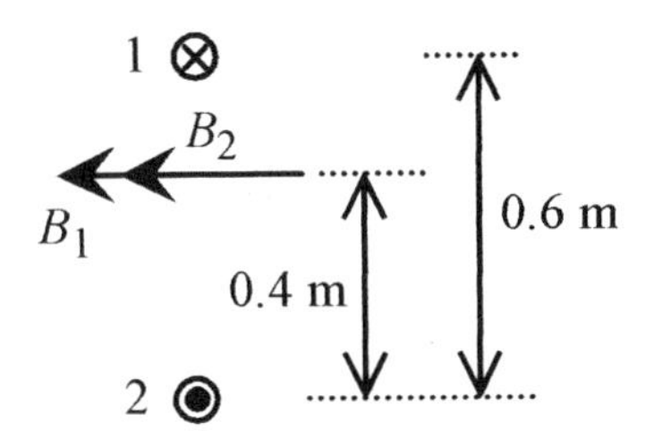

Fig. 35-2

Solution: At the specified position the magnetic field due to wire 1 is

$$B_1 = \frac{\mu_o I}{2\pi r} = \frac{4\pi \times 10^{-7}\,\text{Wb/A}\cdot\text{m}(7.0\,\text{A})}{2\pi \cdot 0.20\,\text{m}} = 7.0 \times 10^{-6}\,\text{T}$$

and directed as shown (right thumb in direction of current with fingers curled). The magnetic field due to wire 2 is

$$B_2 = \frac{\mu_o I}{2\pi r} = \frac{4\pi \times 10^{-7}\,\text{Wb/A}\cdot\text{m}(7.0\,\text{A})}{2\pi \cdot 0.40\,\text{m}} = 3.5 \times 10^{-6}\,\text{T}$$

and directed as shown. The total magnetic field is $10.5 \times 10^{-6}\,\text{T}$.

The Force on Current Carrying Wires

Parallel wires carrying currents I_1 and I_2 produce a force on one other. Take the currents as parallel as shown in Fig. 35-3. The current in wire 1 produces a magnetic field at wire 2 of magnitude $B_1 = \mu_o I_1/2\pi d$. This field is directed down as shown and produces a force on wire 2 of magnitude $F_2 = B_1 I_2 \ell$. These forces are usually written in terms of force per unit length and this is most convenient in this problem. The force per unit length on wire 2 produced by the magnetic field associated with I_1 is

$$\frac{F_2}{\ell} = B_1 I_2 = \frac{\mu_o I_1 I_2}{2\pi d}$$

The force on wire 2 due to the magnetic field produced by wire 1 is toward wire 1. Go through the $\boldsymbol{v} \times \boldsymbol{B}$ operation until this is clear in your mind. If you encounter a test problem concerning forces on wires inevitably one part of the question will concern the direction and you don't want to get something this simple wrong because you didn't practice the $\boldsymbol{v} \times \boldsymbol{B}$ operation.

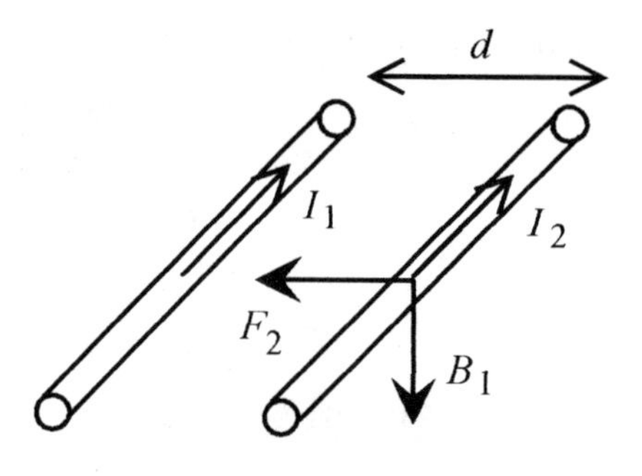

Fig. 35-3

This is the experimental setup that is used to define the ampere. A current produces a magnetic field that produces a force on another current. Using Ampere's law and this setup current can be related directly to mechanical force.

35-3 Consider two wires coming out of the page carrying equal, and oppositely directed, currents as shown in Fig. 35-4. The wires are suspended from a common point with $0.40\,\text{m}$ long cords. The wires have a mass per unit length of $30 \times 10^{-3}\,\text{kg/m}$. Calculate the current to produce an angle of 6°.

Solution: This is a force-balance problem. The horizontal component of the tension (force) in the cord must equal the force due to the currents. Look at 1.0m of the wire. The mass of 1.0m of wire is $30 \times 10^{-3}\,\text{kg}$, and the force due to gravity $mg = 0.29\,\text{N}$. This must equal the vertical component of the tension in the cord. From the geometry,

$$\tan 6^o = F_h/F_v \quad \text{or} \quad F_h = 0.29\,\text{N}(\tan 6^o) = 0.031\,\text{N}$$

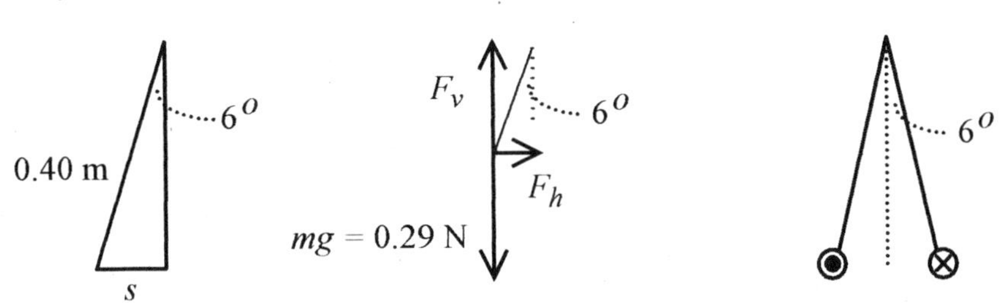

Fig. 35-4

In Fig. 35-4 the separation of the wires is $2s$ where s is defined in the figure through $\sin 6^o = s/0.40\,\text{m}$ or $s = 0.40\,\text{m}\cdot\sin 6^o$. The electric force produced by the current in one wire producing a magnetic field that acts on the current in the other wire is

$$F = BI = \frac{\mu_o I^2}{2\pi(2s)}$$

and this force must equal the mechanical force, or tension, in the cord so

$$0.031\,\text{N} = \frac{4\pi\times 10^{-7}\,\text{T}\cdot\text{m/A}\cdot I^2}{2\pi(2\cdot 0.40\,\text{m}\cdot\sin 6^o)}$$

and solving for I

$$I = \left[\frac{0.40\cdot\sin 6^o(0.031)}{10^{-7}}\text{A}^2\right]^{1/2} = 114\,\text{A}$$

35-4 Calculate the magnetic field 10 cm perpendicular to the bisector of the line connecting two wires 2.0 cm apart and carrying antiparallel currents of 100 A each as shown in Fig. 35-5.

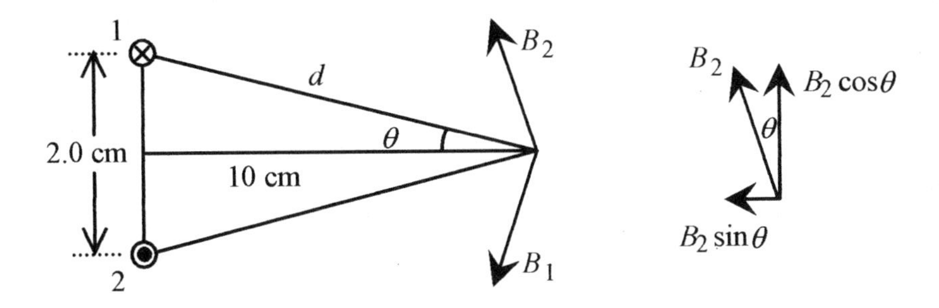

Fig. 35-5

Solution: The magnitude of the magnetic field from each wire is $\mu_o I/2\pi d$. Because of the geometry, the vertical components of B_1 and B_2 add to zero. The horizontal components add to a vector pointing toward the centerline of the wires with magnitude $B = 2B_1 \sin\theta$. The dimension $d = \sqrt{10^2 + 1^2} = \sqrt{101}\,\text{cm}$. The $\sin\theta = 1.0/\sqrt{101}$. The magnitude of the resultant field is

$$B = \frac{2\mu_o I}{2\pi d}\sin\theta = \frac{2\cdot 4\pi\times 10^{-7}\,\text{T}\cdot\text{m/A}(100\,\text{A})}{2\pi\sqrt{101}\times 10^{-2}\,\text{m}}\frac{1.0}{\sqrt{101}} = 4.0\times 10^{-5}\,\text{T}$$

Concentric Currents

In the statement of Ampere's law, $\oint \boldsymbol{B}\cdot d\boldsymbol{l} = \mu_o i$, the i is the current contained within the path of the integral. This is illustrated in the following problem. Take a cylindrical conducting wire of radius 4a and total current I out of the plane of the page. The current is uniformly distributed over the cross-sectional area of the wire.

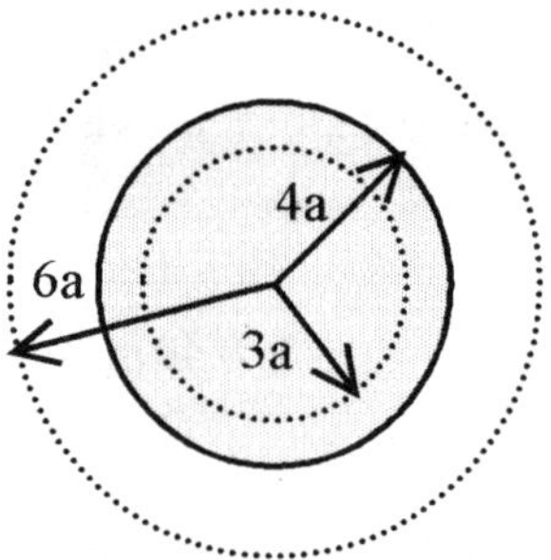

Fig. 35-6

First calculate the magnetic field at $r = 6a$ using $B = \mu_o I/2\pi r$: $B = \mu_o I/12\pi\text{a}$.

Now calculate the magnetic field within the wire at a radius of 3a.

The I in this case is the fractional part of I contained within the 3a radius or $\dfrac{\pi(3\text{a})^2}{\pi(4\text{a})^2}I = \dfrac{9}{16}I$.

Applying Ampere's law $B = \dfrac{\mu_o}{6\pi\text{a}}\dfrac{9I}{16} = \dfrac{3\mu_o I}{32\pi\text{a}}$

Another variation of this problem is concentric conductors with equal and opposite currents as shown in Fig. 35-7. Remember, the magnetic field over a circular path (Ampere's law) is related to the net current within the path.

The inner conductor is solid and has radius a. The outer conductor is hollow, having an inner radius b and an outer radius c. The currents are numerically equal but in opposite directions. Look at the magnetic field over several regions.

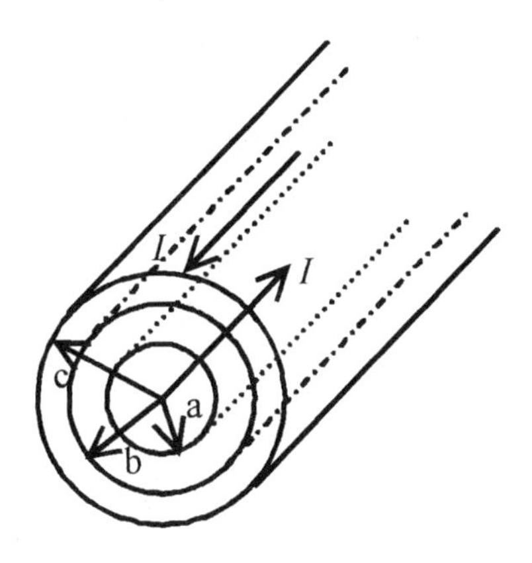

Fig. 35-7

$$0 < r < \mathrm{a} \quad B(2\pi r) = \mu_o I_{\text{inside}} = \frac{\mu_o(\pi r^2)I}{\pi \mathrm{a}^2} \qquad B = \frac{\mu_o r I}{2\pi \mathrm{a}^2}$$

$$\mathrm{a} < r < \mathrm{b} \quad B(2\pi r) = \mu_o I_{\text{inside}} \qquad B = \frac{\mu_o I}{2\pi r}$$

$$\mathrm{b} < r < \mathrm{c} \quad B(2\pi r) = \mu_o I_{\text{inside}} = \mu_o I - \mu_o I \frac{\pi(r^2 - \mathrm{b}^2)}{\pi(\mathrm{c}^2 - \mathrm{b}^2)} \qquad B = \frac{\mu_o I}{2\pi r}\left[1 - \frac{r^2 - \mathrm{b}^2}{\mathrm{c}^2 - \mathrm{b}^2}\right]$$

$$\mathrm{c} < r \quad B(2\pi r) = \mu_o I_{\text{inside}} \qquad B = 0$$

Average Field Outside a Wire

Another application of Ampere's law involves the calculus concept of average value of a function.

(C) **35-5** Calculate the average value of the magnetic field from 0.50 to $0.10\,\mathrm{m}$ radially out from a long wire carrying $100\,\mathrm{A}$.

Solution: The magnetic field falls off radially according to $B = \mu_o I/2\pi r$. The average value of the function over this interval is the integral of the function (the area under the curve of B vs r) divided by the radial distance. See the Introduction, Mathematical Background, for the definition of the average value of a function.

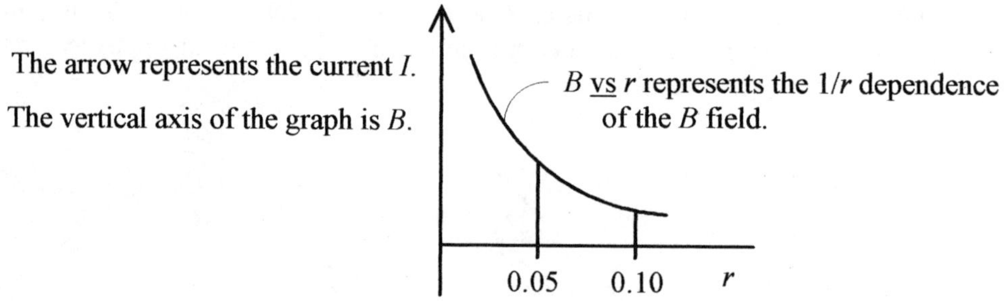

The arrow represents the current I.

The vertical axis of the graph is B.

B vs r represents the $1/r$ dependence of the B field.

0.05 0.10 r

Figure 35-8

$$B_{avg} = \frac{1}{0.10 - 0.050} \frac{\mu_o I}{2\pi} \int_{0.050}^{0.10} \frac{1}{r} dr = \frac{\mu_o I}{0.050 \cdot 2\pi} \ln r \Big|_{0.050}^{0.10} = \frac{\mu_o I}{0.10 \cdot \pi} \ln 2$$

$$B_{avg} = \frac{4\pi \times 10^{-7}\,\mathrm{T \cdot m/A}(100\,\mathrm{A})}{0.10\pi\mathrm{m}} \ln 2 = 2.8 \times 10^{-4}\,\mathrm{T}$$

The Solenoid

A popular device that is based on the application of Ampere's law is the **solenoid**, a tightly wound coil producing a nearly uniform magnetic field along the axis of the coil. The coil is shown in Fig. 35-9.

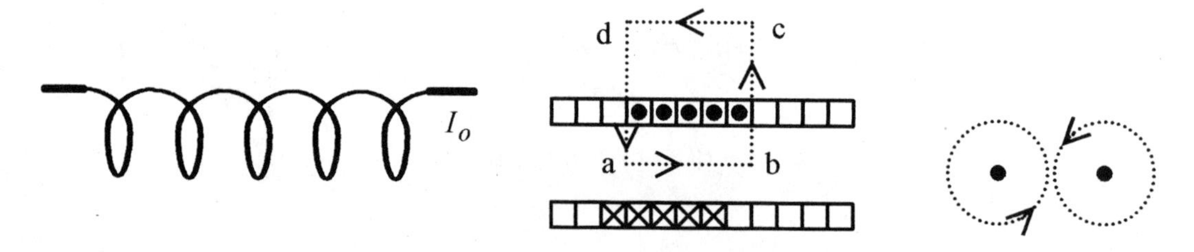

Fig. 35-9

For tightly wound coils the windings take on the form of a sheet of current, and a cross section looks as shown in the middle drawing of the figure. The magnetic field around each coil of wire is such that the magnetic field is concentrated and nearly uniform inside the coil.

C The integral around the path $a \to b \to c \to d \to a$ is conveniently done in four parts.

$$\oint \boldsymbol{B} \cdot d\boldsymbol{l} = \int_a^b Bdl + \int_b^c Bdl + \int_c^d Bdl + \int_d^a Bdl$$

The drawing on the right of Fig. 35-9 shows the magnetic fields of adjacent wires. Between the wires, the fields add to zero so the integrals along the paths $b \to c$ and $d \to a$ are zero. The path from $c \to d$ can be moved sufficiently far away so that the integral along this path approaches zero. The only non-zero integral is along the center line of the solenoid, the $a \to b$ path. Therefore, if the path length is taken as L then the Ampere's law becomes

$$BL = \mu_o I_o NL \quad \text{or} \quad B = \mu_o I_o N \tag{35-3}$$

where N is the number of turns per length, so $I_o NL$ is the total current within the path.

35-6 Calculate the magnetic field inside a solenoid of 300 turns per centimeter carrying a current of $0.050\,\mathrm{A}$.

Solution: $B = \mu_o I_o N = (4\pi \times 10^{-7}\,\mathrm{T \cdot m/A})(0.050\,\mathrm{A})(300/10^{-2}\,\mathrm{m}) = 1.9 \times 10^{-3}\,\mathrm{T}$

The Toroid

A **toroid** is a solenoid wrapped so as to form a circle. A section through a toroid is shown in Fig. 35-10.

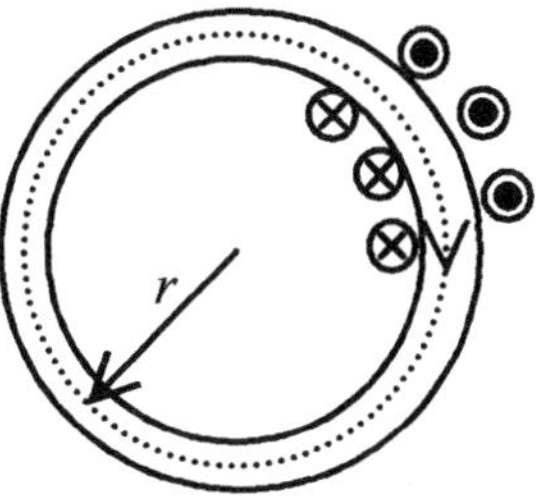

Fig. 35-10

The application of Ampere's law is very straight forward. The path for the integral is along the center line of the toroid where the magnetic field is constant, so

$$\oint \boldsymbol{B} \cdot d\boldsymbol{l} = \mu_o I \quad \text{reduces to} \quad B(2\pi r) = \mu_o n I_o$$

where n is the total number of turns in the toroid. But $n/2\pi r$ is the turns per unit length. Call this N, as before, so the field in a toroid is

$$B = \mu_o N I_o \tag{35-4}$$

35-7 Calculate the magnetic field in a toroid of 600 turns, current of $0.76\,\mathrm{A}$ and center line radius $0.50\,\mathrm{m}$.

Solution: $B = \mu_o N I_o = 4\pi \times 10^{-7}\,\mathrm{T \cdot m/A} \left(\dfrac{600}{2\pi \cdot 0.50\,\mathrm{m}} \right) 0.76\,\mathrm{A} = 1.8 \times 10^{-4}\,\mathrm{T}$

CHAPTER 36

BIOT-SAVART LAW

Ampere's law allows convenient calculation of the magnetic field surrounding a straight current-carrying wire. While the law is written in calculus terms the application is quite easy and involves little formal calculus for most problems. The Biot-Savart law is a more general law and allows calculation of the magnetic field in the vicinity of curved wires. Application of the Biot-Savart law involves considerable calculus. If you are interested only in the application of the formula for the magnetic field on the axis of a current carrying coil you may want to skip directly to the problems involved with these calculations. For the physics student with a reasonable background in calculus the Biot-Savart law is very good for learning how to interpret a differential law written in vector format.

The magnetic field due to a moving charge or a current is given by the Biot-Savart law.

$$\boldsymbol{B} = \frac{\mu_o}{4\pi} q \frac{\boldsymbol{v} \times \hat{\boldsymbol{r}}}{r^2} \quad \text{or} \quad d\boldsymbol{B} = \frac{\mu_o}{4\pi} I \frac{d\boldsymbol{l} \times \hat{\boldsymbol{r}}}{r^2} \tag{36-1}$$

These are vector equations. They allow calculation of the magnitude of the magnetic field and show the direction of that field. The equation on the left applies to moving charges and states that the magnetic field at a point r has magnitude $\mu_o qv/4\pi r^2$ and direction determined by the cross product of $\boldsymbol{v}$ and the unit vector $\hat{\boldsymbol{r}}$. The equation on the right applies to currents in wires and states that the contribution of a length of wire dl to the magnetic field is $\mu_o I\,dl/4\pi r^2$ and the direction is the direction determined by the cross product of $d\boldsymbol{l}$ and the unit vector $\hat{\boldsymbol{r}}$. The following problem will help in visualizing the cross product.

(C) **36-1** Calculate the magnetic field a distance x radially out from a wire carrying a current I.

Solution: Figure 36-1 shows the geometry for calculating the magnetic field due to an element of current in the wire. Use the differential form of the Biot-Savart law

$$d\boldsymbol{B} = \frac{\mu_o I}{4\pi} \frac{d\boldsymbol{l} \times \hat{\boldsymbol{r}}}{r^2} = \frac{\mu_o I}{4\pi} \frac{dl \sin\theta}{r^2}$$

and integrate. The vector dl is in the direction of I, and the angle θ is the angle between the vectors dl and $\boldsymbol{r}$. Cross dl (in the direction of I) with $\boldsymbol{r}$ to see that $\boldsymbol{B}$, at the point P, is into the paper.

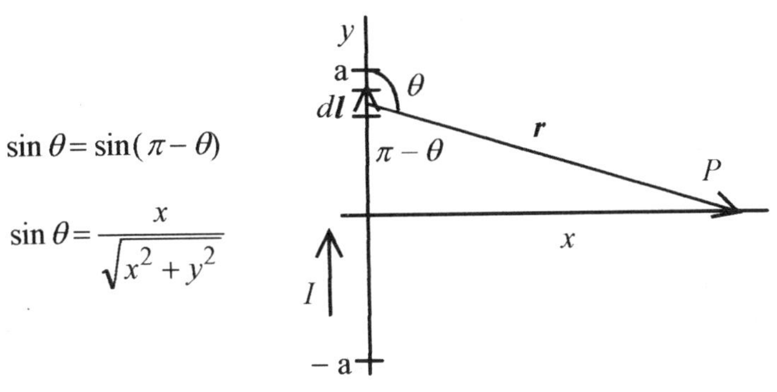

Fig. 36-1

$$B = \frac{\mu_o}{4\pi}\int I\frac{x\,dy}{(x^2+y^2)^{3/2}}$$

x is a constant

Integrate from $-a$ to a and then let a go to infinity to find the expression for the magnetic field due to a long straight wire. This specific integral was done in problem 25-5 in Chapter 25, Electric Field. Follow along the steps in problem 25-5 and work through this problem to confirm that the magnetic field outside a long current-carrying wire is $\mu_o I/2\pi r$, in conformity to Ampere's law.

C **36-2** Calculate the magnetic field on the axis of a circular loop of current.

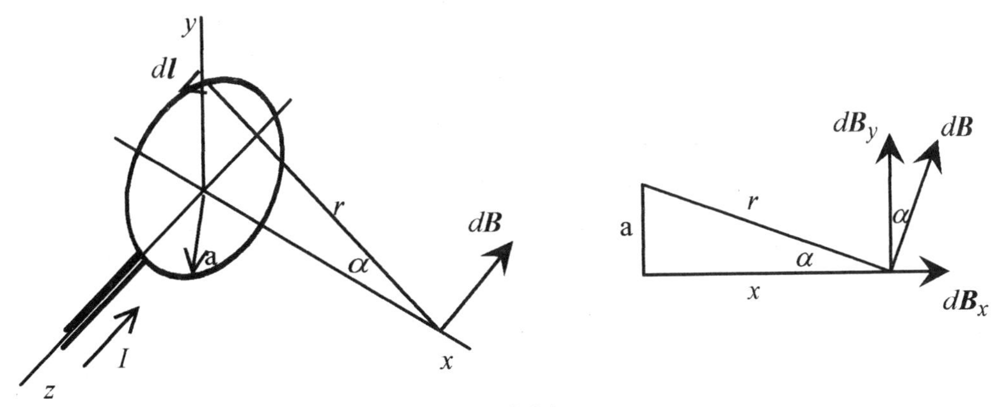

Fig. 36-2

Solution: The Biot-Savart law is most convenient for solving this problem. The cross product of dl and $\hat{r}$ gives the direction of the field along the axis as being at right angles to r and in the plane defined by x and r. The loop of current and dl are in the y-z plane. Start with the law in differential form (equation 36-1) and write the x and y components of the field.

$$dB_x = dB \sin\alpha = \frac{\mu_o I}{4\pi} \frac{d\ell}{x^2+y^2} \frac{\mathrm{a}}{(x^2+y^2)^{1/2}}$$

$$dB_y = dB \cos\alpha = \frac{\mu_o I}{4\pi} \frac{d\ell}{x^2+y^2} \frac{x}{(x^2+y^2)^{1/2}}$$

For every differential increment of length on the loop there is another differential increment of length across a diameter also producing a contribution to the field. Looking at the components of these fields, the x components add while the y components add to zero leaving only the x component that contributes to the resultant field. The field then reduces to the integral of the x component.

$$B_x = \frac{\mu_o I}{4\pi} \int \frac{\mathrm{a}\, d\ell}{(x^2+y^2)^{3/2}} = \frac{\mu_o I}{4\pi} \frac{\mathrm{a} \cdot 2\pi \mathrm{a}}{(x^2+y^2)^{3/2}} = \frac{\mu_o I \mathrm{a}^2}{2(x^2+\mathrm{a}^2)^{3/2}}$$

At the center of the loop, $x = 0$ and the expression for the field reduces to

$$B_x = \frac{\mu_o I}{2\mathrm{a}} \qquad \textbf{(36-2)}$$

For a coil of N loops the current is multiplied by N.

36-3 What is the magnetic field at the center of a circular coil of radius 5.0×10^{-2} m carrying a current of 0.25 A and having 40 turns?

Solution: The field is perpendicular to the plane of the loop at the center and has value

$$B = \frac{\mu_o NI}{2\mathrm{a}} = \frac{4\pi \times 10^{-7}\,\mathrm{T \cdot m/A}(40)0.25\,\mathrm{A}}{2 \cdot 5.0 \times 10^{-2}\,\mathrm{m}} = 1.2 \times 10^{-4}\,\mathrm{T}$$

36-4 For problem 36-3, what is the field 5.0×10^{-1} m distance along the axis?

$$B_x = \frac{\mu_o NI\mathrm{a}^2}{2(x^2+\mathrm{a}^2)^{3/2}} = \frac{4\pi \times 10^{-7}\,\mathrm{T \cdot m/A}(40)0.25\,\mathrm{A} \cdot 25 \times 10^{-4}\,\mathrm{m}^2}{2(25 \times 10^{-2} + 25 \times 10^{-4})^{3/2}\,\mathrm{m}^2} = 1.2 \times 10^{-7}\,\mathrm{T}$$

36-5 A $3.0\,\mu\text{C}$ charge at $x = 0$, $y = 1.0\,\text{m}$ has a velocity $v_x = 5.0\times10^7\,\text{m/s}$. A $-4.0\,\mu\text{C}$ charge at $x = 2.0\,\text{m}$, $y = 0$ has a velocity $v_y = 8.0\times10^7\,\text{m/s}$. What is the magnetic field at the origin due to the motion of these charges?

Solution: For the first charge the $\hat{r}_1$ vector points down (from the charge to the origin) and $v\times\hat{r}_1$ shows the magnetic field at the origin as into the page. The magnitude of this field is

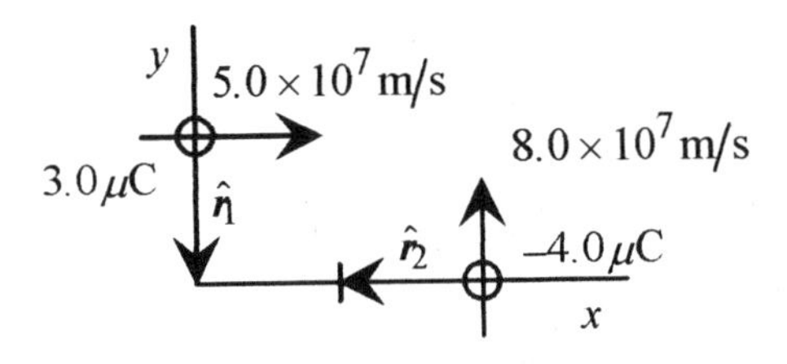

Fig. 36-3

$$B = \frac{\mu_o}{4\pi}q\frac{v}{r_1^2} = \frac{4\pi\times10^{-7}\,\text{T}\cdot\text{m/A}(3.0\times10^{-6}\,\text{C})5.0\times10^7\,\text{m/s}}{4\pi(1.0\text{m}^2)} = 1.5\times10^{-5}\,\text{T}$$

For the second charge the $\hat{r}_2$ vector points to the left (from the charge toward the origin) and $v\times\hat{r}_2$ is a vector pointing out of the page, but the negative charge makes the field point into the page. This magnetic field has magnitude

$$B = \frac{\mu_o}{4\pi}q\frac{v}{r_2^2} = \frac{4\pi\times10^{-7}\,\text{T}\cdot\text{m/A}(3.0\times10^{-6}\,\text{C})8.0\times10^7\,\text{m/s}}{4\pi(4.0\text{m}^2)} = 6.0\times10^{-6}\,\text{T}$$

The total magnetic field is $2.1\times10^{-5}\,\text{T}$ directed into the page.

(C) **36-6** What is the magnetic field at the center of a semicircular piece of wire with radius $0.20\,\text{m}$ and carrying $150\,\text{A}$ of current?

Solution: Start with equation 36-1 in differential form, and referring to Fig. 36-4, form $d\boldsymbol{l}\times\hat{\boldsymbol{r}}$ that shows $\boldsymbol{B}$ into the page at the center of the semicircle. Integrating equation 36-1 produces

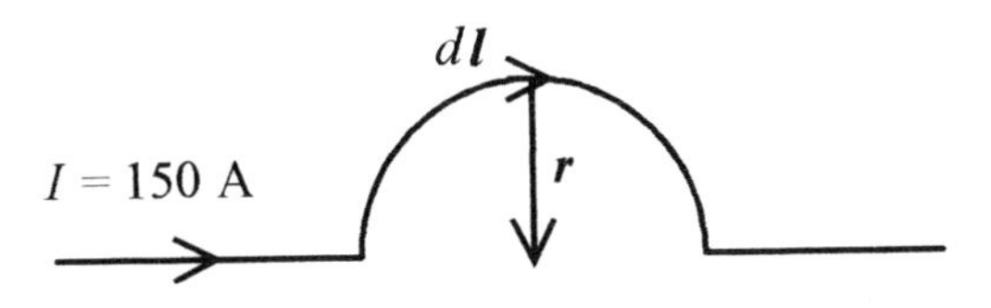

Fig. 36-4

$$B = \frac{\mu_0}{4\pi} I \int \frac{d\boldsymbol{l} \times \hat{\boldsymbol{r}}}{r^2} = \frac{\mu_0}{4\pi} \frac{I}{r^2} \int dl = \frac{\mu_0}{4\pi} \frac{I}{r^2} \pi r = \frac{\mu_0 I}{4r} = \frac{4\pi \times 10^{-7}\,\text{T}\cdot\text{m/A}(150\,\text{A})}{4(0.20\,\text{m})} = 2.4 \times 10^{-4}\,\text{T}$$

The connecting wires are on radii out from the center of the semicircle, so $d\boldsymbol{l} \times \hat{\boldsymbol{r}} = 0$ and they contribute no magnetic field at the center of the semicircle.

CHAPTER 37

FARADAY'S LAW

There are two basic experiments demonstrating Faraday's law. The first experiment involves passing a magnet through a loop of wire. If a magnet is passed through a loop of wire connected to a galvanometer as shown in Fig. 37-1 then three things are observed:

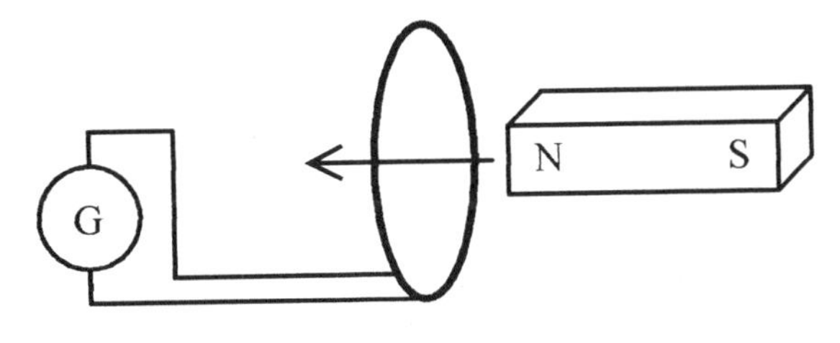

Fig. 37-1

1) a current is observed in the loop when the magnet is moving,

2) the direction of the current depends on the direction of the magnet,

3) the magnitude of the deflection (current) is proportional to the strength of the magnet and its velocity.

The second Faraday experiment involves two loops of wire. If the current in the primary loop (the one with the battery) is <u>changed</u>, then there is a current in the secondary loop (the one with the galvanometer).

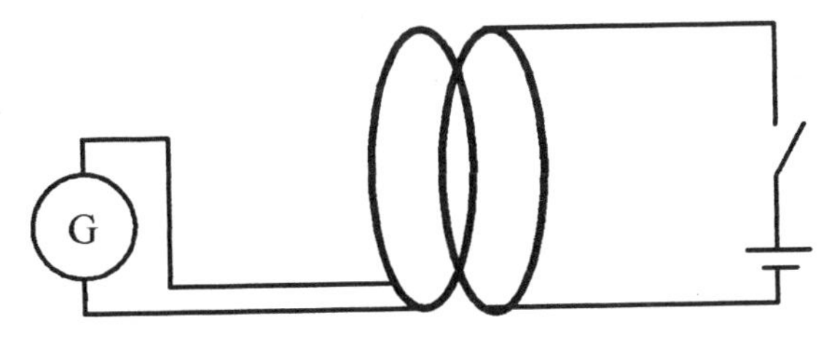

Fig. 37-2

If the switch connecting the battery to the loop is opened and closed three things are observed:

1) a current in the galvanometer loop is observed when the switch is closed,

2) a current in the opposite direction is observed when the switch is opened,

3) no current in the galvanometer loop is observed when there is steady or zero current in the battery loop.

These two experiments are the basis for a statement of **Faraday's law**. An ***emf*** and current is induced in a loop when there is a change in magnetic field in the loop. The magnitude of the induced ***emf*** is proportional to the rate of change in magnetism.

The amount of magnetism is conveniently envisioned as lines of flux Φ. For a constant magnetic field, and area normal to the field, flux is related to magnetic field through

$$\Phi = BA \qquad \textbf{(37-1)}$$

The magnetic field can be envisioned as a collection of (magnetic) lines (of flux) with a higher magnetic field being associated with a higher density of flux lines (more lines of flux per square meter). Picture a square frame one meter on a side oriented normal to the direction of a magnetic field. A stronger (higher Tesla) field is viewed as one with more lines passing through this frame. There are no physical lines of flux. This is a convenient construct that allows us to visualize magnetic fields.

(C) The formal definition of **flux** involves an integral that looks difficult, but in concept it is relatively easy to understand and in practice is often done by inspection.

$$\Phi = \int \boldsymbol{B} \cdot d\boldsymbol{s} \qquad \textbf{(37-2)}$$

The vector ***ds*** is a vector normal to a surface. ***B*** is the vector representing the field passing through this surface. If ***B*** and ***ds*** are pointing in the same direction the integral reduces to equation 37-1. Even when an integral is required most of the time ***B*** is at a fixed angle with respect to ***ds*** so the integral involves the area and an angle without any formal integration.

The unit of flux is Tesla-meter square which has a special name, Weber. One Weber's worth of flux passing through a square meter produces a field strength of one Tesla.

The **Faraday's law** specifically states that the ***emf*** induced in a closed loop is the change in flux per time.

$$\boldsymbol{emf} = -\frac{\Delta\Phi}{\Delta t} \quad \text{or} \quad \boldsymbol{emf} = -\frac{d\Phi}{dt} \qquad \textbf{(37-3)}$$

The delta form of this statement implies that the ***emf*** is the average ***emf*** over the time interval. The minus sign is a reminder of the convention for determining the direction of the ***emf***. For multiple turns of a coil the ***emf*** is just the number of turns times this change in flux over time. The direction of the induced ***emf*** is such as to produce a magnetic field in opposition to the field that created the ***emf***. This will be discussed in detail in the context of a problem. Now look at several means for generating ***emf***'s in wires.

This induced ***emf*** is different from the ***emf*** produced by a battery. In the case of a battery, a terminal voltage can be measured and the ***emf*** of the battery can be associated with the battery raising the potential of charges passing through it. In the case of a loop of wire there are no terminals, yet there is a current so we say that this current must be due to an electromotive force, a force that makes electric charges move. While there are no terminals to measure voltage in the loop, the current in the loop due to magnetic induction is just as real as the current produced by a battery.

37-1 Place a 20 turn coil of radius $0.050\,\mathrm{m}$ inside a solenoid where the magnetic field is changed from $40\times10^{-3}\,\mathrm{T}$ in one direction to $40\times10^{-3}\,\mathrm{T}$ in the other direction in $60\times10^{-3}\,\mathrm{s}$. Find the induced ***emf***.

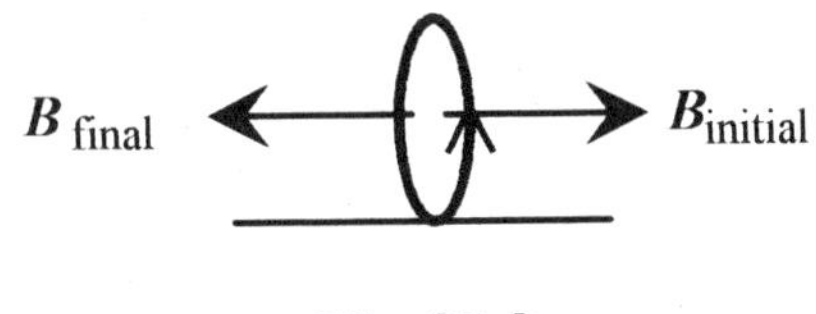

Fig. 37-3

Solution: Inside the coil the flux is $\Phi = BA = 40\times10^{-3}\,\mathrm{T}\,\pi(0.05\,\mathrm{m})^2 = 3.1\times10^{-4}\,\mathrm{Wb}$.

The change in flux is $\Delta\Phi = 6.2\times10^{-4}\,\mathrm{Wb}$, and this occurs over $60\times10^{-3}\,\mathrm{s}$ in the coil of 20 turns so the induced ***emf*** is

$$\boldsymbol{emf} = -N\frac{\Delta\Phi}{\Delta t} = -20\frac{6.2\times10^{-4}\,\mathrm{Wb}}{60\times10^{-3}\,\mathrm{s}} = -0.21\,\mathrm{V}$$

Lenz's Law

To determine the direction of the induced current look at the coil and the direction of the initial and final magnetic fields through the coil. The field originally pointing to the right collapses to zero and then grows to the left to its final value. The current is physically constrained by the wire to go only in one of two directions. If the current in the coil were such that the field produced by this current grew in the same direction as the field that initiated the current then the current would continue to grow (because the field would continue to change in the same direction). This would mean that any time a field changed in a loop of wire the current in the loop would grow without limit. This is clearly contrary to nature. **Lenz's law,** applied to this situation, states that an induced current is in a direction so as to produce a field that opposes the changing field that initiated the current. This will be shown in subsequent problems.

37-2 A circular loop $0.040\,\text{m}$ in radius is perpendicular to a magnetic field of $0.50\,\text{T}$. The loop is stretched, across a diameter, so that its area goes to zero in $0.20\,\text{s}$. What is the induced voltage and direction of the current through the resistor?

Solution: The flux through the loop is $\Phi = BA = 0.50\,\text{T}\cdot\pi(0.04\,\text{m})^2 = 2.5\times10^{-3}\,\text{Wb}$.

The *emf* is $emf = -\dfrac{\Delta\Phi}{\Delta t} = -\dfrac{2.5\times10^{-3}\,\text{Wb}}{0.20\,\text{s}} = -0.012\,\text{V}$.

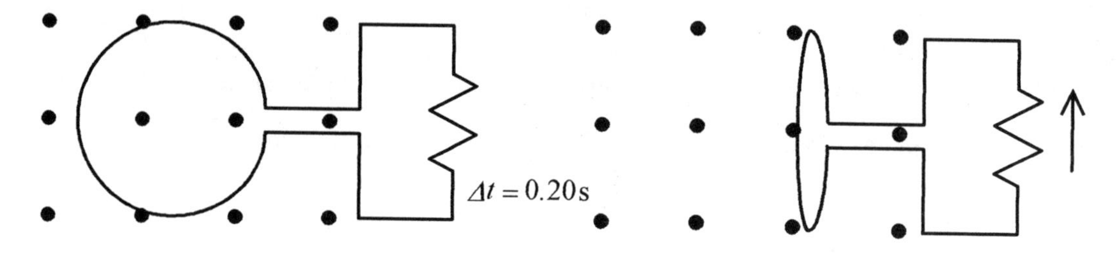

Fig. 37-4

The current is such as to produce a magnetic field out of the page (or in opposition to the collapsing magnetic field). The current is counterclockwise in the stretched loop and up on the resistor.

37-3 Take a 50 turn rectangular coil of dimensions $0.10\,\text{m}$ by $0.20\,\text{m}$ and rotate it from a position perpendicular to a field of $0.50\,\text{T}$ to parallel to the field in $0.10\,\text{s}$ and calculate the induced *emf*.

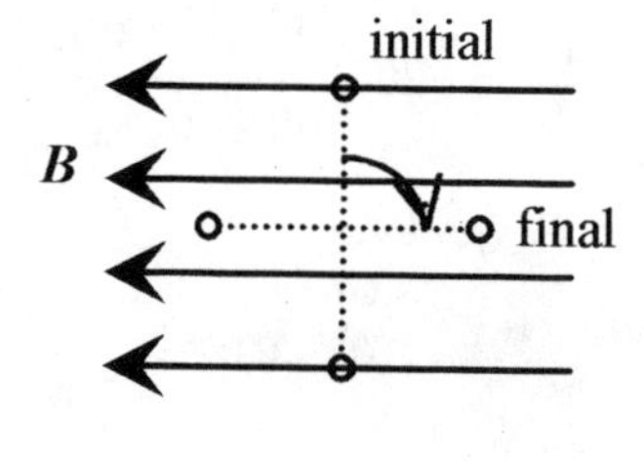

Fig. 37-5

Solution: The flux $\Phi = BA$. In addition, $\Delta\Phi = BA$ since the flux goes from maximum to zero.

$$\Delta\Phi = BA = 0.50\,\text{T}\cdot2.0\times10^{-2}\,\text{m}^2 = 1.0\times10^{-2}\,\text{T}\cdot\text{m}^2$$

The coil is 50 turns, so the induced *emf* is $emf = -N\dfrac{\Delta\Phi}{\Delta t} = -50\dfrac{1.0\times10^{-2}\,\text{T}\cdot\text{m}^2}{0.10\,\text{s}} = -5.0\,\text{V}$.

37-4 Now instead of flipping the coil from perpendicular to parallel, rotate it with uniform angular velocity.

Solution: The expression for the flux, assuming the rotation is clockwise is

$$\Phi = BA\cos\omega t \qquad \text{where} \qquad \omega = \frac{\Delta\theta}{\Delta t} = \frac{\pi/2}{0.10\,\mathrm{s}}.$$

The coil rotates through an angle of $\pi/2$ in $0.10\,\mathrm{s}$ so $\Phi = 0.50\,\mathrm{T}\cdot 2.0\times10^{-2}\,\mathrm{m}^2\cos(\pi t/0.20\,\mathrm{s})$

The instantaneous ***emf*** is

$$\textbf{\textit{emf}} = -N\frac{d\Phi}{dt} = -50\cdot 1.0\times10^{-2}\,\mathrm{T}\cdot\mathrm{m}^2\frac{\pi}{0.20\,\mathrm{s}}\sin\frac{\pi t}{0.20\,\mathrm{s}} = -7.85\,\mathrm{V}\sin\frac{\pi t}{0.20\,\mathrm{s}}$$

The units in this calculation are helpful in understanding the ***emf*** generated in a loop. A $\mathrm{T}\cdot\mathrm{m}^2$ is a Weber, a measure of the number of lines of flux. Webers per second is the number of flux lines per second that are changed within a loop. One Weber's worth of flux lines per second generates an ***emf*** of one Volt in the loop.

(C) **37-5** Now find the average ***emf*** induced over this time. This is the average value of the instantaneous ***emf*** function over this one-quarter cycle.

Solution: This requires the concept of the average value of a function from integral calculus. Review this procedure in the Introduction, Mathematical Background, if necessary. This integral is

$$\textbf{\textit{emf}}\big|_{avg} = -\frac{1}{\pi/2}7.85\,\mathrm{V}\int_0^{\pi/2}\sin\theta\,d\theta = -5.0\,\mathrm{V}\cos\theta\Big|_0^{\pi/2} = -5.0\,\mathrm{V}$$

The integral is done over θ for convenience rather than over ωt and a time interval. The average value of the first quarter cycle of the sine function is the same as the average value for the first half cycle of the sine function and the same as the absolute value of the average value of the entire sine function. This is an excellent example of a problem where the average ***emf*** can be calculated using either $\Delta\Phi/\Delta t$ (problem 37-3) or the average value of the function for ***emf***.

(C) **37-6** For the coil and magnetic field situation of Fig. 37-5 consider continuous rotation at some angular velocity ω. The maximum flux through the coil is Φ_o.

Solution: The flux is $\Phi = \Phi_o\cos\omega t$.

The time rate of change of flux is $\frac{d\Phi}{dt} = -\Phi_0 \omega \sin \omega t$.

The instantaneous ***emf*** is ***emf*** $= -N\frac{d\Phi}{dt} = N\Phi\omega \sin \omega t$.

These three quantities are shown in Fig. 37-6.

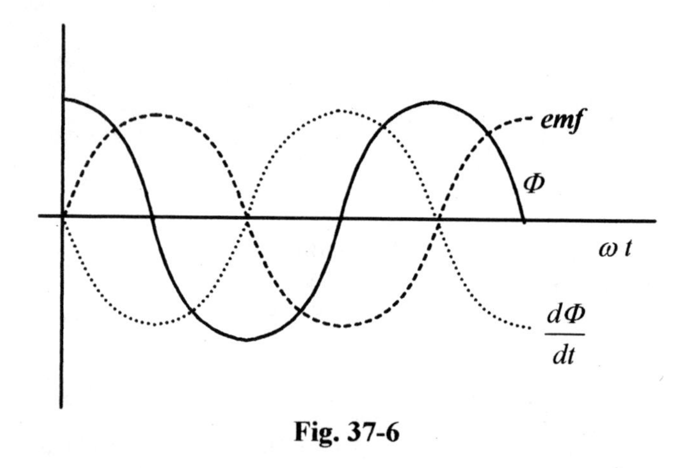

Fig. 37-6

Consider a rectangular loop of wire being pulled through a magnetic field. The field is strength ***B***, width of the loop *L*, length of the loop in the field *x*, and velocity with which the loop is being pulled through the field *v*.

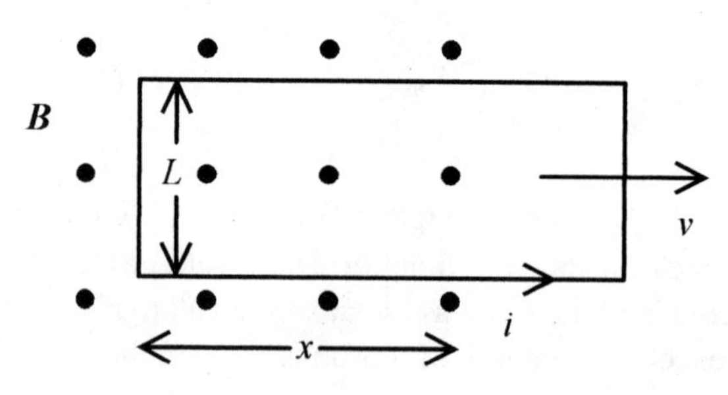

Fig. 37-7

The total flux within the loop is $\Phi = BLx$

The induced ***emf*** is $\quad$ ***emf*** $= -\frac{\Delta\Phi}{\Delta t} = -BL\frac{\Delta x}{\Delta t} = -BLv$

emf $= -\frac{d\Phi}{dt} = -BL\frac{dx}{dt} = -BLv$

As the loop is being pulled through the field there is a decrease in the number of lines of magnetic flux pointing out of the page. Therefore the induced current will be in a direction so as to create more flux out of the page. This requires a counterclockwise current around the loop. Interestingly this current produces forces on the sides of the loop. Apply the right-hand rule to the forces and note that the force on the bottom section of the wire is down, while the force on the top section of the wire is up. These two forces are equal and opposite and so add to zero. The force on the vertical section of the wire is to the left in opposition to the force producing the velocity. The force on the wire due to the induced current is BiL. The work performed by this force is (in analog with mechanics) force times distance or $F \cdot x = BiLx$.
The power (again in analog with mechanics) is force times velocity or $P = F \cdot v = BiLv$.

37-7 Pull a loop of width $0.20\,\mathrm{m}$, length $0.80\,\mathrm{m}$, and resistance $200\,\Omega$ through a magnetic field of $0.40\,\mathrm{T}$ at $0.20\,\mathrm{m/s}$. First find the induced ***emf*** and current. Then find the force necessary to pull the loop, the work performed, and the power, the rate of doing the work.

Solution: The induced ***emf*** is $\boldsymbol{emf} = -BLv = -0.40\,\mathrm{T} \cdot 0.20\,\mathrm{m} \cdot 0.20\,\mathrm{m/s} = -1.6 \times 10^{-2}\,\mathrm{V}$.

The induced current (in the direction shown on Fig. 37-7) is

$$i = \frac{\boldsymbol{emf}}{R} = \frac{1.6 \times 10^{-2}\,\mathrm{V}}{200\,\Omega} = 8.0 \times 10^{-5}\,\mathrm{A}$$

The force necessary to pull the wire out of the field is

$$F = BiL = 0.40\,\mathrm{T} \cdot 8.0 \times 10^{-5}\,\mathrm{A} \cdot 0.20\,\mathrm{m} = 6.4 \times 10^{-6}\,\mathrm{N}$$

The work performed in completely removing the loop is this force times the length of the loop.

$$W = F \cdot x = 6.4 \times 10^{-6}\,\mathrm{N} \cdot 0.80\,\mathrm{m} = 5.1 \times 10^{-6}\,\mathrm{J}$$

The power delivered to the loop is $P = F \cdot v = 6.4 \times 10^{-6}\,\mathrm{N} \cdot 0.20\,\mathrm{m/s} = 1.3 \times 10^{-6}\,\mathrm{W}$.

37-8 A variation of the problem done above is one where the loop is replaced by a "U"-shaped piece of wire with a sliding piece along the arms of the "U." Take a $0.60\,\mathrm{T}$ field and a "U"-shaped piece with width 0.30 m. The entire circuit has resistance of 20 Ω and the sliding bar is moving to the right at $6.0\,\mathrm{m/s}$.

Solution: The ***emf*** generated in the wire and moving rod is

$$\boldsymbol{emf} = -BLv = -0.60\,\mathrm{T} \cdot 0.30\,\mathrm{m} \cdot 6.0\,\mathrm{m/s} = -1.1\,\mathrm{V}$$

The current in the loop is $i = emf/R = 1.1\text{V}/20\Omega = 0.054\text{A}$.

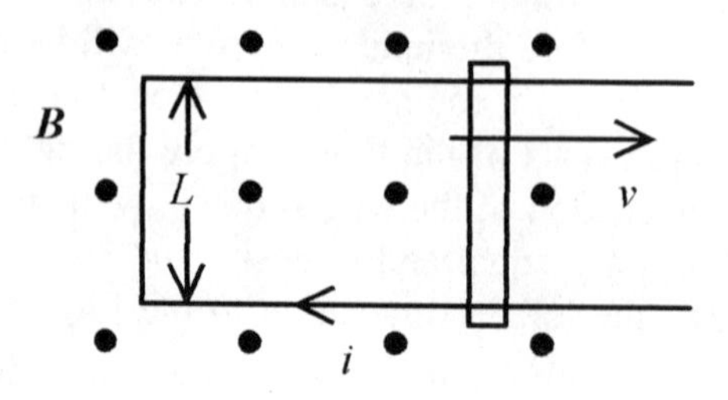

Fig. 37-8

When the bar is moving to the right, the number of lines of flux (amount) is increasing out of the page, so the current is clockwise so as to produce a field pointing into the page, i.e., a field opposing the increasing field, within the "loop" causing the current.

The work performed in moving the bar is

$$W = Fv = BiLv = 0.60\text{T}\cdot 0.054\text{A}\cdot 0.30\text{m}\cdot 6.0\text{m/s} = 0.058\text{J}$$

Compare this with the Joule heat production $H = i^2R = (0.054\text{A})^2(20\Omega) = 0.058\text{J}$.

(C) Calculation of the total flux through a rectangular frame placed with its long side parallel to a current-carrying wire is very helpful in understanding how to apply calculus to a physics problem.

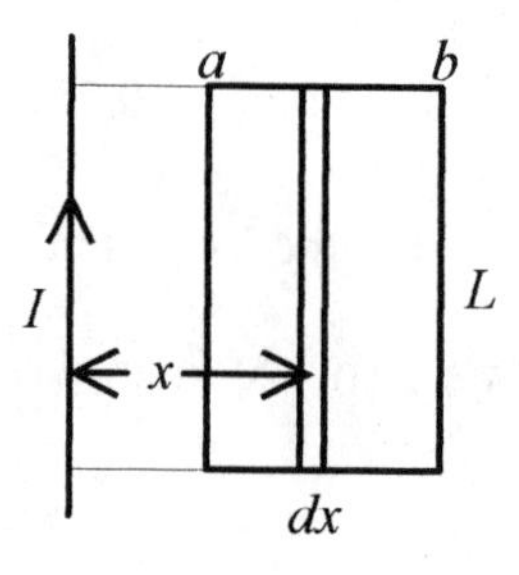

Fig. 37-9

The geometry is shown in Fig. 37-9. What is desired is the total flux in the rectangle of width $b-a$ and height L. The magnetic field at any point, x, a radial distance away from the current, is $B = \mu_0 I/2\pi x$.

The incremental flux over the strip of width dx is the flux at the point x times the incremental area

$$d\Phi = \frac{\mu_0 I}{2\pi x} L\,dx$$

The total flux over the frame is the integral of $d\Phi$ from a to b.

$$\Phi = \frac{\mu_o I L}{2\pi}\int_a^b \frac{dx}{x} = \frac{\mu_o I L}{2\pi}\ln x\Big|_a^b = \frac{\mu_o I L}{2\pi}\ln\frac{b}{a}$$

37-9 For the situation described above the current in the wire increases from zero to $20\,\mathrm{A}$ in $0.10\,\mathrm{s}$. Find the induced ***emf*** over a rectangle with $a = 0.20\,\mathrm{cm}$, $b = 0.60$ cm, and $L = 1.0$ cm.

Solution: ***emf*** $= \Delta\Phi/\Delta t$ and $\Delta\Phi$ is the total flux, so

$$\textit{emf} = \frac{\mu_o I L}{2\pi}\ln\frac{b}{a} = \frac{4\pi\times 10^{-7}\,\mathrm{T\cdot m^2/A}\cdot 20\mathrm{A}\cdot 1.0\times 10^{-2}\,\mathrm{m}}{2\pi(0.10\mathrm{s})}\ln 3 = 4.4\times 10^{-7}\,\mathrm{V}$$

Time Varying Magnetic Fields

There is another aspect of Faraday's law associated with time varying magnetic fields. Place a conducting loop of wire in a magnetic field that varies with time and make the field increase out of the page so that the flux contained within the loop increases as shown in Fig. 37-10.

The ***emf*** generated in the loop is $d\Phi_B/dt$. In the wire there must be an electric field. In equations involving time varying magnetic fields the magnetic flux is often indicated with Φ_B to avoid possible confusion with electric flux. Because of the symmetry ***emf*** $= \oint \mathbf{E}\cdot d\mathbf{l} = 2\pi r E$.

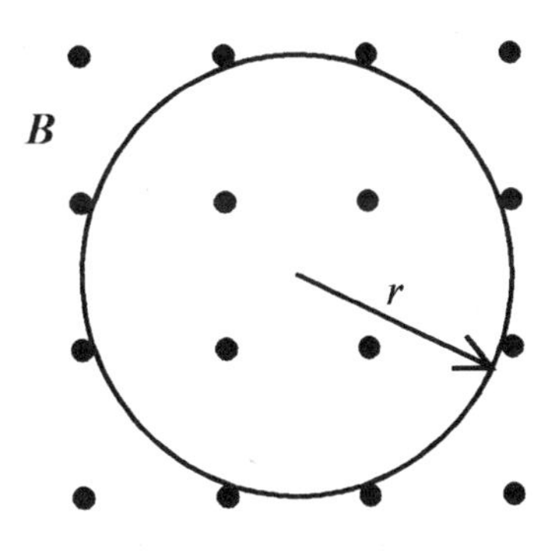

Fig. 37-10

In this instance potential has no meaning (cannot be defined). Further, a physical wire is not necessary for the existence of the electric field caused by this changing flux. With the ***emf*** defined by the changing flux and the integral of the electric field, we can make the identification

$$-\frac{d\Phi_B}{dt} = \oint E \cdot dl$$

and remembering that $\Phi_B = BA$ we have as a solution

$$-\pi r^2 \frac{dB}{dt} = 2\pi r E \quad \text{or} \quad E = -\frac{r}{2}\frac{dB}{dt}$$

37-10 Calculate how an electron placed at 0.30 m radius would be accelerated in a changing magnetic field of $50 \times 10^{-6}\,\text{T/s}$.

Solution: The force on the electron would be $eE = ma$ so

$$a = \frac{e}{m}\frac{r}{2}\frac{dB}{dt} = \frac{1.6 \times 10^{-19}\,\text{C}}{9.1 \times 10^{-31}\,\text{kg}} \frac{0.30\,\text{m}}{2} \frac{50 \times 10^{-6}\,\text{T}}{\text{s}} = 1.3 \times 10^{6}\,\text{m/s}^2$$

Accelerating devices using this principle are called synchrotrons.

CHAPTER 38

INDUCTANCE

If two coils are placed nearby, the current in one sets up a magnetic field or flux through the other. If the current in the one is changing, then there is a changing magnetic field and flux in the other producing an induced ***emf*** and current. In equation-like statements:

Current in one produces flux in the other.

Changing current in one produces an ***emf*** in the other.

Self Inductance

Further, a changing current in an isolated coil produces an ***emf*** in itself. This is called a self-induced ***emf***. The direction of the self-induced ***emf*** is so as to produce a current that opposes the current that set it up.

Calculation of the self-induced ***emf*** follows from a consideration of the geometry of the coil. Consider a close packed coil with a flux, Φ, passing through each of N turns of the coil. The number of turns times the flux is called the flux linkages, $N\Phi$. The self-induced ***emf*** is proportional to the time rate of change of these flux linkages. In the language of calculus, the self-induced ***emf*** is proportional to d/dt of $N\Phi$, so

$$emf = -\frac{\Delta(N\Phi)}{\Delta t} \quad \text{or} \quad emf = -\frac{d(N\Phi)}{dt} \tag{38-1}$$

(C) The quantity $N\Phi$, the flux linkages, is proportional to the current, so make $N\Phi = Li$ where L is a constant that depends on how much flux is linked to the coil. Rewriting equation 38-1 with the defined constant L

$$emf = -L\frac{di}{dt} \tag{38-2}$$

The unit of inductance is the Henry which, from equation 38-2, must have units of $\text{V}\cdot\text{s/A}$.

The self inductance, or inductance as it is often called, can be calculated for a solenoid. Consider a solenoid of length ℓ having N turns or n turns per unit length. Inside the solenoid the flux is BA, so the number of flux linkages is $N\Phi = (n\ell)BA$. For a solenoid the magnetic field is $\mu_o ni$, so

$$N\Phi = (n\ell)BA = \mu_o n^2 \ell i A$$

The quantity $N\Phi$ is equal to the constant L times i so

$$L = \mu_o n^2 \ell A \quad \textbf{(38-3)}$$

The inductance of a solenoid is entirely geometry dependent. Adding a magnetic material to enhance the flux linkages only makes the inductance also dependent on material.

38-1 Calculate the inductance of a solenoid of radius $0.0040\,\mathrm{m}$, length $0.015\,\mathrm{m}$, and 1.0×10^4 turns per meter.

Solution:

$$L = \mu_o n^2 \ell A = (4\pi \times 10^{-7}\,\mathrm{T \cdot m/A})(1.0 \times 10^8\,1/\mathrm{m}^2)(0.015\,\mathrm{m})\pi(0.0040\,\mathrm{m})^2 = 9.5 \times 10^{-5}\,\mathrm{H}$$

The geometric factors n, ℓ, and A (equation 38-3) are difficult to calculate in a closely wound coil, so the measurement of inductance is often done operationally using ***emf*** $= L(di/dt)$. Most inductors, or choke coils as they are sometimes called, are not solenoids but short (in length) tightly wound coils. In this geometry it is very difficult to calculate the inductance based on geometry, so this operational method is employed.

38-2 For a coil of 300 turns and inductance $12 \times 10^{-3}\,\mathrm{H}$, find the flux through the coil when the current is $5.0 \times 10^{-3}\,\mathrm{A}$.

Solution: Use the relationship between flux linkages and inductance $N\Phi = Li$.

$$\Phi = \frac{Li}{N} = \frac{12 \times 10^{-3}\,\mathrm{H} \cdot 5.0 \times 10^{-3}\,\mathrm{A}}{300} = 2.0 \times 10^{-7}\,\mathrm{Wb}$$

Mutual Inductance

If there are coils within, or adjacent to, other coils, then we can define mutual inductance, the linking of one coil with another via the flux. If the flux is changing, then the current in the one coil is linked to the current in the other coil.

(C) The simplest case to consider is one solenoid-like coil inside another solenoid-like coil. The voltage induced in the second coil is proportional, via the flux linkage, to the changing current in the first coil. The mathematical statement is

$$emf_2 = M\frac{di_1}{dt}$$

The reverse is also true. The voltage induced in the first coil is proportional to the changing current in the second coil.

$$emf_1 = M\frac{di_2}{dt}$$

The constants are the same. The geometry is the same. These "geometric coupling constants" are again labeled L, the inductance.

38-3 Consider a coil within a coil. The larger coil is 3.0 cm in radius, 10 cm in length with 1200 turns. The smaller coil is 2.0 cm in radius, 4.0 cm in length with 50 turns. Calculate the inductance.

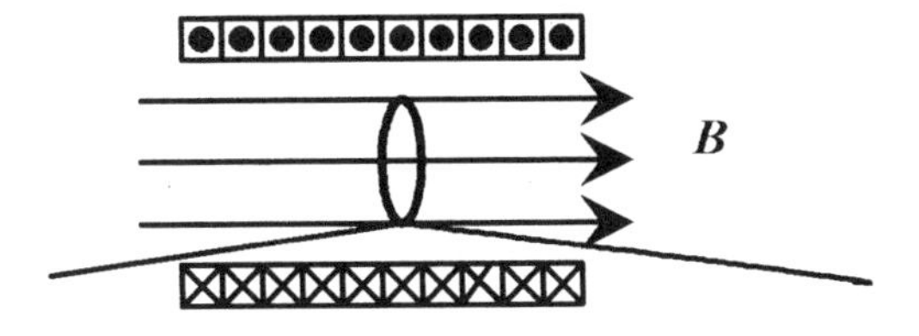

Fig. 38-1

Solution: The current in the large solenoid sets up a magnetic field within the coil of $B = \mu_o Ni/\ell$. Assume the flux is uniform across the inside of the large coil so that the flux through the small coil is this B times the area of the small coil $\Phi_2 = BA_2$. The 1 subscript refers to the large coil and 2 refers to the smaller coil. The mutual inductance, in accord with the definition of self inductance is the number of flux linkages divided by the current

$$M \text{ or } L = \frac{N_2\Phi_2}{i_1} = \frac{N_2}{i_1}BA_2 = \frac{N_2}{i_1}\frac{\mu_o N_1 i_1}{\ell}A_2 = \mu_o\frac{N_1}{\ell}N_2A_2$$

The inductance is totally geometry dependent. For this particular combination of coils

$$L = \mu_o\frac{N_1}{\ell}N_2A_2 = 4\pi\times10^{-7}\,\mathrm{T\cdot m/A}\frac{1200}{0.10\,\mathrm{m}}(50)\pi(4.0\times10^{-4}\,\mathrm{m}^2) = 9.5\times10^{-4}\,\mathrm{H}$$

38-4 For the situation of problem 38-3 find the voltage induced in the second coil as a result of a steady change in current of $0.50\,\text{A/s}$.

Solution:

$$emf = L\frac{\Delta i}{\Delta t} = 9.5\times10^{-4}\,\text{H}\frac{0.50\,\text{A}}{\text{s}} = 4.7\times10^{-4}\,\text{V}$$

(C) **38-5** Two coils wound together as a single package have an inductance of $6.0\times10^{-3}\,\text{H}$. Find the voltage induced in the second coil when the current is changing in the first at the instantaneous rate of $20\times10^{-3}\,\text{A/s}$.

Solution:

$$emf = L(di/dt) = 6.0\times10^{-3}\,\text{H}\cdot20\times10^{-3}\,\text{A/s} = 1.2\times10^{-4}\,\text{V}$$

Power and Energy Storage

(C) If the ***emf*** generated in an inductor is $L(di/dt)$ and the power in an electric circuit is voltage times current, then the general expression for the power in an inductor is

$$P = Li\frac{di}{dt} \qquad \textbf{(38-4)}$$

The energy stored in a conductor is the integral over time of the power (remember power is work over time), or in differential form $dU = Lidi$, so the total energy stored in a coil (the energy is actually stored in the magnetic field) when the current goes from zero to some value i is the integral of this expression

$$U = \int_0^i Lidi = \frac{Li^2}{2} \qquad \textbf{(38-5)}$$

See Chapter 39, R-L Circuits, for another discussion on this topic.

The energy density in a solenoid is the total energy stored in the solenoid divided by the volume.

$$u_B = \frac{U_B}{A\ell} = \frac{Li^2}{2A\ell}$$

Using $L = \mu_o n^2 \ell A$,

$$u_B = \frac{\mu_o n^2 i^2}{2}$$

Or using $B = \mu_o ni$,

$$u_B = \frac{B^2}{2\mu_o} \qquad \textbf{(38-6)}$$

Compare this with the expression for the energy density in an electric field $u_E = \frac{\varepsilon_o E^2}{2}$.

38-6 Calculate the energy density in the large coil of problem 38-3 for a current of 0.10 A.

Solution:

$$u_B = \frac{\mu_o n^2 i^2}{2} = \frac{(4\pi \times 10^{-7}\,\mathrm{T\cdot m/A})(1200/0.10\,\mathrm{m})^2(0.10\,\mathrm{A})^2}{2} = 0.90\,\mathrm{J/m^3}$$

CHAPTER 39

R-L CIRCUITS

Our first look at R-L circuits is with a simple word description of the behavior of the current in the circuit of Fig. 39-1. After this general description of how the circuit behaves, we will take a closer look at the mathematical description. The important point to keep in mind in this discussion is that the voltage across an inductor is proportional to the rate of change in current.

Increasing Current

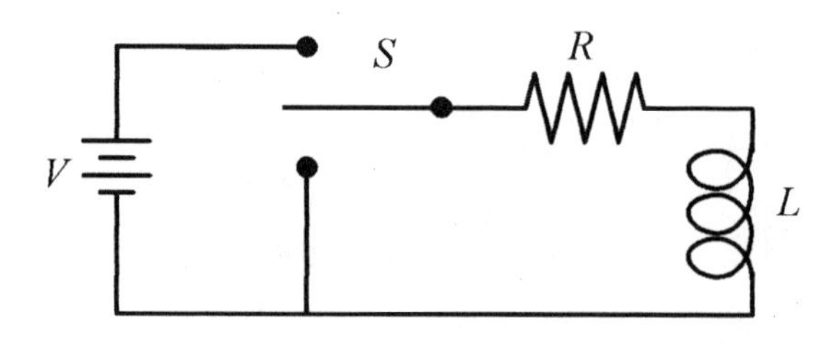

Fig. 39-1

When the switch is closed to place the resistor and inductor in series with the battery, current begins to flow in the circuit. Because of the self-induced ***emf*** in the inductor, that is proportional to the rate of change of current, all of the battery voltage does not instantaneously appear across the resistor. The current in the circuit rises in an exponential manner to (its final value of) V/R according to equation 39-1.

$$i = \frac{V}{R}(1 - e^{-Rt/L}) \qquad \textbf{(39-1)}$$

This function is consistent with our understanding of inductors: as time goes on the current rises to V/R.

(C) When the battery voltage is applied to R and L in series, voltages across these components vary with time. We can, however, write a Kirchhoff-type voltage statement that is valid for all time.

$$V - iR - L\frac{di}{dt} = 0$$

which can be solved for i. The solution is similar to the one for the R-C circuit done in an earlier chapter. A review of that more detailed analysis may be helpful before continuing. Rewrite this equation as

$$\frac{di}{dt} = \frac{V}{L} - i\frac{R}{L} \quad \text{or} \quad \frac{di}{dt} = \left(\frac{V}{R} - i\right)\frac{R}{L}$$

Switching to a more convenient form for integration

$$\int \frac{di}{i - (V/R)} = -\int \frac{R}{L}dt$$

Integrating with a change of variable as before

$$\ln\left(i - \frac{V}{R}\right) = -\frac{R}{L}t + \ln K$$

Switching to exponentials

$$i - (V/R) = Ke^{-Rt/L} \quad \text{or} \quad i = (V/R) + Ke^{-Rt/L}$$

The constant, K, can be evaluated in either of these equations by imposing the initial condition that at $t = 0,\ i = 0,$ so $K = -V/R$ and

$$i = \frac{V}{R}(1 - e^{-Rt/L})$$

The time constant in the circuit is L/R. The rate of change of current in the circuit is

$$\frac{di}{dt} = \frac{V}{L}e^{-Rt/L} \tag{39-2}$$

There are three graphs (Fig. 39-2) that describe how the voltage and current vary in the circuit. The graph of $i = (V/R)(1 - e^{-Rt/L})$ versus t shows how the current in the circuit varies with time. The graph of $v_R = iR = V(1 - e^{-Rt/L})$ versus t shows the exponential rise of voltage across R. The graph of $v_L = L(di/dt) = Ve^{-Rt/L}$ versus t shows the exponential decay of voltage across L.

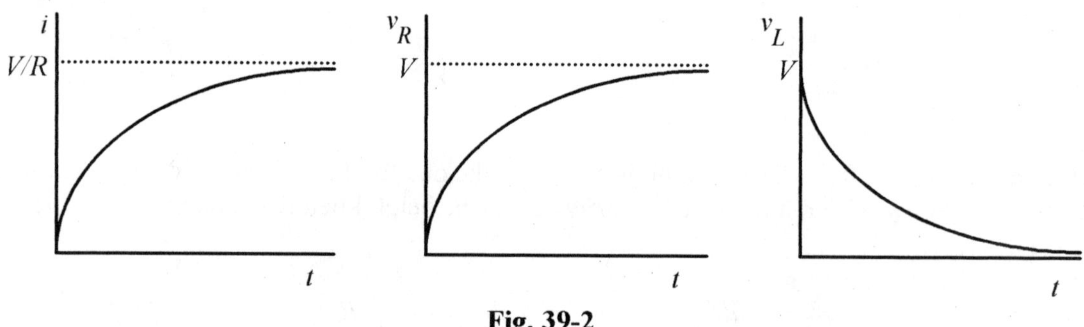

Fig. 39-2

As an exercise show that the Kirchhoff equation $V = iR + L(di/dt)$ is satisfied by substituting the expressions for i and di/dt. The analysis is also verified by the graphs: the voltage across the resistor is growing in a one minus exponential manner while the voltage across the inductor is decaying in an exponential manner.

39-1 Place a $60\,\text{V}$ battery in series with an inductor of $50 \times 10^{-3}\,\text{H}$ and a resistor of $180\,\Omega$. What is the current and the rate of change of current at $t = 0$?

Solution: Using $i = \dfrac{V}{R}(1 - e^{-Rt/L})$ and $\dfrac{di}{dt} = \dfrac{V}{L}e^{-Rt/L}$ it is clear that at $t = 0$, $i = 0$,

but $\dfrac{di}{dt} = \dfrac{V}{L} = \dfrac{60\,\text{V}}{50 \times 10^{-3}\,\text{H}} = 1.2 \times 10^{3}\,\text{A/s}.$

39-2 For the same circuit, what is the time constant?

Solution: The time constant $\tau = \dfrac{L}{R} = \dfrac{50 \times 10^{-3}\,\text{H}}{180\,\Omega} = 2.8 \times 10^{-4}\,\text{s}$

39-3 For the same circuit, what is the rate of current increase at $t = 1.0 \times 10^{-4}\,\text{s}$?

Solution: Using $\dfrac{di}{dt} = \dfrac{V}{L}e^{-t/2.8 \times 10^{-4}\,\text{s}}$ and $t = 1.0 \times 10^{-4}\,\text{s}$,

$$\frac{di}{dt} = \frac{60\,\text{V}}{50 \times 10^{-3}\,\text{H}}e^{-1/2.8} = 840\,\text{A/s}$$

39-4 For the same circuit described in problem 39-3, what is the rate of current increase when i has reached 90% of its final value?

Solution: Translated into mathematics, the question reads, "Find di/dt when $i = 0.90(V/R)$." First find t when $i = 0.90(V/R)$. Set $i = 0.90V/R$ in equation 39-1 and solve for t.

$$0.90(V/R) = (V/R)(1 - e^{-t/2.8\times10^{-4}\,\mathrm{s}}) \quad \text{or} \quad e^{-t/2.8\times10^{-4}\,\mathrm{s}} = 0.10 \quad \text{or} \quad e^{t/2.8\times10^{-4}\,\mathrm{s}} = 10$$

Solve for t by switching to logarithms: $t = 2.8 \times 10^{-4} \ln 10 = 6.4 \times 10^{-4}\,\mathrm{s}$

This is not a common type of problem, so go back over these equations and be sure you know how to manipulate the exponents and logarithms.

Now place this time in equation 39-2.

$$\frac{di}{dt} = \frac{V}{L} e^{-t/2.8\times10^{-4}\,\mathrm{s}} = \frac{60\,\mathrm{V}}{50 \times 10^{-3}\,\mathrm{H}} e^{-6.4\times10^{-4}/2.8\times10^{-4}} = 122\,\mathrm{A/s}$$

39-5 Construct the circuit of Fig. 39-1 with an inductor of 10 H, a resistor of 12 Ω, and a 10 V battery. Calculate the time constant, the rate of change of current, and the voltages across the resistor and inductor at one time constant.

Solution: The time constant is $\tau = \dfrac{L}{R} = \dfrac{10\,\mathrm{H}}{12\,\Omega} = 0.83\,\mathrm{s}$.

The rate of change of current is $\dfrac{di}{dt} = \dfrac{V}{R} e^{-t/0.83\mathrm{s}} = \dfrac{10\,\mathrm{V}}{12\,\Omega} e^{-1} = 0.31\,\mathrm{A/s}$.

The voltage across the resistor is $v_R = iR = V(1 - e^{-t/0.83\mathrm{s}}) = 10(1 - e^{-1})\,\mathrm{V} = 6.3\,\mathrm{V}$.

The voltage across the inductor is $v_L = L\dfrac{di}{dt} = Ve^{-t/0.83\mathrm{s}} = 10e^{-1}\,\mathrm{V} = 3.7\,\mathrm{V}$.

Decreasing Current

Referring back to Fig. 39-1, suppose that the current has reached the steady state condition, the R and L are placed in series, and the battery removed. At the instant the battery is removed there is no voltage across the inductor. The circuit has reached steady state and $(di/dt) = 0$ so $L(di/dt) = v_L$ is zero. There is, however, full battery voltage across the resistor. This voltage drops as the current in the circuit drops in an exponential manner according to equation 39-3.

$$i = \frac{V}{L} e^{-Rt/L} = i_o e^{-Rt/L} \tag{39-3}$$

C In this situation the loop equation is

$$L\frac{di}{dt} + iR = 0 \quad \text{or} \quad \int \frac{di}{i} = -\frac{R}{L}\int dt$$

which has solution

$$\ln i = -\frac{R}{L}t + \ln k \quad \text{or} \quad i = ke^{-Rt/L}$$

The constant can be obtained by imposing the initial conditions. At $t = 0$ (in decay mode), $i = i_o$. The i_o has the value V/R (initially all the battery voltage is across the resistor). With the constant evaluated

$$i = i_o e^{-Rt/L}$$

and

$$\frac{di}{dt} = -\frac{V}{L} e^{-Rt/L} \tag{39-4}$$

The voltage across the resistor is $v_R = iR = Ve^{-Rt/L}$.

The voltage across the inductor is $v_L = L\frac{di}{dt} = -Ve^{-Rt/L}$.

Graphs of i, v_R, and v_L are shown in Fig. 39-3.

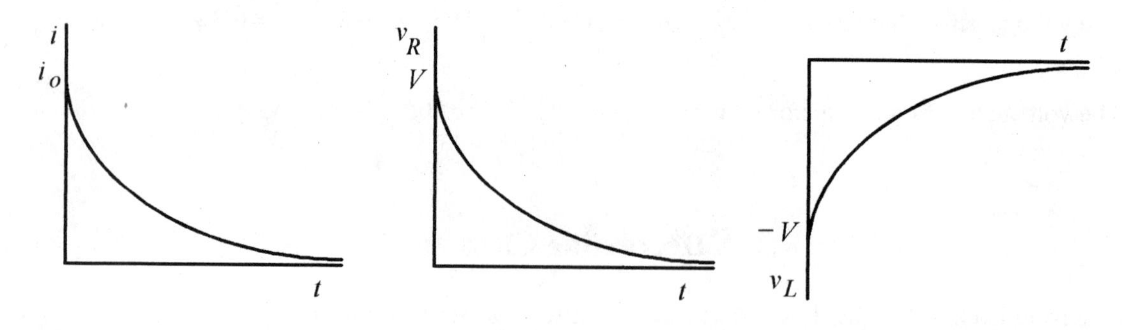

Fig. 39-3

39-6 For the circuit described in problem 39-5 operating in decreasing current mode, calculate the time for the voltage to decay to one-half the steady state value. Then find the voltages across R and L at this time.

Solution: The time constant, calculated earlier, is $0.83\,\mathrm{s}$. The time for the current to drop to one-half the steady state value is found by replacing i in equation 39-3 with $0.50 i_o$.

$$0.50 i_o = i_o e^{-t/0.83\,\mathrm{s}}$$

A more convenient form for logarithms is

$$1/2 = e^{-t/0.83\,\mathrm{s}}$$

Taking logarithms and remembering that $\ln(1/2) = \ln 1 - \ln 2 = 0 - \ln 2 = -\ln 2$,

$$-\ln 2 = -t/0.83\,\mathrm{s} \quad \text{or} \quad t = 0.83 \ln 2\,\mathrm{s} = 0.58\,\mathrm{s}$$

The voltage across the inductor is $v_L = L\dfrac{di}{dt} = -Ve^{-Rt/L} = -10e^{-0.58/0.83}\,\mathrm{V} = -5.0\,\mathrm{V}$.

The voltage across the resistor is $v_R = Ve^{-Rt/L} = 10e^{-0.58/0.83}\,\mathrm{V} = 5.0\,\mathrm{V}$.

The voltages $v_R + v_L$ add to zero as they must since they are the entire circuit!

Energy Storage

The energy stored in an inductor is, from Chapter 38, Inductance,

$$U = \frac{Li^2}{2}$$

In this circuit in the increasing current mode, energy is being stored in the magnetic field associated with the current in the coil. In the decreasing current mode the energy stored in the magnetic field of the coil leaves as the magnetic field and current collapse. This energy is dissipated as heat in the resistor.

39-7 In the circuit described in problem 39-5, what is the maximum energy stored in the (magnetic field of the) coil?

Solution:

$$U = \frac{Li^2}{2} = \frac{10\,\mathrm{H}}{2}\left(\frac{10\,\mathrm{V}}{12\,\Omega}\right)^2 = 3.5\,\mathrm{J}$$

CHAPTER 40

OSCILLATING L-C CIRCUITS

In the next two chapters we will look at L-C and R-L-C circuits from two different points of view. Our view in this chapter is concerned with the transient response of the circuit, that is how the circuit responds to an initially charged capacitor. In the next chapter we will look at how the circuit behaves when it is driven by a sinusoidal voltage source.

L-C Circuits

Charge a capacitor and place it in series with an inductor as shown in Fig. 40-1.

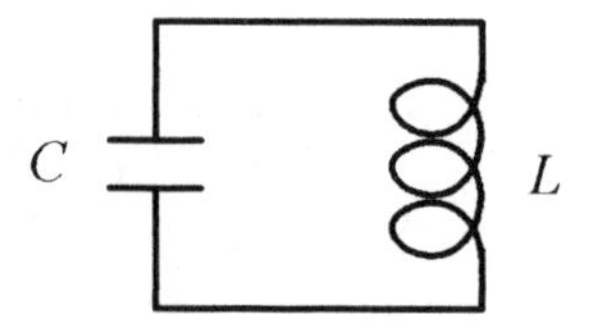

Fig. 40-1

When the charged capacitor is connected to the inductor, the maximum possible voltage is applied across the inductor. This maximum voltage results in the maximum rate of increase in current in the circuit, $v_L = L(di/dt)$. As time goes on, the voltage drops, the rate of change in current drops, but the current increases. Maximum current and zero voltage across the capacitor are coincident. The current continues in the same direction until the capacitor is fully charged in the opposite (from original) direction. In the absence of energy-consuming resistance the voltage (across the capacitor) and current (in the circuit) continue to vary sinusoidally 90° out of phase.

This circuit behaves analogous to the mechanical mass-spring system with the capacitor playing the role of the spring and the inductor playing the role of the mass. Voltage is analogous to the compression of the spring. Energy stored in the spring is equal to $kx^2/2$, and energy stored in the capacitor is equal to $CV^2/2$. Current is analogous to the velocity of the mass. Energy stored in the velocity of the mass is equal to $mv^2/2$, and energy stored in the inductor is equal to $Li^2/2$. Zero spring displacement is coincident with maximum velocity of the mass, as zero voltage is coincident with maximum current.

C The Kirchhoff loop statement for this circuit is

$$\frac{q}{C} + L\frac{di}{dt} = 0 \qquad \textbf{(40-1)}$$

This equation can be rewritten with q the only time dependent variable as

$$\frac{d^2q}{dt^2} = -\frac{1}{LC}q \qquad \textbf{(40-2)}$$

This equation has the same form as the one for the simple harmonic oscillator. (See Chapter 15, Simple Harmonic Motion.) It is solved by asking the question, "What function differentiated twice produces a negative constant times itself?" A sine or cosine function is the obvious answer, so "guess" $q = q_o \cos \omega t$ with

$$\frac{d^2q}{dt^2} = -\omega^2 q_o \cos\omega t$$

and putting this into the Kirchhoff equation $-\omega^2 q_o \cos\omega t = -\frac{1}{LC}q_o \cos\omega t$, we see immediately that the frequency is

$$\omega = \frac{1}{\sqrt{LC}} \quad \text{or} \quad f = \frac{1}{2\pi\sqrt{LC}} \qquad \textbf{(40-3)}$$

The circuit oscillates between maximum electric field in the capacitor and zero current in the circuit, and maximum current in the circuit and zero charge (and field) in the capacitor.

40-1 An L-C circuit consists of a $50\,\text{mH}$ inductor and a $40\,\mu\text{F}$ capacitor. What is the frequency of oscillation of this circuit? If the maximum charge on the capacitor is $100\,\mu\text{C}$, what is the maximum voltage on the components?

Solution: The frequency $f = \dfrac{1}{2\pi\sqrt{LC}} = \dfrac{1}{2\pi\sqrt{50\times10^{-3}\,\text{H}\cdot 4.0\times10^{-5}\,\text{C}}} = 112\,\text{Hz}$

The maximum voltage on either component is the same as the maximum voltage on the capacitor, which comes from the basic definition of capacitance

$$V_{max} = \frac{q_{max}}{C} = \frac{100\,\mu\text{C}}{40\,\mu\text{F}} = 2.5\,\text{V}$$

In an ideal circuit with no resistance, there is no way for energy to enter or leave the system, and the total energy must be passed back and forth between capacitor and inductor. Energy is stored in the electric field of the capacitor and the magnetic field of the inductor. The maximum energy in each component is equal, so

$$\frac{q_{max}^2}{2C} = \frac{Li_{max}^2}{2} \qquad \textbf{(40-4)}$$

40-2 For the circuit described in problem 40-1, what is the maximum energy and the maximum current in the circuit?

Solution: Maximum energy is $\frac{q_{max}^2}{2C} = \frac{(1.0\times 10^{-4}\,\mathrm{C})^2}{2\cdot 4.0\times 10^{-5}\,\mathrm{F}} = 1.25\times 10^{-4}\,\mathrm{J}$.

The maximum current is from $\frac{Li_{max}^2}{2} = 1.25\times 10^{-4}\,\mathrm{J}$, so

$$i_{max} = \left[\frac{2\cdot 1.25\times 10^{-4}\,\mathrm{J}}{50\times 10^{-3}\,\mathrm{H}}\right]^{1/2} = 0.071\mathrm{A}$$

R-L-C Circuits

The introduction of series resistance produces the circuit of Fig. 40-2.

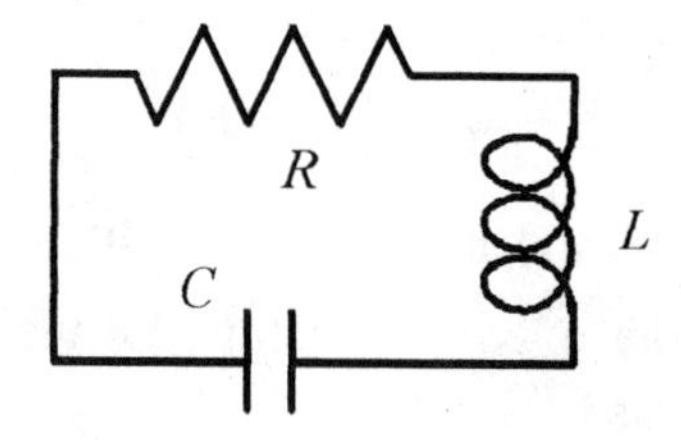

Fig. 40-2

If the capacitor is initially charged (before it is placed in the circuit), oscillation will occur. The total energy in the circuit, however, will not remain constant but decrease at the rate of i^2R, the power dissipated in the resistor as heat.

(C) A Kirchhoff-type voltage equation for the circuit can be written

$$L\frac{di}{dt} + iR + \frac{q}{C} = 0 \qquad \textbf{(40-5)}$$

Again, rewriting with q

$$L\frac{d^2q}{dt^2} + R\frac{dq}{dt} + \frac{q}{C} = 0 \qquad \textbf{(40-6)}$$

The introduction of the iR term makes the equation much more difficult to solve. It is, however, identical in form to the equation for the damped harmonic oscillator found in Chapter 15, Simple Harmonic Motion. The addition of the iR term produces an exponential envelope to the oscillations. The specific solution is

$$q = q_o e^{-Rt/2L}\cos(\omega' t + \phi) \qquad \text{where } \omega' = \sqrt{1/LC - (R/2L)^2} \qquad \textbf{(40-7)}$$

This exponential envelope of a sinusoidal oscillation is what we would expect from this circuit; oscillations that decrease exponentially. While this equation hasn't been derived, it can be justified by taking the appropriate derivatives and substituting back into equation 40-6.

40-3 An R-L-C circuit consists of a $10\,\Omega$ resistor, $200\,\text{mH}$ inductor, and a $10\,\mu\text{F}$ capacitor. The capacitor is initially charged to $10\,\mu\text{C}$. What is the maximum voltage on the capacitor at $0.10\,\text{s}$?

Solution: The maximum voltage on the capacitor corresponds to the maximum charge, and that occurs when the cosine function is a maximum of 1. Therefore the exponential term determines the maximum value of q. Using equation 40-6,

$$q = q_o e^{-Rt/2L} = (10\,\mu\text{C})\exp\left[-\frac{10\,\Omega \cdot 0.10\,\text{s}}{2 \cdot 0.20\,\text{H}}\right] = 10\exp[-2.5]\,\mu\text{C} = 0.82\,\mu\text{C}$$

The notation for e has been switched to exp for convenience.

The voltage at this time is $V = \dfrac{q}{C} = \dfrac{0.82\,\mu\text{C}}{10\,\mu\text{F}} = 0.082\,\text{V}$.

CHAPTER 41

SERIES R-L-C CIRCUITS AND PHASORS

The previous chapter was concerned with the natural oscillations of L-C circuits. This chapter is devoted to the study of R's, L's and C's, individually and in combination, in response to sinusoidal voltages.

Alternating Currents

Problems 37-3 through 37-6 in Chapter 37, Faraday's Law, describe how a loop rotated in a magnetic field produces a sinusoidal voltage and current. This is the basis of alternating current generators. An external agent, such as falling water or steam, is used to rotate the loop of wire in a magnetic field thus generating a sinusoidal, or alternating, voltage and current. This alternating current, ac for short, has two basic advantages over direct current. First, it is easy to increase the voltage for transmission over long distances or decrease the voltage for distribution to individual users or specific applications. (See the "Transformers" section in this chapter.) Second, an alternating current in a loop placed in a magnetic field rotates, providing rotational power for all sorts of machines. See the "A Current-Carrying Loop in a *B* Field" section in Chapter 34, Magnetic Forces. The basic circuit for the study of alternating current in R-L-C circuits is shown in Fig. 41-1.

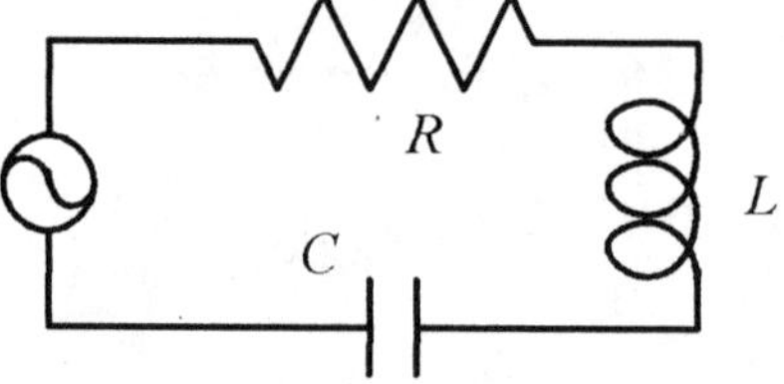

Fig 41-1

The circuit is driven by the ac source. This is in contrast with the circuit of Fig. 40-2 where the capacitor is charged, placed in the circuit, and the transient response studied. The Kirchhoff voltage differential equation for this circuit is like equation 40-5 except that the right-hand side contains a driving voltage $V_0 \cos\omega t$. This dramatically complicates the mathematics beyond the level for most people taking their first physics course. For this reason we take another, less mathematical, view of this circuit but one that is very helpful in understanding how the circuit operates.

The Resistive Circuit

The simplest way to start the study of ac circuits is with a resistor and ac source as shown in Fig. 41-2. The time varying voltage is $v = V_R \cos\omega t$.

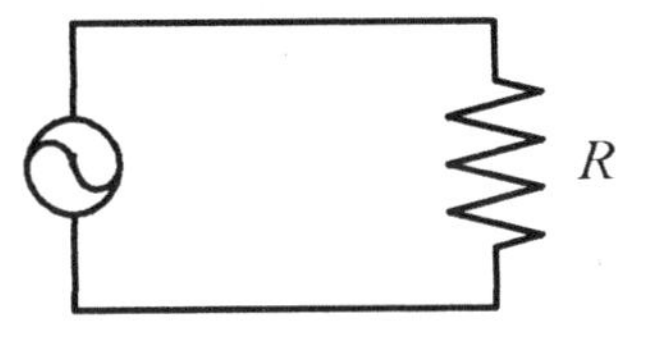

Fig. 41-2

The current tracks (is in phase with) the voltage and is $i = (V_R/R)\cos\omega t = I_R \cos\omega t$.
Fig. 41-3 shows the voltage and current as a function of time and (on the right) what is known as a phasor diagram, an important analysis tool in the study of R-L-C circuits.

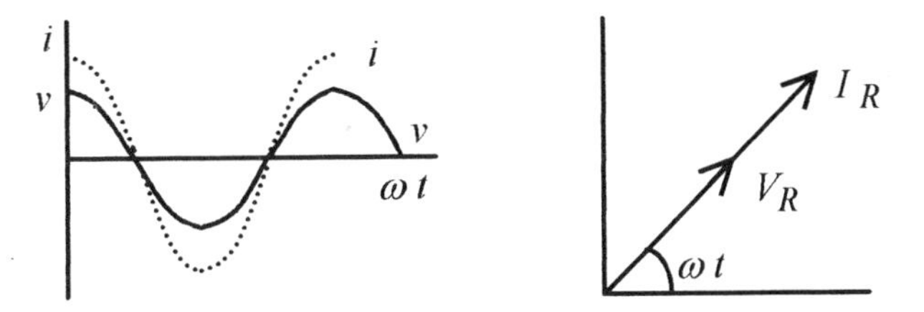

Fig. 41-3

The individual vector-like arrows are called phasors. The diagram is started by drawing a phasor I_R at some arbitrary angle ωt. The phasor rotates counterclockwise with the length proportional to I_R and the projection on the horizontal axis proportional to the instantaneous current. An increase in ωt toward the right on the graph corresponds to an increase in ωt in the counterclockwise direction on the phasor diagram. When an entire cycle of a sine wave is completed the phasor will have rotated through 360°. Next the phasor representing the voltage is drawn at the same arbitrary angle ωt (in a resistor the instantaneous voltage and current are in phase). The length is proportional to V_R, and the projection along the horizontal axis is proportional to the instantaneous voltage across R.

The Capacitive Circuit

Figure 41-4 shows a capacitive circuit driven by an ac source.

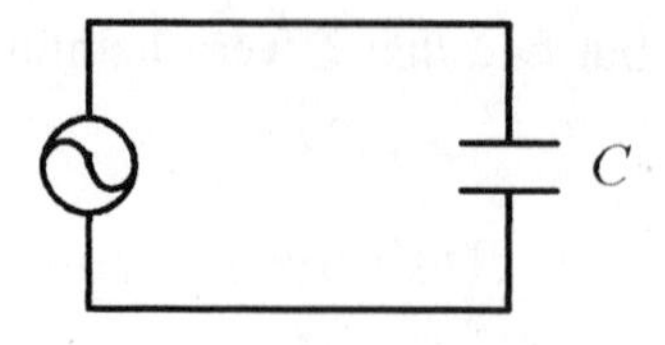

Fig. 41-4

The time varying voltage is $v = V_C \cos\omega t$. The phase relation between this v and i_C is different than for a resistor. When an alternating voltage is applied to a capacitor the current alternates (flows in one direction, then in the opposite direction) but does not track with (is not in phase with) the voltage across the capacitor. When the voltage reaches a maximum the capacitor is fully charged and the current is zero! When the voltage reaches a maximum in the other direction the capacitor is again fully, but oppositely, charged and the current is again zero. The current then must be a maximum when the (alternating) voltage is passing through zero.

The charge is in phase with the voltage $q = Cv = CV_C \cos\omega t$.

The current is $i_C = (dq/dt) = -\omega CV_C \sin\omega t$

The voltage and current are plotted as a function of time in Fig. 41-5 along with the phasor diagram.

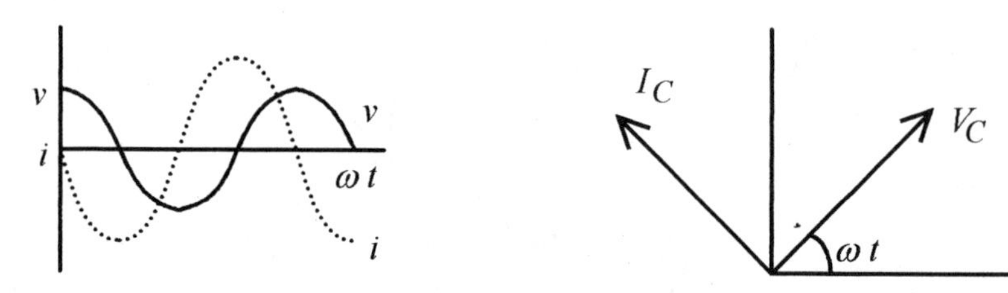

Fig. 41-5

The phasor diagram is drawn starting with V_C at an arbitrary angle ωt. (Starting with I_C produces the same result.) The hard part in drawing the I_C phasor is to figure out how to orient it with respect to V_C. The easiest way to do this is to look at the graph of v and i versus time and ask the question, "Which quantity leads the other and by how much?" By looking at adjacent peaks note that i reaches its maximum 90° before v. Therefore we say "i leads v by 90° in a capacitive circuit." The I_C phasor is 90° ahead (rotated 90° further counterclockwise) of V_C.

The maximum current is $I_C = \omega CV_C$ or $V_C = I_C(1/\omega C) = I_C X_C$.

The $1/\omega C$ term plays the role of resistance and is called capacitive reactance X_C.

41-1 A $20\,\mu\text{F}$ capacitor is connected to a variable frequency ac source with maximum voltage $30\,\text{V}$. What is the capacitive reactance at 60 Hz, 600 Hz, and 60 kHz?

Solution: $X_C\big|_{60} = \dfrac{1}{\omega C} = \dfrac{10^6\,\text{s}}{2\pi\cdot 60\cdot 20\,\text{F}} = 130\,\Omega \quad X_C\big|_{600} = 13\,\Omega \quad X_C\big|_{60\text{k}} = 0.13\,\Omega$

The Inductive Circuit

Figure 41-6 shows an inductor driven by an ac source. The time varying voltage is $v = V_L \cos\omega t$. Again the phase relationship between v and i_L is different from either the resistor or capacitor. The maximum voltage across an inductor is proportional to the rate of change of current. Therefore the maximum voltage corresponds not to maximum current but to maximum rate of change of current. A quick look at a sine curve indicates that the maximum rate of change (slope) is when the curve crosses the axis, so we expect the current to be 90° out of phase with the voltage.

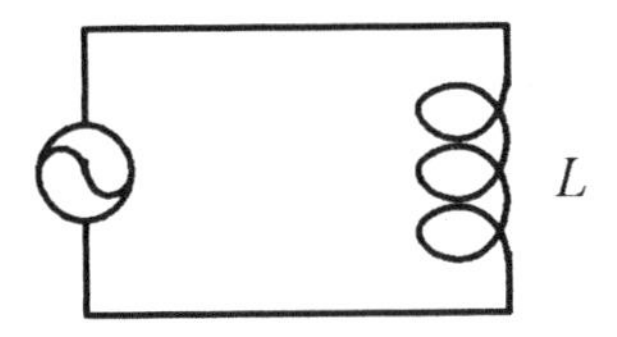

Fig 41-6

The Kirchhoff-type voltage statement for this circuit is $V\cos\omega t = L(di/dt)$

This statement is easily integrated $\int V\cos\omega t\,dt = \int L\,di$ to $i = (V/\omega L)\sin\omega t$.

The voltage and current are plotted as a function of time in Fig. 41-7 along with the phasor diagram.

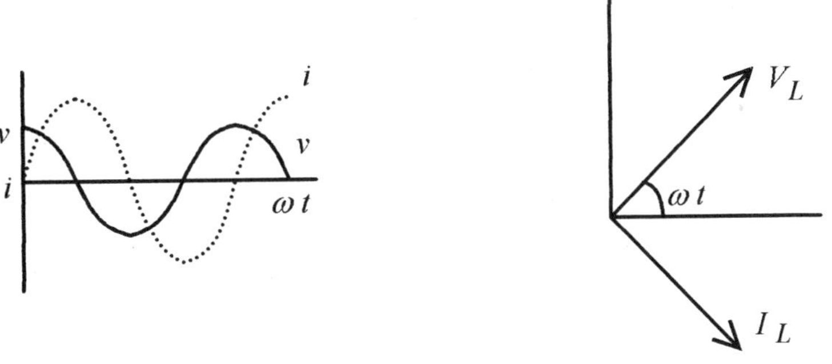

Fig. 41-7

The phasor diagram is drawn by starting with V_L. Now look at the adjacent peaks in the graph of v and i versus ωt and note that v reaches its peak earlier in time than i_L. Therefore we say "v leads i by 90° in an inductive circuit." Notice how, as the phasors rotate at this fixed 90° difference, the voltage phasor traces out the cosine function on the horizontal axis and the current phasor traces out the sine function.

The maximum current is $I_L = V_L/\omega L = V_L/X_L$ or $V_L = I_L X_L$.

The ωL term plays the role of resistance and is called inductive reactance X_L.

41-2 A $120\,\text{mH}$ inductor is connected to a variable frequency ac source with maximum voltage $10\,\text{V}$. What is the inductive reactance at 100 Hz, and 1.0 MHz?

Solution:

$$X_L\big|_{100} = \omega L = 2\pi\cdot 100\,\text{Hz}\cdot 0.12\,\text{H} = 75\,\Omega$$

$$X_L\big|_{1.0\text{M}} = 2\pi\cdot 1.0\times 10^6\,\text{Hz}\cdot 0.12\,\text{H} = 7.5\times 10^5\,\Omega$$

The basic relation $v = L(di/dt)$ shows that $[\text{H} = \text{V}\cdot\text{s/A}]$ making X_L have the units of Ω.

41-3 At what frequency do a $65\,\text{mH}$ inductor and a $20\,\mu\text{F}$ capacitor have the same reactance?

Solution: When $X_L = X_C$, or $\omega L = \dfrac{1}{\omega C}$, The frequency then is

$$\omega = \frac{1}{\sqrt{LC}} \quad \text{or} \quad f = \frac{1}{2\pi\sqrt{LC}} \qquad \textbf{(41-1)}$$

This is the condition for resonance or oscillation in the L-C circuit!

$$f = \frac{1}{2\pi\sqrt{65\times 10^{-3}\,\text{H}\cdot 20\times 10^{-6}\,\text{F}}} = 140\,\text{Hz}$$

41-4 An ac source of $100\,\text{Hz}$ and maximum voltage of $20\,\text{V}$ is connected to a $70\,\text{mH}$ inductor. What is the maximum current? When the current is maximum, what is the voltage of the source?

Solution: First calculate the inductive reactance $X_L = \omega L = 2\pi\cdot 100\,\text{Hz}\cdot 0.070\,\text{H} = 44\,\Omega$.

The maximum current is $I_L\big|_{\max} = V_L/X_L = 20\text{V}/44\Omega = 0.45\text{A}$.

Look at the graphs in Fig. 41-7 and note that when the current is a maximum, the voltage is zero.

41-5 For the circuit described in problem 41-4, what is the current when the source voltage is 12 V and increasing?

Solution: Look at the graphs of voltage and current, and voltage versus time as shown in Fig. 41-8 and note where on the graph the voltage is 12 V and increasing. This point (12 V and increasing) is between 270° and 360°.

The curve between 270° and 360° is a sine curve so the angle $\beta = \sin^{-1}(12/20) = 37^o$.

Therefore the point where the voltage is 12 V and increasing is $270^o + 37^o = 307^o$.

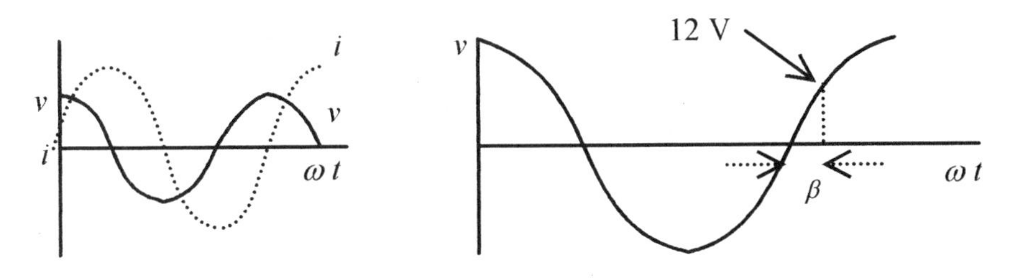

Fig. 41-8

The current at this point is $i_L\big|_{12\text{V}+} = I_L \sin\omega t = (0.45\text{A})\sin 307^o = -0.36\text{A}$.

This solution using the graphs may appear a bit cumbersome to you. The visual aspect of using the graphs, however, is a considerable help in keeping track of where you are as you proceed through the problem. It is easy to make the mistake of finding the arcsin of 12/20 and putting 37° into the current equation, completely missing that this is the point where i is decreasing, not increasing. After you have done a few problems like this you may be able to shorten the procedure.

The R-L-C Circuit

The phasor diagrams are most helpful in understanding R-L-C circuits as shown in Fig. 41-1. There are two important points to keep in mind in the analysis of these circuits. First, the sum of the instantaneous voltages must equal the source voltage $V\cos\omega t = v_R + v_L + v_C$. Second, since there is only one current path, the current is everywhere the same. Voltages on the various components have different phase relationships, but the current is the same everywhere in the circuit. The phasor diagram for a typical R-L-C circuit is shown in Fig. 41-9. Do not try to take this in all at once. Follow along the steps in the construction of the diagram.

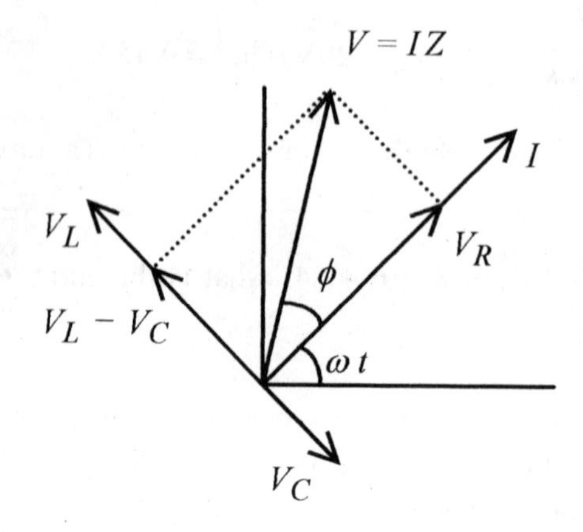

Fig. 41-9

Place the I phasor at the arbitrary angle ωt.

Place the V_R phasor over I. The voltage and current in the resistor are in phase.

Add $V_L = I\,X_L$ leading V_R by 90°.

Add $V_C = I\,X_C$ lagging V_R by 90°.

On an axis perpendicular to V_R and I, V_L and V_C, add in a vector manner to produce $V_L - V_C$. In this example $V_L > V_C$.

If the load is resistive, voltage is in phase with current. If the load is entirely inductive or entirely capacitive the voltage is 90° out of phase with current. In this situation, with all elements present, the voltage is the vector-like sum of V_R and $V_L - V_C$.

In equation form

$$V^2 = V_R^2 + (V_L - V_C)^2 = I^2 R^2 + (I\,X_L - I\,X_C)^2 = I^2\left[R^2 + (X_L - X_C)^2\right] \tag{41-2}$$

or

$$V = I\sqrt{R^2 + (X_L - X_C)^2}$$

This suggests another resistance-like expression

$$Z = \sqrt{R^2 + (X_L - X_C)^2} \tag{41-3}$$

which is called impedance. We now have the numeric relations between voltage, current , and the vaues of R, L, and C.

The phase relation between V and I is seen from the phasor diagram as

$$\tan\phi = \frac{V_L - V_C}{V_R} \qquad \textbf{(41-4)}$$

To obtain a better picture of what is going on here imagine measuring the ac voltages of the source, resistor, capacitor, and inductor, and the current in the circuit. The voltages across the resistor, capacitor, and inductor do not add up to the source voltage! They are not in phase! These voltages will satisfy equation 41-2. The source voltage divided by the impedance, equation 41-3, will equal the current. Finally the phase angle between the source voltage and current comes from equation 41-4.

41-6 An R-L-C circuit with $R = 200\Omega$, $L = 0.40\,\mathrm{H}$, and $\mathrm{C} = 3.0\,\mu\mathrm{F}$ is driven by an ac source of $20\,\mathrm{V}$ maximum and frequency $100\,\mathrm{Hz}$. Find the reactances, impedance, and maximum current and voltages across each of the components.

Solution: R-L-C circuit problems can be confusing. The key to successfully solving them is to follow a logical path through the problem. The current is everywhere the same and is determined by the source voltage and the impedance. The impedance is determined by the resistance and the reactances, and the reactances are frequency dependent. As you proceed through this problem be aware of the logic in the calculations. The schematic of the circuit is shown in Fig. 41-1.

$$X_L = \omega L = 2\pi \cdot 100\,\mathrm{Hz} \cdot 0.40\,\mathrm{H} = 251\Omega$$

$$X_C = \frac{1}{\omega C} = \frac{10^6}{2\pi \cdot 100\,\mathrm{Hz} \cdot 3.0\,\mathrm{F}} = 530\Omega$$

$$(X_L - X_C)^2 = (251\Omega - 530\Omega)^2 = 77800\Omega^2$$

$$Z = \sqrt{R^2 + (X_L - X_C)^2} = \sqrt{(200)^2 + 77800}\,\Omega = 343\Omega$$

$$I = V/Z = 20\,\mathrm{V}/343\Omega = 0.058\,\mathrm{A}$$

$$V_R = 0.058\,\mathrm{A} \cdot 200\Omega = 11.6\,\mathrm{V}$$

$$V_L = 0.058\,\mathrm{A} \cdot 251\Omega = 14.6\,\mathrm{V}$$

$$V_C = 0.058\,\mathrm{A} \cdot 530\Omega = 30.7\,\mathrm{V}$$

41-7 For the circuit described in problem 41-6 construct the phasor diagram and find the angle of the source voltage with respect to the current.

Solution: The phasor diagram is shown in Fig. 41-10. Again, follow the logic in the construction

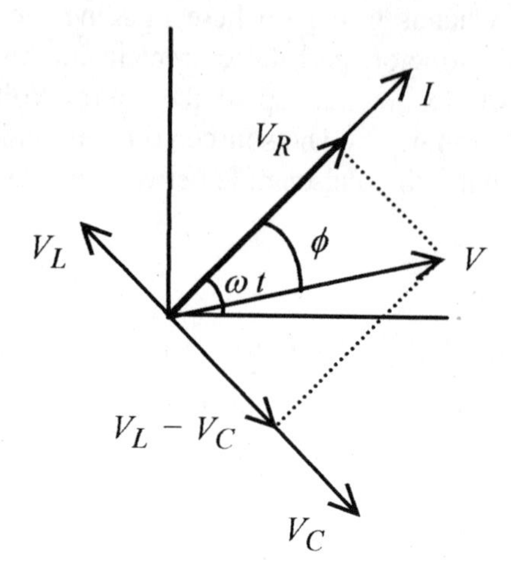

Fig 41-10

of the diagram. Draw the phasor representing $I = 0.058\,\mathrm{A}$ at an arbitrary angle ωt. Draw the phasor representing $V_R = 11.6\,\mathrm{V}$ in the same direction as I. Draw $V_L = 14.6\,\mathrm{V}$ leading V_R by 90°. Add $V_C = 30.7\,\mathrm{V}$ lagging V_R by 90°. Complete the rectangle with side $V_L - V_C$ and V_R. Since $V_C > V_L$, this vector points in the same direction as V_C. The source voltage V is the diagonal of this rectangle. The phase angle is determined from equation 41-4.

$$\tan\phi = \frac{V_L - V_C}{V_R} = \frac{14.6 - 30.7}{11.6} \quad \text{or} \quad \phi = -54^o$$

The load is resistive and capacitive and (from the diagram) the source voltage lags the current by 54°.

After you are clear on the calculations go back over these last two problems and concentrate on the logic. The biggest pitfall in R-L-C circuit problems is losing your way!

41-8 An R-L-C circuit consists of a $300\,\Omega$ resistor, $0.15\,\mathrm{H}$ inductor, and $4.5\,\mu\mathrm{F}$ capacitor driven by an ac source of 800 rad/s. Find the phase angle of the source voltage with respect to the current.

Solution: Notice that the maximum voltage is not given in the problem, yet a phase diagram is to be drawn. The problem is written this way to emphasize that it is not necessary to know the maximum source voltage to find the phase angle. The phase diagrams so far have used voltage.

The voltage across each component is the resistance or reactance times a constant, the current. Thus the diagram could as well be drawn with resistance and reactances as well as voltage.

Calculate the reactances: $X_L = \omega L = 800\,\text{rad/s} \cdot 0.15\,\text{H} = 120\,\Omega$

$$X_C = \frac{1}{\omega C} = \frac{1}{(800\,\text{rad/s})4.5 \times 10^{-6}\,\text{F}} = 278\,\Omega$$

Now draw the phasor diagram as shown in Fig. 41-11. Draw I at some arbitrary angle ωt.

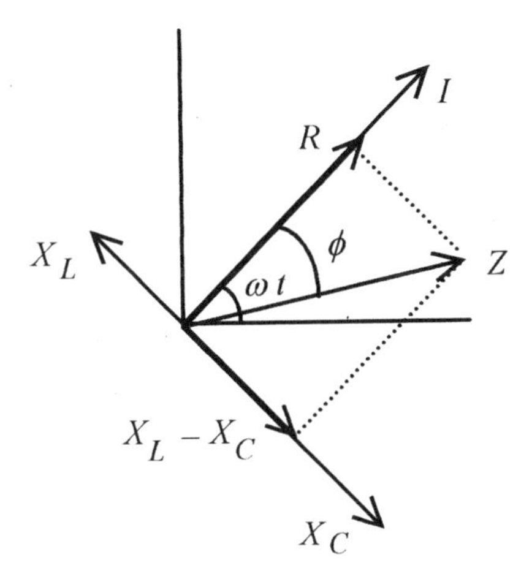

Fig. 41-11

Draw R along I proportional to 300 Ω. Draw X_L leading R by 90° proportional to 120 Ω. Draw X_C lagging R by 90° proportional to 278 Ω. Draw $X_L - X_C = -158\ \Omega$. Complete the rectangle formed by R and $X_L - X_C$ and draw the diagonal, which is proportional to Z. The phase angle is

$$\tan\phi = \frac{X_L - X_C}{R} = \frac{-158}{300} \quad \text{or} \quad \phi = -28^o$$

The load is resistive and capacitive, and the phase angle is -28° with the source voltage lagging the current.

Power in ac Circuits

The instantaneous power in the resistive circuit of Fig. 41-2 is

$$p = vi = V\cos\omega t\, I\cos\omega t = VI\cos^2\omega t$$

The average power is the average value of the $\cos^2$ function over one cycle. The $\sin^2$ function and $\cos^2$ functions have the same shape (area under the curve), and $\sin^2\theta + \cos^2\theta = 1$. The only way for $\sin^2\theta$ to equal $\cos^2\theta$ and their sum to equal 1 is for $\cos^2\theta$ to equal $1/2$. Therefore the average value of the cosine squared function over one cycle is $1/2$, and the average power is

$$P = \frac{VI}{2} = \frac{V}{\sqrt{2}}\frac{I}{\sqrt{2}}$$

The most convenient associations are shown as $V/\sqrt{2}$ and $I/\sqrt{2}$.

These values of $V/\sqrt{2}$ and $I/\sqrt{2}$ used to compute the average power are equivalent to V and I used to compute power in a dc circuit. DC voltmeters and ammeters measure V and I with the product being power, $P = VI$. AC voltmeters and ammeters must measure $V/\sqrt{2}$ and $I/\sqrt{2}$, the time average of these quantities, so power calculations in ac and dc will be the same. The $V/\sqrt{2}$ and $I/\sqrt{2}$ measurements are called the rms (root mean square) values of voltage and current.

$$V_{rms} = \frac{V_{max}}{\sqrt{2}} \qquad I_{rms} = \frac{I_{max}}{\sqrt{2}}$$

41-9 For the circuit described in problem 41-6 calculate the power loss in each of the components.

Solution: There is no energy loss in an inductor. Energy goes in to build up the magnetic field and is released in the collapse of the magnetic field. Look at the curves of Fig. 41-7. The product vi averaged over one cycle adds to zero. A similar argument can be made for a capacitor.

For the resistor the voltage is $v = (11.6\,\text{V})\cos\omega t$ and the current is $i = (0.058\,\text{A})\cos\omega t$, so the power is $p = 11.6 \cdot 0.058\,\text{W}\cos^2\omega t$. The average power is

$$P = 11.6 \cdot 0.058(1/2)\text{W} = 0.34\,\text{W}$$

Using the rms voltages and currents

$$P = \frac{11.6\,\text{V}}{\sqrt{2}}\frac{0.058\,\text{A}}{\sqrt{2}} = 0.34\,\text{W}$$

Transformers

A transformer consists of two coils (called primary and secondary) wound one over the other with usually a soft iron core to enhance the magnetic field or an arrangement with two coils wound on a soft iron core as illustrated in Fig. 41-12. Based on Faraday's law the voltage induced in a coil is proportional to the number of windings.

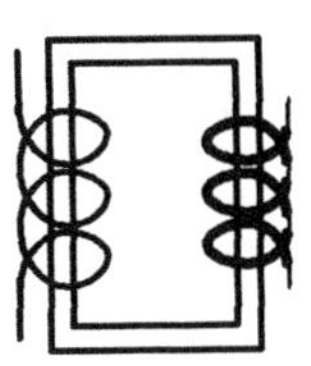

Fig. 41-12

$$\frac{V_p}{N_p} = \frac{V_s}{N_s}$$

By varying the relative number of windings we can make either a step-up or step-down (voltage) transformer. The relative currents in the primary and secondary are determined with a simple statement that the power in equals the power out.

$$V_p I_p = V_s I_s = \frac{N_s}{N_p} V_p I_s \quad \text{or} \quad I_p = \frac{N_s}{N_p} I_s$$

41-10 A power distribution transformer steps voltage down from 8.5kV to 120V (both rms values). If the 120V side of the transformer supplies 500A to a resistive load, what current is taken from the primary (high) side of the transformer and what is the turns ratio?

Solution: First find the turns ratio $\frac{8.5\,\text{kV}}{N_p} = \frac{120\,\text{V}}{N_s}$ or $\frac{N_p}{N_s} = \frac{8500}{120} = 71.$

The current is $I_p = \frac{N_s}{N_p} I_s = \frac{1}{71} 500\,\text{A} = 7.0\,\text{A}.$

CHAPTER 42

MAXWELL'S EQUATIONS

The discussion of Maxwell's equations starts with four basic laws of electromagnetism, each of which has been described in a previous chapter. These laws can be described in words, but a full appreciation of them requires a good calculus background. Whether you have sufficient calculus background to fully appreciate the scope of these equations or not, our discussion, which will use both words and equations, should be helpful to you.

Four Basic Laws of Electromagnetism

The first basic law is Gauss' law for electricity.

$$\varepsilon_o \oint \boldsymbol{E} \cdot d\boldsymbol{s} = q \qquad \textbf{(42-1)}$$

This law states that the summation of the electric vector over a closed surface times the vector representing that surface must add up to the total charge contained. Visualize a sphere, or irregular volume if you like, containing charge. On each element of this surface there is an electric field vector and a vector $d\boldsymbol{s}$ normal to the surface representing the element of surface. The dot product of these two vectors summed over the surface must equal the charge contained.

The second basic law is Gauss' law for magnetism.

$$\oint \boldsymbol{B} \cdot d\boldsymbol{s} = 0 \qquad \textbf{(42-2)}$$

The integral in this law is the same form as Gauss' law for electricity. Again visualize a surface but with $\boldsymbol{B}$ representing the magnetic field on the surface with the dot product of $\boldsymbol{B}$ and $d\boldsymbol{s}$ being summed over the surface. In this case, however, the summation (integral) is equal to zero!

The third basic law is Faraday's law of induction.

$$\oint \boldsymbol{E} \cdot d\boldsymbol{l} = -\frac{d\Phi_B}{dt} \qquad \textbf{(42-3)}$$

This law states that magnetic flux, this construct visualized as magnetic field lines, changing with time through a closed area equals the sum of the dot product of the electric field and the vector representing the loop (around the area). Visualize a loop of wire normal to a changing magnetic field. An ***emf*** and current is induced in the wire when the flux (of magnetic field) through the loop changes.

The fourth basic law is Ampere's law.

$$\oint \boldsymbol{B} \cdot d\boldsymbol{l} = \mu_o i \qquad \textbf{(42-4)}$$

The sum of the dot product of the magnetic field and a vector representing an element of a loop around the current is equal to the current contained. A simple visual picture is a wire carrying a current with a circle symmetric about the wire and in a plane normal to the current. The field is constant on the circle, so this integral reduces to the constant field times the length of the circle. This integral is proportional to the current.

These four equations are the basic laws of electromagnetism. They are unique among electric laws in that they are valid for both stationary and rapidly moving charges. Now let's take a look at these equations from a symmetry point of view remembering that ε_o and μ_o operationally serve only to define a system of units.

Notice that equations 42-1 and 42-2 are symmetric in that they are surface integrals, but the first is equal to charge and the second to zero. This difference implies that there are discrete charges but no discrete magnetic poles. Equation 42-4 describes a current of charges, but there is no corresponding current of magnetic monopoles evident in equation 42-3. This is as we would expect, since based on equation 42-2 there are no magnetic monopoles.

Looking at equations 42-3 and 42-4 another asymmetry is the lack of a time varying electric field term in equation 42-4. If there are no magnetic monopoles then we expect no term corresponding to a flow of monopoles in equation 42-3, but the question of a time varying electric field term in equation 42-4 is not so easily dismissed. Time varying electric fields do exist.

Look at the consequence of a $d\Phi_E/dt$ term in equation 42-4. To be dimensionally correct $d\Phi_E/dt$ would have to be multiplied by a constant with the same units as $\mu_o\varepsilon_o$. Note that $\varepsilon_o(d\Phi_E/dt)$ has the units of current. Adding the term $\mu_o\varepsilon_o(d\Phi_E/dt)$ produces Ampere's law as modified by Maxwell.

$$\oint \boldsymbol{B} \cdot d\boldsymbol{l} = \mu_o i + \mu_o \varepsilon_o \frac{d\Phi_E}{dt} \qquad \textbf{(42-5)}$$

This completes our discussion leading to Maxwell's equations as summarized below.

$$\varepsilon_o \oint \boldsymbol{E} \cdot d\boldsymbol{s} = q$$

$$\oint \boldsymbol{B} \cdot d\boldsymbol{s} = 0$$

$$\oint E \cdot dl = -\frac{d\Phi_B}{dt}$$

$$\oint \mathbf{B} \cdot dl = \mu_o i + \mu_o \varepsilon_o \frac{d\Phi_E}{dt}$$

The Displacement Current

The most convenient way to study the time varying electric field term is with a parallel plate capacitor. Figure 42-1 shows a parallel plate capacitor with current to the plate and a growing electric field producing a magnetic field. Note the direction of the magnetic field for the displacement current is the same as for the physical (or real) current.

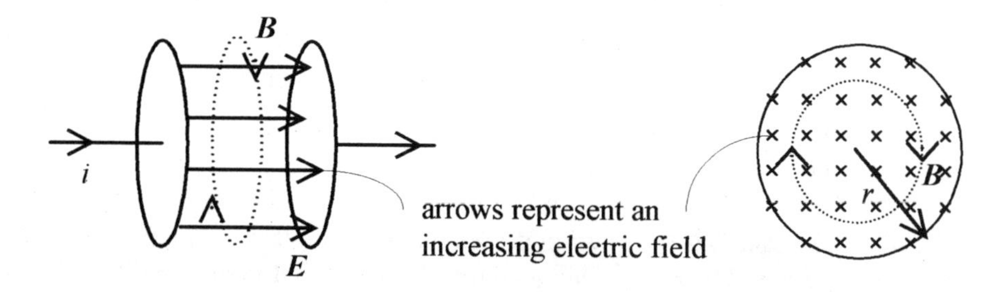

Fig. 42-1

For the connecting wire, Ampere's law is valid, $\oint \mathbf{B} \cdot dl = \mu_o i$.

For the space between the parallel plates the physical current is zero, so the Maxwell added term in Ampere's law is valid and $\oint \mathbf{B} \cdot dl = \mu_o \varepsilon_o \frac{d\Phi_E}{dt}$.

Both integrals can be evaluated over a circular path with radius equal to the radius of the parallel plates. $\oint \mathbf{B} \cdot dl = B(2\pi r)$. The electric flux is simply EA, so $\mu_o \varepsilon_o \frac{d\Phi_E}{dt} = \mu_o \varepsilon_o A \frac{dE}{dt}$ and equating the physical current to the displacement current

$$B(2\pi r) = \mu_o \varepsilon_o (\pi r^2) \frac{dE}{dt} \quad \text{or} \quad B = \mu_o \varepsilon_o \frac{r}{2} \frac{dE}{dt}$$

This is the magnetic field at the edge of the capacitor. This is a small field and a transient one. The dE/dt cannot be increased rapidly for a long time! A magnetic field due to this "additional term" in Ampere's law was discovered over 60 years after it was first predicted by Maxwell.

The quantity $\varepsilon_o \dfrac{d\Phi_E}{dt} = \varepsilon_o A \dfrac{dE}{dt} = i_d$ is known as the **displacement current** and the Ampere-Maxwell equation is sometimes written as

$$\oint \boldsymbol{B} \cdot d\boldsymbol{l} = \mu_o(i + i_d) \qquad \textbf{(42-6)}$$

In the case of the parallel plate capacitor the displacement current is equal to the physical current to the plates.

42-1 Find the displacement current and the magnetic field in a parallel plate capacitor with circular plates of radius $5.0\,\text{cm}$ at an instant when E is changing at $1.0 \times 10^{12}\,\text{V/m}\cdot\text{s}$.

Solution: The displacement current is

$$i_d = \varepsilon_o A \frac{dE}{dt} = 8.8 \times 10^{-12}\,\text{C}^2/\text{N}\cdot\text{m}^2(\pi \cdot 25 \times 10^{-4}\,\text{m}^2)1.0 \times 10^{12}\,\text{V/m}\cdot\text{s} = 0.070\,\text{A}$$

The magnetic field is

$$B = \mu_o \varepsilon_o \frac{r}{2}\frac{dE}{dt} = (4\pi \times 10^{-7}\,\text{T}\cdot\text{m/A})(8.8 \times 10^{-12}\,\text{C}^2/\text{N}\cdot\text{m}^2)(2.5 \times 10^{-2}\,\text{m})(1.0 \times 10^{12}\,\text{V/m}\cdot\text{s}) = 2.8 \times 10^{-7}\,\text{T}$$

42-2 Find the maximum magnetic field at the edge of a parallel plate capacitor with radius $6.0\,\text{cm}$ and separation $2.0\,\text{cm}$ connected to a sinusoidal voltage source of $1000\,\text{Hz}$ and $200\,\text{V}$ maximum.

Solution: The basic equation is $B = \mu_o \varepsilon_o \dfrac{r}{2}\dfrac{dE}{dt}$, but for this case $\dfrac{dE}{dt} = \dfrac{1}{d}\dfrac{dV}{dt}$ so

$$B\big|_{max} = \mu_o \varepsilon_o \frac{r}{2}\frac{1}{d}\frac{dV}{dt}\bigg|_{max}$$

The voltage is $v = V_o \sin\omega t$, and $\dfrac{dV}{dt} = V_o \omega \cos\omega t$ so $\dfrac{dV}{dt}\bigg|_{max} = V_o \omega$, and

$$B\big|_{max} = \mu_o \varepsilon_o \frac{r}{2d} V_o \omega = (4\pi \times 10^{-7})(8.8 \times 10^{-12})\frac{6.0}{2 \cdot 2.0}(200)(2\pi \cdot 1000)\,\text{T} = 2.1 \times 10^{-11}\,\text{T}$$

CHAPTER 43

ELECTROMAGNETIC WAVES

This chapter, especially in the middle part, is very calculus intensive. Depending on your course you may want to be selective of the sections you read. If your course does not deal with the derivations of the speed and energy transport in an electromagnetic wave, you may want to pass over or skim the sections on "The Speed of the Wave" and "Energy Transport in Electromagnetic Waves." While the understanding of how electromagnetic waves are propagated is fairly complicated, many of the problems are quite simple. Because the understanding of how waves propagate is so important, and fairly difficult to understand, we will develop the ideas underpinning electromagnetic wave propagation in some detail.

Generating Waves

There is a broad spectrum of electromagnetic waves generated in a variety of ways. For convenience in discussion we will, however, concentrate on waves generated from an oscillating L-C circuit. The circuit shown in Fig. 43-1 is an L-C circuit with some additions.

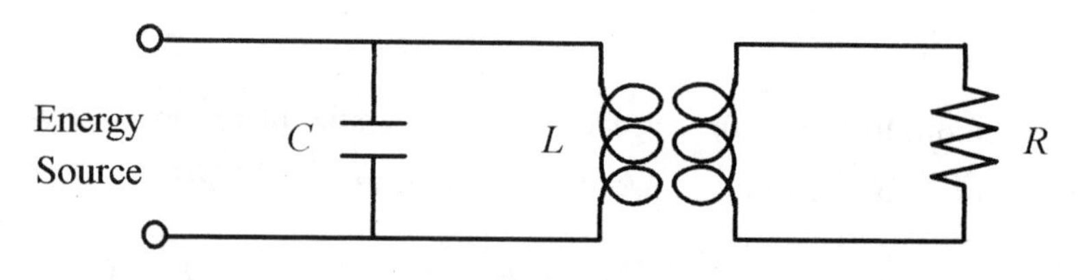

Fig. 43-1

The L-C circuit has components that produce resonance with a frequency suitable for a radio transmitter (a typical radio frequency of 300 kHz corresponds to a wavelength of 1.0 m). The energy source keeps the circuit oscillating. The inductor consists of a primary and secondary, with the secondary connected to a resistor. At the natural resonance frequency of the L-C circuit, the voltage across the resistor is sinusoidal. If a wire (antenna) is substituted for this resistor, then a sinusoidal voltage appears across the antenna. This is not a complete description of a radio transmitter; it is intended only to illustrate that it is possible to generate a sinusoidal voltage and current in a wire (antenna) of finite resistance.

"Can the oscillating electric field and current be detected a distance away from the antenna?" "Of course they can." A current in a wire produces a magnetic field, and a changing current produces a changing magnetic field (Ampere's law). A changing magnetic field means flux is changing in space, and according to Faraday's law, changing magnetic flux produces an electric field.

"Do these oscillating fields proceed out as waves?" "Yes, they do." If you were floating on an inflatable raft in a pool and you sensed a sinusoidal up and down motion of the raft, you would say a (water) wave passed by. If you were viewing a small length of stretched rope and the small segment you were viewing executed a sinusoidal up and down motion, you would say a sine wave traveled down the rope.

Likewise if you observed a sinusoidally varying electric field accompanied at right angles by a sinusoidally varying magnetic field, you would say an electromagnetic wave passed by.

Electromagnetic waves are on firm theoretical ground. The above discussion presents only a plausible argument for the existence of electromagnetic waves.

The Speed of the Waves

Start with an antenna with oscillating current. Any wave that propagates out from the antenna has velocity c.

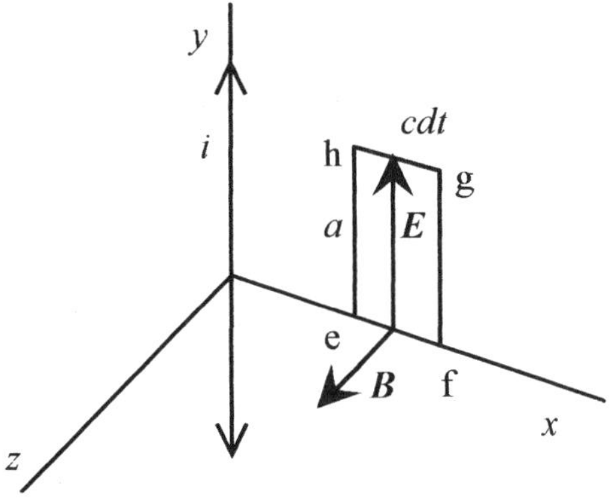

Fig. 43-2

Consider some point along x, an axis oriented radially out from y, the direction of the oscillating current. The oscillating current produces a sinusoidally varying magnetic field at right angles to x and in the plane (x-z) normal to y. Construct the rectangle e-f-g-h with height a and width cdt. This is a "snapshot in space" of the wave passing by the fixed point on x. When the current oscillations start, the oscillating magnetic field progresses in the positive x direction taking a finite amount of time to reach the place where the rectangle is constructed. At the instant of this "snapshot" the leading edge of the wave, as manifest by the $\boldsymbol{B}$ vector, has progressed nearly all of the way through the rectangle.

Now apply Faraday's law $\oint \boldsymbol{E} \cdot d\boldsymbol{l} = -(d\Phi_B/dt)$ to this rectangle.

The differential change in flux over this rectangle is $d\Phi_B = Bacdt$, so $(d\Phi_B/dt) = Bac$.

Faraday's law, $\oint \boldsymbol{E}\cdot d\boldsymbol{l}$, requires $\boldsymbol{E}\cdot d\boldsymbol{l}$ be integrated around a region where Φ_B is changing.

In the "Time Varying Magnetic Fields" section of Chapter 37, Faraday's Law, the area was bounded by a circle. In this situation a rectangle is more convenient. Along e-f and g-h, $\boldsymbol{E}$ is perpendicular to $d\boldsymbol{l}$ and along f-g, $\boldsymbol{E}$ is zero (remember the wave front is passing through the rectangular region), so the only contribution is from h-e. If, as is shown, $\boldsymbol{B}$ is in the positive x direction, $\boldsymbol{E}$ must be in the positive y direction. Positive E along h-e will produce a clockwise current around the rectangle and a magnetic field in the negative z-direction within the rectangle satisfying Lenz's law. This verifies the relationship of $\boldsymbol{E}$ and $\boldsymbol{B}$ as shown in Fig. 43-2. Integrating counterclockwise around the rectangle, the only contribution to the integral is $-Ea$. Faraday's law applied to this rectangle yields

$$-Ea = -Bac \quad \text{or} \quad E = cB \tag{43-1}$$

The speed of the wave is the ratio of the oscillating electric and magnetic fields.

43-1 A radio signal at a certain point has a measured maximum electric field of $5.0\times10^{-3}\ \text{V/m}$. What is the maximum magnetic field?

Solution: The magnetic field maximum is at right angles to the electric field and has magnitude

$$B = \frac{E}{c} = \frac{5.0\times10^{-3}\ \text{V/m}}{3.0\times10^{8}\ \text{m/s}} = 1.7\times10^{-11}\,\text{T}$$

Now apply Ampere's law using only the displacement term. $\oint \boldsymbol{B}\cdot d\boldsymbol{l} = \mu_o\varepsilon_o\dfrac{d\Phi_E}{dt}$

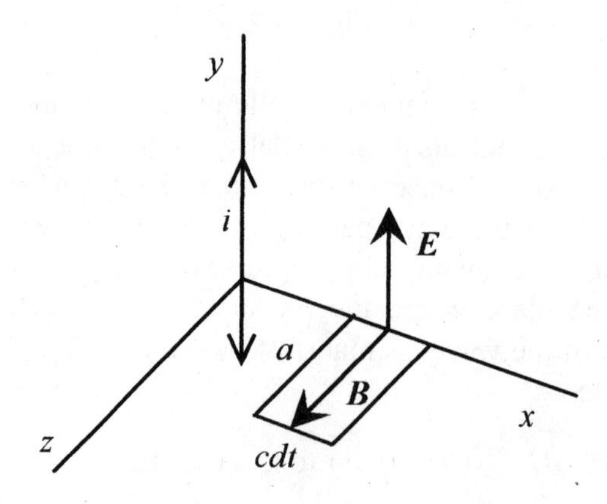

Fig. 43-3

Figure 43-3 shows the relationship of $\boldsymbol{E}$ and $\boldsymbol{B}$. The procedure is much the same as before. The only contribution to the integral is along one side of the rectangle, so the integral becomes Ba (again integrating counterclockwise).

The differential flux is $d\Phi_E = Eacdt$.

Ampere's law then is $Ba = \mu_o \varepsilon_o Eac$, or $B = \mu_o \varepsilon_o Ec$.

Substituting $E = cB$, $B = \mu_o \varepsilon_o c^2 B$, so $\mu_o \varepsilon_o = 1/c^2$, or

$$c = \frac{1}{\sqrt{\mu_o \varepsilon_o}} \qquad \textbf{(43-2)}$$

The velocity of electromagnetic waves (including light) is derivable from static constants!

There are three conclusions to this short discussion.

1. The electric and magnetic fields are related by the velocity; $E = cB$.
2. The velocity is $1/\sqrt{\mu_o \varepsilon_o}$.
3. The wave moves in a direction determined by $\boldsymbol{E} \times \boldsymbol{B}$.

43-2 How long does it take for an electromagnetic wave to travel to the moon, 3.8×10^8 m away?

Solution:
$$t = \frac{d}{c} = \frac{3.8 \times 10^8\,\text{m}}{3.0 \times 10^8\,\text{m/s}} = 1.3\,s$$

43-3 How far does light travel in 1.0 ns $(1.0 \times 10^{-9}\,\text{s})$?

Solution: $d = ct = (3.0 \times 10^8\,\text{m/s})(1.0 \times 10^{-9}\,\text{s}) = 0.30\,\text{m} \approx 1\,\text{ft}$

Energy Transport in Electromagnetic Waves

When electric and magnetic fields are absorbed in material objects, energy is transferred from the wave to the material. This is true for radio waves impinging on antennas, infrared waves absorbed in water molecules, or microwaves into your lunch.

The discussion of energy transport starts with the energy density expressions from static electric and magnetic fields. (See the "Energy Storage" section in Chapter 28, Capacitance, and the "Power and Energy Storage" section in Chapter 38, Inductance.) The total energy density for a region of space where $\boldsymbol{E}$ and $\boldsymbol{B}$ fields are present is

$$u = \frac{\varepsilon_o E^2}{2} + \frac{B^2}{2\mu_o} \qquad \textbf{(43-3)}$$

In the electromagnetic wave $E = cB$ and $E^2 = c^2 B^2$ so $B^2 = \mu_o \varepsilon_o E^2$ allowing rewriting

$$u = \frac{\varepsilon_o E^2}{2} + \frac{\varepsilon_o E^2}{2} = \varepsilon_o E^2 \qquad \textbf{(43-4)}$$

This last expression shows that while the ***B*** field is numerically smaller than the ***E*** field, the energy density associated with both fields is the same.

Referring to Fig. 43-4, the differential amount of energy in the volume with cross-sectional area A and length cdt is

$$dU = udV = \varepsilon_o E^2 Acdt$$

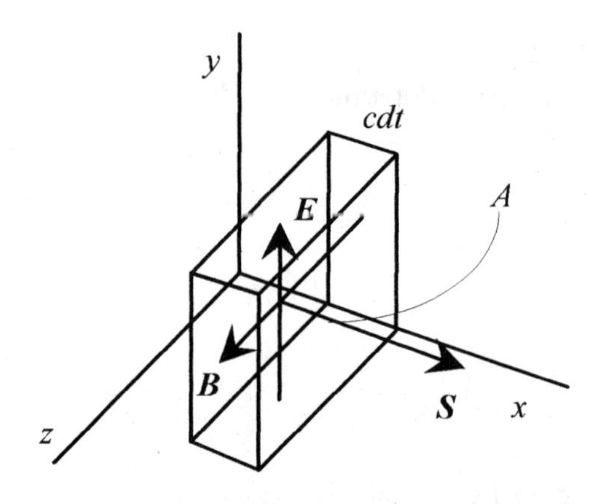

Fig. 43-4

the expression dU/dt is the energy transported per time over the area A. The more useful expression is the energy flowing across an area per unit of time

$$S = \frac{1}{A}\frac{dU}{dt} = \varepsilon_o c E^2 = \sqrt{\frac{\varepsilon_o}{\mu_o}} E^2 = \frac{EB}{\mu_o} \qquad \textbf{(43-5)}$$

The alternative forms come from $E = cB$ and $c = 1/\sqrt{\mu_o \varepsilon_o}$. The units of S are energy per unit time per unit area or power per unit area.

The Poynting Vector

The magnitude and direction of the energy flow rate is defined by the **Poynting Vector**.

$$\boldsymbol{S} = \frac{1}{\mu_o} \boldsymbol{E} \times \boldsymbol{B} \qquad \textbf{(43-6)}$$

This vector, like the flow rate, varies with time. The average value of the magnitude of the Poynting vector is the **intensity**. The average value of the magnitude of $\boldsymbol{S}$ is

$$S = \frac{EB}{\mu_o} = \frac{E_{max} B_{max}}{\mu_o} \sin^2 \omega t$$

The average value of $\sin^2 \omega t$ is $1/2$, so

$$S = \frac{E_{max} B_{max}}{2\mu_o} = \frac{E_{max}^2}{2\mu_o c} = \frac{1}{2}\sqrt{\frac{\varepsilon_o}{\mu_o}} E_{max}^2 = \frac{1}{2}\varepsilon_o c E_{max}^2 \qquad \textbf{(43-7)}$$

43-4 Take the intensity of sunlight at the earth's surface as $600\,\mathrm{W/m^2}$. Assuming 100% collection efficiency, how much energy is collected on a $1.0\,\mathrm{m^2}$ panel exposed to this amount of sunlight for 10 hours?

Solution: The key to getting problems involving energy, power, and intensity correct is to watch the units closely. In this case see how the units dictate how to make the calculation.

$$E = \frac{600\,\mathrm{W}}{\mathrm{m^2}} \frac{\mathrm{J/s}}{\mathrm{W}} \cdot 1.0\,\mathrm{m^2} \cdot 10\,\mathrm{hr} \frac{60\,\mathrm{min}}{\mathrm{hr}} \frac{60\,\mathrm{s}}{\mathrm{min}} = 2.2 \times 10^7\,\mathrm{J}$$

Another way to do this problem is to find the energy in $\mathrm{W \cdot hr}$ rather than Joules.

$$E = \frac{600\,\mathrm{W}}{\mathrm{m^2}} 1.0\,\mathrm{m^2} \cdot 10\,\mathrm{hr} = 6.0\,\mathrm{kW \cdot hr}$$

This is a strange looking unit, but it gives us a feel for the amount of energy that can be collected this way. This $6000\,\mathrm{W \cdot hr}$ would light sixty $100\,\mathrm{W}$ light bulbs for an hour or run a $1.0\,\mathrm{kW}$ microwave oven for 6 hours.

43-5 The maximum value of electric field $2.0\,\mathrm{m}$ away from a spherically symmetric source is $1.8\,\mathrm{V/m}$. Find the maximum value of magnetic field, average intensity, and power output of the source.

Solution: The magnetic field is from $B = \frac{E}{c} = \frac{1.8\,\text{V/m}}{3.0\times10^{8}\,\text{m/s}} = 6.0\times10^{-9}\,\text{T}$.

The average intensity is from

$$S = \frac{1}{2}\varepsilon_o c E_{max}^2 = \frac{1}{2} 8.8\times10^{-12}\frac{\text{C}^2}{\text{N}\cdot\text{m}^2} 3.0\times10^{8}\frac{\text{m}}{\text{s}}\left(\frac{1.8\,\text{V}}{\text{m}}\right)^2 = 4.3\times10^{-3}\frac{\text{W}}{\text{m}^2}$$

As an exercise work through the units in this problem.

The total power is the power per unit area at 2.0 m radius times the total area.

$$P = 4.3\times10^{-3}\,\text{W/m}^2\cdot 4\pi(2.0\text{m})^2 = 0.21\text{W}$$

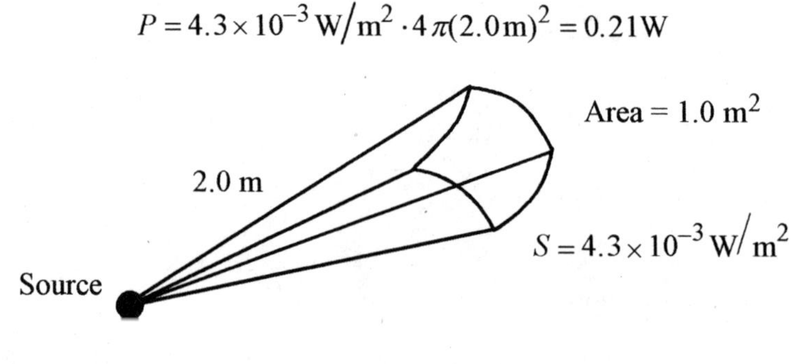

Fig. 43-5

43-6 What is the maximum electric field strength and intensity of a 500W spherically symmetric light source at 10m radius?

Solution: The area of a sphere 10m in radius is $A = 4\pi(10\text{m})^2 = 1.2\times10^3\,\text{m}^2$.
The entire 500W is delivered to this area. The power, intensity, and area are related by

$$S = \frac{P}{A} = \frac{500\,\text{W}}{1.2\times10^3\,\text{m}^2} = 0.40\,\text{W/m}^2$$

The electric field intensity comes from

$$E_{\text{max}} = [2\mu_o cS]^{1/2} = \left[2(4\pi\times10^{-7}\,\text{T}\cdot\text{m/A})(3.0\times10^{8}\,\text{m/s})(0.40\,\text{m/s})\right]^{1/2} = 17\,\text{V/m}$$

Radiation Pressure

When an electromagnetic wave is absorbed by a material, charges are accelerated by the electric field transverse to the direction of the wave. The electric and magnetic field vectors are shown in Fig. 43-4. The electric field vector is pointing in the $+y$ direction, and this is the direction a positive charge would accelerate. A charge moving in the $+y$ direction acted on by a magnetic field in the $+z$ direction has a $(\boldsymbol{v} \times \boldsymbol{B})$ force in the $+x$ direction, the direction of the wave. The Poynting vector always points in the direction the wave is traveling, so the net force on any charges in the material is in the direction of the wave.

This force can be associated with a momentum. This momentum is

$$p = \frac{U}{c}$$

This is the momentum absorbed by a non-reflecting surface. For a completely reflecting surface the momentum transfer is twice this value. Think of a tennis ball striking a surface and sticking or striking a surface and rebounding with the incident velocity.

43-7 A light beam with energy flux $20\,\mathrm{W/cm^2}$ falls normal onto a completely reflecting surface of $2.0\,\mathrm{cm^2}$. What is the force on this reflector?

Solution: The average force is $F = \Delta p/\Delta t$ and $\Delta p = 2\Delta U/c$, but $\Delta U = SA\Delta t$, so

$$F = \frac{2S}{c}A = \frac{2(20\,\mathrm{W/cm^2})}{3.0\times10^8\,\mathrm{m/s}}2.0\,\mathrm{cm^2} = 2.7\times10^{-7}\,\mathrm{N}$$

CHAPTER 44

REFLECTION, REFRACTION, AND POLARIZATION

In the study of reflection and refraction it is very convenient to depict light as rays. A light ray represents the path of a thin beam (ray) of light. This ray construct is very convenient in understanding reflection and refraction as well as images formed by mirrors and lenses.

Reflection and Refraction

When light strikes a surface such as an air-glass or air-water interface, part of the incident light is reflected and part is refracted. Reflection of light at any interface follows a very simple law. The angle of incidence equals the angel of reflection with these angles measured from the normal to the surface. Incident and reflected rays are shown in Fig. 44-1.

$$\theta_1 = \theta_1' \qquad \textbf{(44-1)}$$

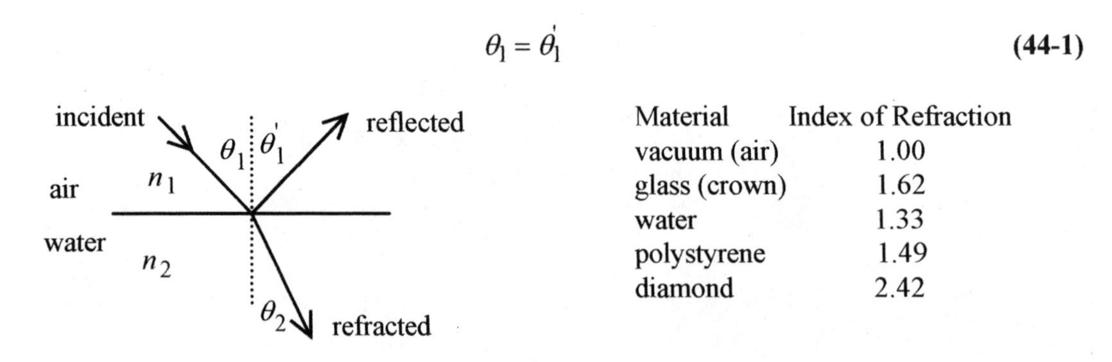

Material	Index of Refraction
vacuum (air)	1.00
glass (crown)	1.62
water	1.33
polystyrene	1.49
diamond	2.42

Fig. 44-1

44-1 A ray of light is incident on a reflecting surface at 70° with respect to the surface. What is the angle of the reflected ray?

Solution: The 70° with respect to the surface is 20° with respect to the normal so the reflected beam is 20° from the normal.

The refracted (or bent) ray is the one that enters the new medium. The angular relationship between the incident and refracted ray is given by **Snell's law**

$$n_1 \sin\theta_1 = n_2 \sin\theta_2 \qquad \textbf{(44-2)}$$

where the n_1 and n_2 are the indexes of refraction of the two media. The index of refraction is a property of the material. Figure 44-1 contains some index of refraction values.

In addition to describing refraction, the index of refraction is also the ratio of the velocity of light in vacuum to the velocity of light in the indexed medium.

$$n = c/v \qquad \textbf{(44-3)}$$

44-2 What are the angles of reflection and refraction for a light ray in air incident on glass at an angle of 35°?

Solution: The angle of incidence is measured from the normal so the angle of reflection is 35° The angle of refraction is from Snell's law $n_1 \sin\theta_1 = n_2 \sin\theta_2$

$$1.00\sin 35^o = 1.62\sin\theta_2 \quad \text{or} \quad \theta_2 = 21^o$$

In passing from a medium of lower index of refraction to one with higher index of refraction the light bends toward the normal and vice versa. This is best illustrated in problem 44-3.

44-3 A light ray in water is incident on the water-air interface at an angle of 45°. What is the angle of reflection in the water and refraction into the air?

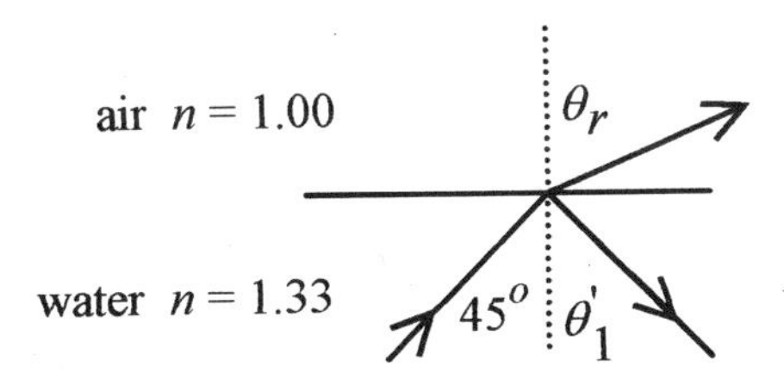

Fig 44-2

Solution: The problem is diagrammed in Fig. 44-2. The incident angle equals the reflected angle, so $\theta_1' = 45^o$. The refracted angle is from Snell's law

$$1.33\sin 45^o = 1.00\sin\theta_r \quad \text{or} \quad \theta_r = 70^o$$

This last problem suggests that there is an angle of incidence where there is no refracted ray; the refracted ray follows the surface. Beyond this **critical angle,** as it is called, all light will be reflected back into the water. Fiber-optic cable is made of light transmitting material with an index greater than one, so that light injected at the end of the cable will be internally reflected many times over its entire length.

44-4 Find the critical angle for a fiber optic material with index 1.50.

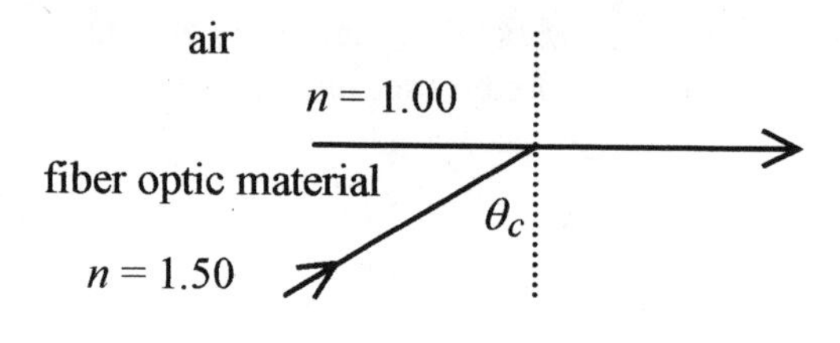

Fig 44-3

Solution: The critical angle is the one where the angle of refraction is 90°. Applying Snell's law

$$1.50\sin\theta_c = 1.00\sin 90^o \quad \text{or} \quad \theta_c = 42^o$$

Any beam that strikes the fiber-air interface at an angle greater than 42° will be internally reflected down the fiber.

Figure 44-4 shows a ray of light incident on a equilateral triangular glass prism at an angle of 54°.

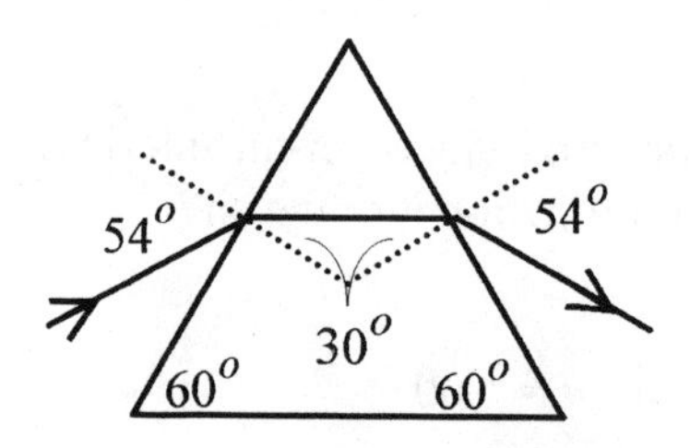

Fig. 44-4

Applying Snell's law to find the refracted ray angle yields $1.00\sin 54^o = 1.62\sin\theta_r$, or $\theta_r = 30^o$. This means that the ray travels parallel to the base inside the prism and is refracted at 54° on exit.

Now change the problem slightly by having the ray incident parallel to the base of a prism with the base angles 70°.

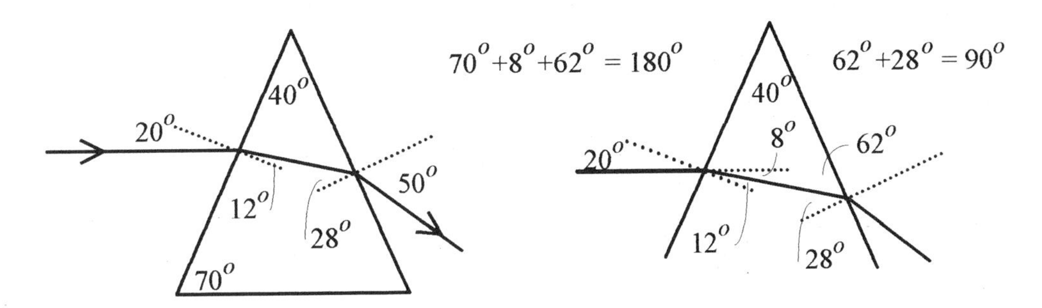

Fig. 44-5

From the geometry the angle of incidence for the ray entering the prism is 20°, and applying Snell's law $1.00 \sin 20^o = 1.62 \sin \theta_r$, $\theta_r = 12^o$. Again from the geometry, the angle of incidence for the ray leaving the prism is 28°, and applying Snell's law $1.62 \sin 28^o = 1.00 \sin \theta_r$, $\theta_r = 50^o$. A prism bends light incident parallel to its base toward the base.

If a collection of parallel rays were incident on the face of this prism they would all be refracted toward a line extending from the base. Further, if the incident and exit faces were curved, all parallel rays could be focused at one point. Two curved-face prisms placed base-to-base are a lens of sorts focusing parallel rays to a single point.

The index of refraction, or wave speed, depends slightly on the wavelength of the light. A ray of sunlight passing through a prism exhibits the **dispersion** effect, whereby different wavelength components of the sunlight are refracted at different angles producing a rainbow of colors from red to violet.

44-5 Light is incident at an angle of 35° on a slab of glass 2.0cm thick. Part of the light travels through the glass and part along its original direction beside the glass. Find the path of the two rays, the velocity of the ray in the glass, and the time difference between the two when they emerge from the opposite side of the glass. The geometry is shown in Fig. 44-6.

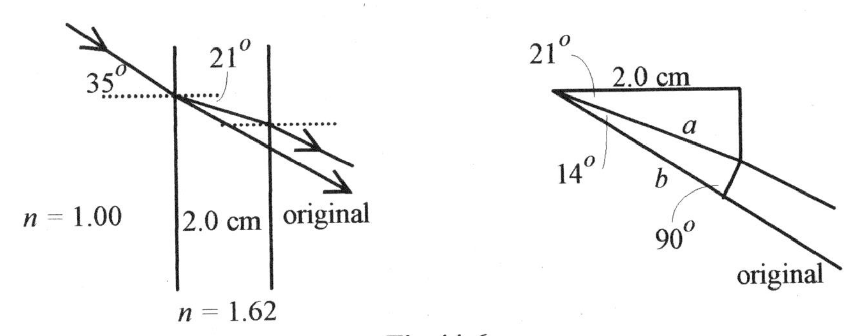

Fig 44-6

Solution: Apply Snell's law to find the path through the slab. $1.00\sin 35^o = 1.62\sin\theta_r$, or $\theta_r = 21^o$. Because of symmetry, the ray through the slab exits parallel to the original ray. The velocity of the ray in the glass is from $n = c/v$, or $v = c/n = 3.0\times 10^8\,\text{m/s}/1.62 = 1.85\times 10^8\,\text{m/s}$. Now that the velocities are known the time difference requires finding the distance the rays travel. The distance a (the path through the slab) comes from $\cos 21^o = 2.0\,\text{cm}/a$, or $a = 2.14\,\text{cm}$. Construct a line perpendicular to the two rays at the point where the refracted one leaves the slab. The distance b in the original ray direction comes from $\cos 14^o = b/2.14\,\text{cm}$, or $b = 2.08\,\text{cm}$.

The time for the refracted beam is $\tau_r = 0.0214\,\text{m}/3.0\times 10^8\,\text{m/s} = 7.1\times 10^{-11}\,\text{s}$.

The time for the unaltered beam is $\tau = 0.0208\,\text{m}/3.0\times 10^8\,\text{m/s} = 6.9\times 10^{-11}\,\text{s}$.

The time delay is $2.0\times 10^{-12}\,\text{s}$.

Polarization

Light from a light bulb or the sun is circularly polarized. This means that if you could look at the electric vectors there would be no preferred direction. The electric vectors would be randomly oriented in space. Polarization is described in terms of electric vectors rather than magnetic vectors. A radio transmitting antenna that is vertical produces waves with vertical electric vectors, and the radiation from this antenna is vertically polarized.

Polarization is usually studied in the context of light. There are two ways to polarize light. It can be polarized by passing through a material that contains molecules all oriented in the same direction that absorb (in this one direction) only the electric vector in the light. These materials are called **polaroids**. Polaroid sun glasses are made of a polaroid that absorbs the electric vector in one direction.

Light, and other electromagnetic waves are more or less polarized by reflection, depending on the angle of incidence and the index of refraction. When unpolarized or circularly polarized light is incident from air onto water the reflected beam is of lowered intensity and polarized with the electric vector parallel to the reflecting surface. The refracted beam is transmitted with a diminished intensity and is partially polarized by the reduction of the electric vector in the plane of the reflecting surface. The angle of incidence for complete polarization occurs when the angle between the reflected and refracted beams is 90°. The geometry of the situation is shown in Fig. 44-7.

The criterion for complete polarization is that $\theta_r + \theta_p = 90^o$. Snell's law requires that $n_{inc}\sin\theta_p = n_{refr}\sin\theta_r$, but $\theta_r = 90^o - \theta_p$ and $\sin\theta_r = \sin(90^o - \theta_p) = \cos\theta_p$, so $n_{inc}\sin\theta_p = n_{refr}\cos\theta_p$

or

$$\tan\theta_p = \frac{n_{refr}}{n_{inc}} \qquad \textbf{(44-4)}$$

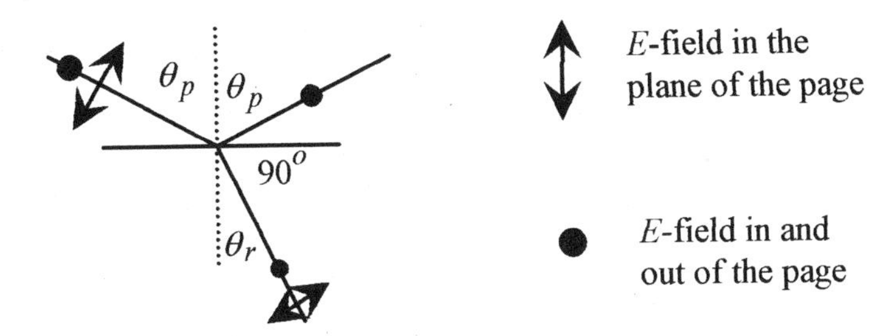

Fig. 44-7

This polarization law is known as **Brewster's law,** and the angle of polarization is called the **Brewster angle**. In working problems be careful not to invert the fraction of the indexes of refraction.

44-6 What is the polarization angle for light reflected from water?

Solution:

$$\tan\theta_p = \frac{n_{refr}}{n_{inc}} = \frac{1.33}{1} \quad \text{or} \quad \theta_p = 53^o$$

44-7 A beam of unpolarized light in water is incident on what is reported to be diamond. By means of polaroids the angle (of incidence) for complete polarization is determined to be 61°. Is this diamond?

Solution: For light in water reflected from diamond the Brewster angle should be

$$\tan\theta_p = \frac{n_{refr}}{n_{inc}} = \frac{2.42}{1.33} = 1.81 \quad \text{or} \quad \theta_p = 61^o$$

From the indexes of refraction given in Fig. 44-1 we conclude that this is diamond.

CHAPTER 45

MIRRORS AND LENSES

The subject of mirrors and lenses is difficult to treat briefly first because for a thorough discussion there needs to be considerable attention to sign convention and second because different instructors and authors approach the subject with slight, but very significant differences that impact the working of problems.

The only way to become proficient in working with lenses and mirrors is to place the sign conventions appropriate to your course on a card in front of you and do problems drawing the rays and working the calculations. Be sure that you know the sign conventions and practice doing sample problems before taking any tests on mirrors or lenses. Signs will be your major (perhaps only) source of error. Sample sign conventions for mirrors and lenses are given below.

Sign Conventions for Mirrors

o is positive if object is in front of the mirror
o is negative if object is in back of the mirror

i is positive if the image is in front of the mirror
i is negative if the image is in back of the mirror

f is positive if the center of curvature is in front of the mirror
f is negative if the center of curvature is in back of the mirror

Sign Conventions for Lenses

o is positive if the object is in front of the lens
o is negative if the object is in back of the lens

i is positive if the image is in back of the lens
i is negative if the image is in front of the lens

f is positive for a converging lens
f is negative for a diverging lens

Mirrors

An object placed in front of a plane mirror appears to an observer to be behind the mirror.

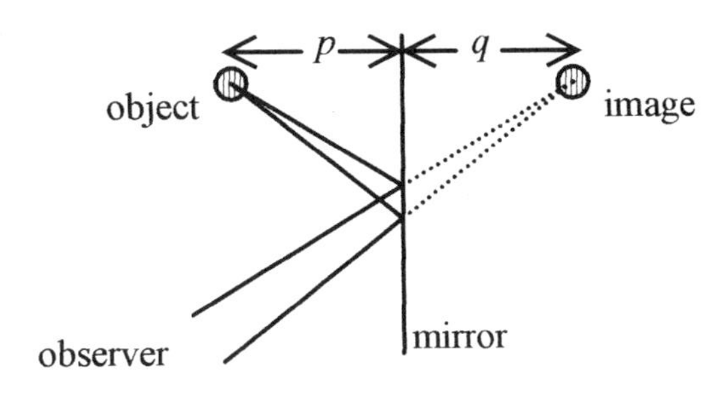

Fig. 45-1

Because of the law of reflection, divergent rays intercepted by the observer on reflection from the mirror appear to come from behind the mirror. The object distance, p, is numerically equal to the image distance, q. The image is called a **virtual image** because the light does not physically come from the image. A **real image** is one where the light comes from or passes through the image.

45-1 A light source is 4.0cm in front of a plane mirror. Where does an observer looking into the mirror see the image and is it real or virtual?

Solution: The image is 4.0cm behind the mirror. It is a virtual image because light does not pass through this image point.

A concave (converging) spherical mirror as shown in Fig. 45-2 can be analyzed by rays.

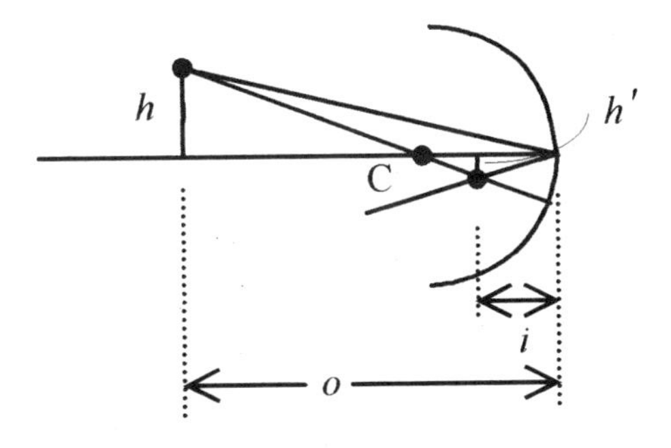

Fig 45-2

An object of height h placed at o, the object distance from the concave mirror, will produce a smaller image, h' at i, the image distance, according to the formula

$$\frac{1}{o}+\frac{1}{i}=\frac{2}{R}=\frac{1}{f} \qquad \textbf{(45-1)}$$

where R is the radius of curvature and $f(=R/2)$ is the focal length. The magnification is

$$M=\frac{h'}{h}=-\frac{i}{o}=-\frac{R-i}{o-R} \qquad \textbf{(45-2)}$$

The minus sign in the definition indicates that the image is inverted.

Draw a ray from the top of the object through the center of curvature (C in Fig. 45-2). Next draw a ray to the point where the principal axis (the horizontal line through C) intersects the mirror reflecting this ray back to intersect the one drawn previously. The intersection of these rays defines the top of the object. Drawing these rays requires experience. Set up several situations and draw the ray diagrams to become familiar with the procedure.

45-2 For a spherical concave mirror of 12 cm radius of curvature describe the image of a 2.0 cm height object placed 20 cm on the center line of the mirror according to Fig. 45-2.

Solution: The image is inverted. The ray diagrams show this. Using the radius of curvature and the object distance find the image distance from

$$\frac{1}{20\,\text{cm}}+\frac{1}{i}=\frac{2}{12\,\text{cm}} \quad \text{or} \quad i=8.6\,\text{cm}$$

The magnification is from $M=-\frac{8.6}{20}=-\frac{12-8.6}{20-12}=-0.43$.

The height of the image is from $\frac{h'}{2.0\,\text{cm}}=-0.43$, or $h'=-0.86\,\text{cm}$.

The image is 8.6 cm from the mirror, inverted (minus sign) and real (rays pass through image).

When the object is at infinity (very far away) the mirror equation reduces to $1/i=2/R$, and we can say that the rays from infinity are focused at $R/2$. This defines the focal length as $f=R/2$.

A convex (diverging) spherical mirror is illustrated in Fig. 45-3.

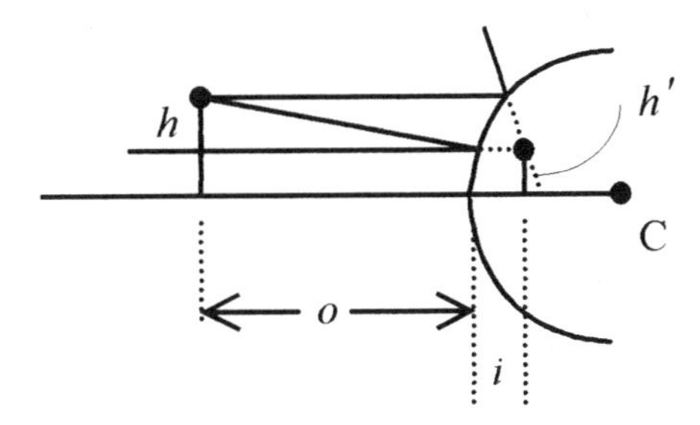

Fig. 45-3

Objects placed in front of a convex mirror appear to come from behind the mirror, and they are smaller. First draw a ray from the top of the object parallel to the principal axis and reflect it from the mirror. This ray appears to come from the focus (behind the mirror). Next draw a ray so as to produce a reflected ray parallel to the principal axis of the mirror. The extension of this ray intersects the extension of the first one locating the top of the image.

The equations for concave mirrors also work for convex mirrors if a sign convention is adopted. Lengths where the light moves (to the left of the mirrors in Figs. 45-2 and 45-3) are positive, and lengths on the other side of the mirror (to the right of the mirror in Fig. 45-3) are negative. Lengths are measured (positive and negative) from the intersection of the principal axis and the mirror. These positive and negative regions are often referred to as the front and the back sides of the mirrors.

45-3 For a spherical convex mirror of 14 cm radius of curvature, describe the image of a 2.5 cm object placed 30 cm out on the principal axis of the mirror.

Solution: Here is where we get into the signs. The focus and the image are on one side of the mirror and the object is on the other side. Therefore we take the focus as negative and expect the image distance to be negative. The image distance comes from equation 45-1

$$\frac{1}{30\,\text{cm}} + \frac{1}{i} = -\frac{1}{7\,\text{cm}} \quad \text{or} \quad i = -5.7\,\text{cm}$$

The magnification comes from equation 45-2 $M = -\dfrac{-5.7}{30} = \dfrac{-14+5.7}{30+14} = 0.19$.

The height of the image is from $\dfrac{h'}{2.5\,\text{cm}} = 0.19$, or $h' = 0.48\,\text{cm}$.

The minus sign for the image distance indicates that the image is behind the mirror, or on the same side as all the other minus signs. The plus sign for the magnification indicates that the image is erect (not inverted). The image is virtual.

Lenses

There are two types of thin lenses, converging and diverging, as shown in Fig. 45-4.

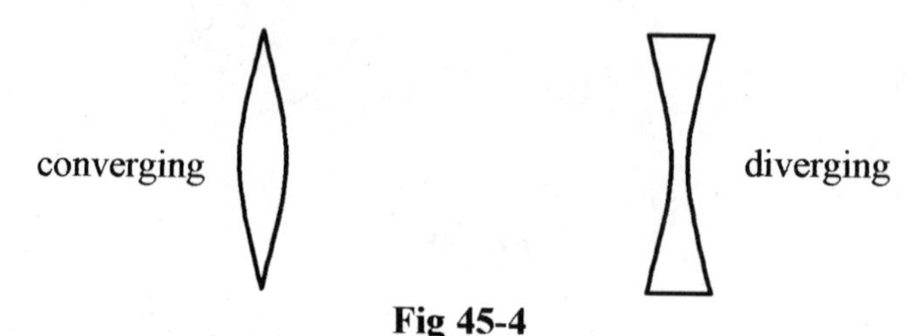

Fig 45-4

The converging lens converges parallel rays to a point called the focus while a diverging lens refracts rays to make them appear as to come from a focus. The sign conventions become more involved for lenses than for mirrors. Rather than set out a sign convention, we will handle the signs in the context of each problem. The relationship between image distance, object distance and focal length is the same as for mirrors (equation 45-1).

45-4 A converging lens of focal length 8.0 cm forms an image of an object placed 20 cm in front of the lens. Describe the image.

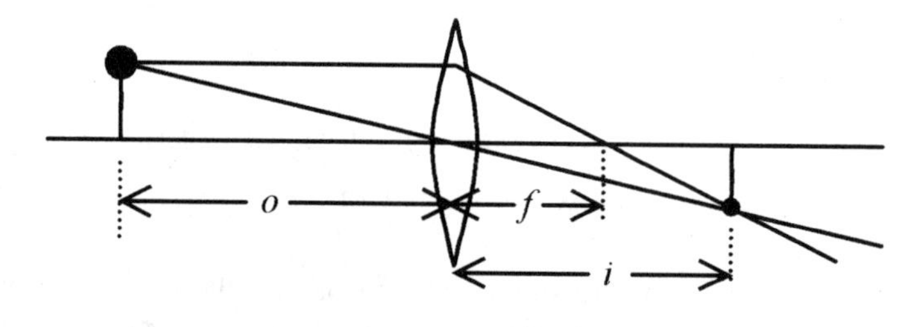

Fig. 45-5

Solution: Draw a ray from the top of the object parallel to the axis then through the focus. Next draw a ray from the top of the object through the center of the lens to intersect the first ray. This locates the top of the object. Use equation 45-1 to find the image distance.

$$\frac{1}{20\,\text{cm}} + \frac{1}{i} = \frac{1}{8.0\,\text{cm}} \quad \text{or} \quad i = 13\,\text{cm}$$

The magnification is $(-)$image distance over object distance or $M = -\frac{i}{o} = -\frac{13}{20} = -0.67$.

The image is located 13 cm on the side of the lens opposite the object with magnification 0.67. It is inverted (minus sign) and real (rays pass through image).

45-5 A diverging lens has a 14 cm focal length. Describe the image of a 4.0 cm object placed 40 cm from the lens.

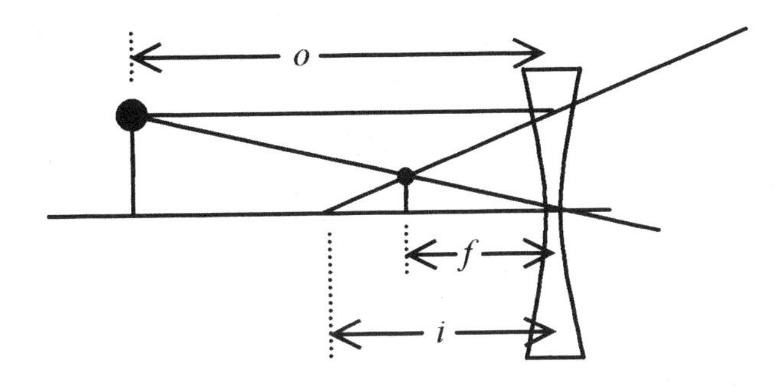

Fig. 45-6

Solution: Draw a ray from the top of the object to the lens parallel to the principal axis and refract it back to the focus. Next draw a line from the top of the object through the center of the lens. The intersection of these rays locates the top of the object.
The image distance is from equation 45-1.

$$\frac{1}{40\,\text{cm}} + \frac{1}{i} = -\frac{1}{14\,\text{cm}} \quad \text{or} \quad i = -10\,\text{cm}$$

The negative sign for the focal length is because this is a diverging lens. The image distance is negative because it is on the same side of the lens as the object (opposite to the converging lens). The magnification is $10/40 = 0.25$.

The image is $0.25 \times 4.0\,\text{cm} = 1.0\,\text{cm}$ high, erect, virtual, and appears to come from a point 10 cm from the lens on the same side as the object.

Go back over the problems in this chapter paying particular attention to the signs. As an exercise change the numbers in these problems and work them through until the sign conventions are clear in your mind.

CHAPTER 46

DIFFRACTION AND INTERFERENCE

The wave nature of light is used to explain the several diffraction and interference phenomena discussed in this chapter.

Double Slit Diffraction

Figure 46-1 shows the setup for the classic Young's double slit experiment first performed around 1800.

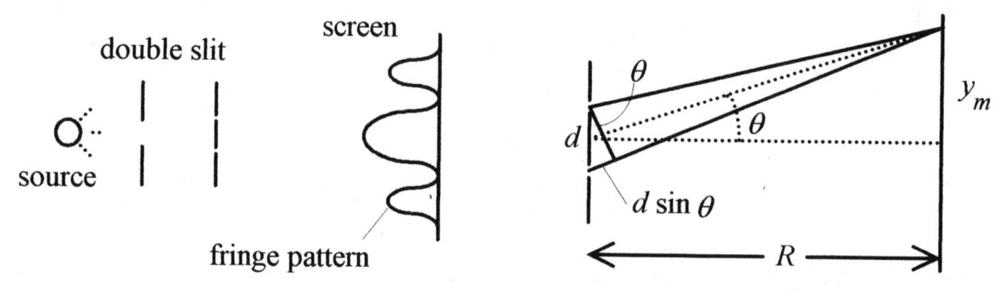

Fig. 46-1

Monochromatic light passes through a sufficiently narrow slit to produce a coherent source for the double slits. The coherent (in phase) light spreads out from each of the slits creating an interference pattern on the screen. The geometry for the interference is shown in Fig. 46-1. Destructive interference occurs when the path length differs by one-half wavelength and constructive interference occurs when the path length is an integral number of wavelengths. From the figure we can write the criteria for constructive interference as

$$d \sin\theta = m\lambda \quad (m \text{ is an integer}) \tag{46-1}$$

The distance from the central maxima (opposite the center of the two slits) to the first bright fringe on the screen is related to the angle via $\tan\theta = y_m/R$. For small angles, which is the case with the Young's experiment, $\sin\theta \approx \tan\theta = \theta$, so equation 46-1 can be rewritten as

$$\frac{d\,y_m}{R} = m\lambda \tag{46-2}$$

Historically this experiment was used to measure the wavelength of light.

46-1 A Young's experiment is set up with slit separation 2.0×10^{-4} m, screen distance 0.80 m, and distance from central maxima to second bright fringe 3.9×10^{-3} m. What is the wavelength of the source?

Solution: $$\lambda = \frac{d y_m}{mR} = \frac{2.0 \times 10^{-4}\,\text{m} \cdot 3.9 \times 10^{-3}\,\text{m}}{2 \cdot 0.80\,\text{m}} = 488\,\text{nm}$$

46-2 In a different Young's experiment with 488 nm light the second dark fringe is 1.2×10^{-3} m away from the center of the central bright fringe. The screen is 1.2 m from the slits. What is the separation of the slits?

Solution: This second dark fringe corresponds to $3/2$ of a wavelength in path difference. One wavelength path difference is $2/3$ of the 1.2×10^{-3} m or 8.0×10^{-4} m. This number now fits with equation 46-2 (one wavelength path difference), and solving for d

$$d = \frac{m\lambda R}{y_m} = \frac{1 \cdot 488 \times 10^{-9}\,\text{m} \cdot 1.2\,\text{m}}{8.0 \times 10^{-4}\,\text{m}} = 7.3 \times 10^{-4}\,\text{m}$$

Single Slit Diffraction

The intensity distribution pattern for single slit diffraction is similar to the pattern for double slit diffraction. The geometry and analysis, however, are quite different. Coherent light incident on a narrow slit (100 to 1000 wavelengths wide) produces a fringe pattern as shown in Fig. 46-2.

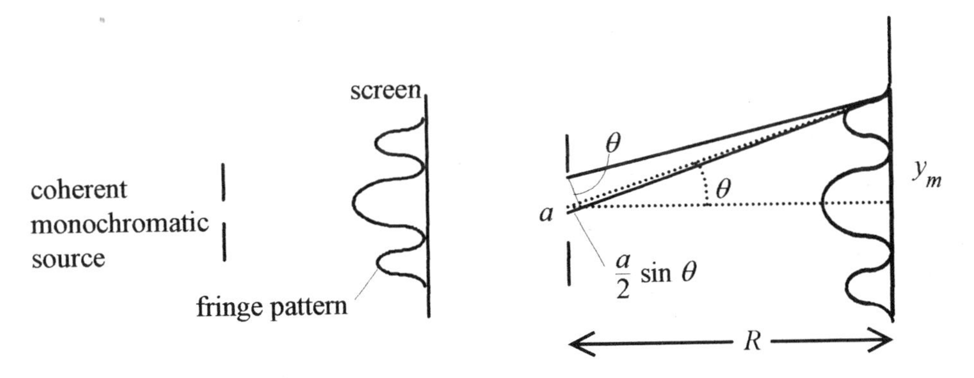

Fig. 46-2

In the analysis of single slit, or Fraunhofer, diffraction we will use the dark fringes. The slit of width a is viewed as a region where interference occurs between light at the top of the slit and the middle of the slit. The particular situation for the first dark fringe obtains for corresponding pairs of source points separated by $a/2$ across the width of the slit. The criterion for dark fringes is

$$\frac{a}{2}\sin\theta = \frac{m\lambda}{2} \quad (m \text{ integer}) \qquad \textbf{(46-3)}$$

Again $\sin\theta$ is approximately the same as $\tan\theta$, which is approximately the same as θ. So rewrite equation 46-3 with this approximation as

$$\frac{a}{2}\frac{y_m}{R} = \frac{m\lambda}{2} \qquad \textbf{(46-4)}$$

where m corresponds to the number of the dark fringes away from the central maxima.

46-3 Monochromatic light is incident on a 4.0×10^{-4} m wide slit producing a Fraunhofer diffraction pattern on a screen 1.5m away. The distance from the central maxima to the second dark fringe is 3.4×10^{-3} m. What is the wavelength of the light?

Solution: Reworking equation 46-4 yields $\lambda = \frac{a y_m}{mR} = \frac{4.0 \times 10^{-4}\,\text{m} \cdot 3.4 \times 10^{-3}\,\text{m}}{2 \cdot 1.5\text{m}} = 453\text{nm}$.

46-4 Light from a He-Ne laser of 633nm is incident on a 2.0×10^{-4} m wide slit. What is the width of the central maxima (the distance between the dark fringes on either side of the central maxima) on a screen 2.0m away.

Solution: Rearranging equation 46-4, the distance to the first dark fringe is

$$y_1 = \frac{1 \cdot \lambda R}{a} = \frac{1 \cdot 633 \times 10^{-9}\,\text{m}(2.0\text{m})}{2.0 \times 10^{-4}\,\text{m}} = 6.3 \times 10^{-3}\,\text{m}$$

The width of the central maxima is two times this value or 1.3×10^{-2} m.

The Diffraction Grating

A diffraction grating consists of multiple slits as shown in Fig. 46-3. When light is incident on this arrangement constructive interference occurs when the path difference between adjacent slits is $m\lambda$. Increasing the number of slits produces two effects: the intensity of each interference maximum increases, and the width of each interference maximum decreases. This makes the grating a very convenient tool for the study of the various component wavelengths (spectra) of gases.

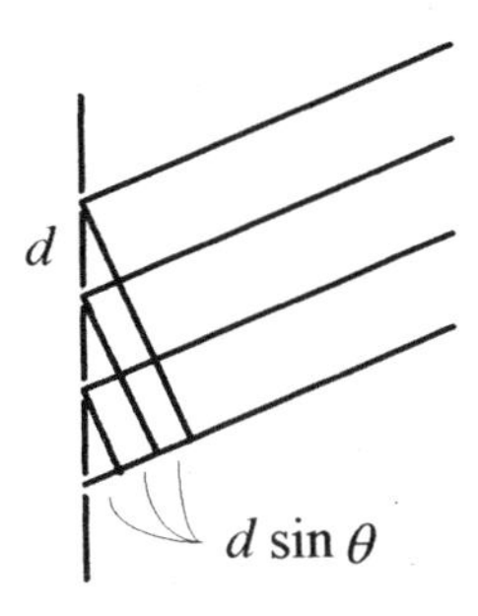

Fig. 46-3

46-5 A grating with 2000 lines per cm is illuminated with a hydrogen gas discharge tube. Two of the hydrogen lines are at 410 and 434 nm. What is the first order spacing of these lines on a screen 1.0 m from the grating. The experimental arrangement is shown in Fig. 46-4.

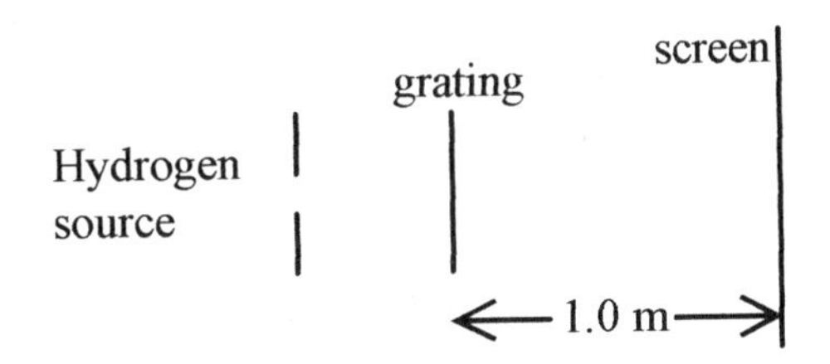

Fig. 46-4

Solution: The phrase "first order" refers to the first maxima.

First find the spacing $1.0\times10^{-2}\,\mathrm{m}/2000\text{ lines} = 5.0\times10^{-6}\,\mathrm{m}$.

For the 410 nm line $y_1 = \dfrac{1\cdot\lambda R}{d} = \lambda\dfrac{R}{d} = 410\times10^{-9}\,\mathrm{m}\dfrac{1.0\,\mathrm{m}}{5.0\times10^{-6}\,\mathrm{m}} = 0.082\,\mathrm{m}$.

For the 434 nm line $y_1 = \dfrac{1\cdot\lambda R}{d} = \lambda\dfrac{R}{d} = 434\times10^{-9}\,\mathrm{m}\dfrac{1.0\,\mathrm{m}}{5.0\times10^{-6}\,\mathrm{m}} = 0.087\,\mathrm{m}$.

The lines are separated by 5 mm on the screen.

CHAPTER 47

SPECIAL RELATIVITY

The special theory of relativity is based on two postulates.

1. The laws of physics are the same in all inertial reference frames.

Inertial frames are reference (coordinate) frames moving at constant velocity with respect to one another; that is, they are not accelerating.

2. The speed of light is the same in all inertial frames.

These two postulates lead to very interesting and highly significant conclusions concerning simultaneity and how we measure the fundamental quantities of length, mass, and time. These discussions, though very interesting, are inappropriate for a problems book, so we will go directly to the consequences of special relativity as it impacts our understanding of physics.

The concepts and calculations of special relativity require a change in how we view the world. Many of the things we study in special relativity may go against our intuition, but remember, intuition is often wrong and even if it were correct we have no experience to base our intuition on when dealing with particles approaching the speed of light where relativistic effects are observable. How many of Galileo's colleagues intuitively "knew" that when he dropped those two different sized balls from the Tower at Pisa the heavier one would reach the ground first? Success in understanding Special Relativity requires first that you rid your mind of intuitive knowledge based on what could be called "low velocity experience." As you study Special Relativity you will encounter situations where you will be challenged to look at the postulates and change your view of the world.

Time Dilation

Time intervals are different in different (moving) inertial frames. Place a light source, mirror, and detector in a moving vehicle as shown in Fig. 47-1. An observer in the moving vehicle measures the time for a light pulse to move from the source to the mirror and back to the detector as the distance traveled divided by the velocity of light, so $\Delta t_o = 2d/c$. The zero subscript indicates that this time is measured by an observer in the same frame where the event is taking place. This is also called the proper time.

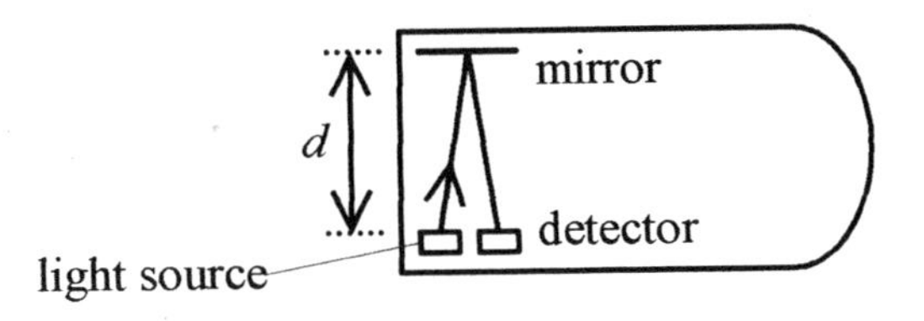

Fig 47-1

An observer in another inertial frame observes the vehicle moving at a velocity v (see Fig. 47-2) and over the time interval of the event observes that the detector has traveled a distance $v\Delta t$.

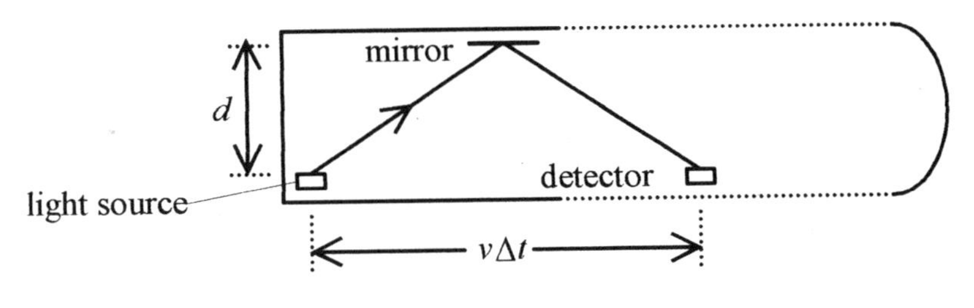

Fig. 47-2

The total distance the light has traveled is $2\sqrt{d^2 + (v\Delta t/2)^2}$.

The time interval is $\Delta t = \frac{2}{c}\sqrt{d^2 + (v\Delta t/2)^2}$.

Remember, c is the same to all observers regardless of inertial frame. The two times can be related with $d = c\Delta t_o/2$ so that

$$\Delta t = \frac{2}{c}\sqrt{(c\Delta t_o/2)^2 + (v\Delta t/2)^2} \quad \text{or} \quad \Delta t = \frac{\Delta t_o}{\sqrt{1 - v^2/c^2}} \qquad \textbf{(47-1)}$$

The external observer of an event taking place in a moving reference frame, the one who measures Δt, always sees an event as taking longer than the stationary observer, the one who measures Δt_o, in the frame moving with the event.

47-1 You are in a railroad car moving at constant velocity of zero with respect to the surface of the earth. What do you measure as the time it takes for a coin to drop 1.0 m to the floor of the railroad car? What does an observer in another railroad car traveling at a constant 25 m/s measure for this time?

Solution: You apply $s = \frac{1}{2}gt^2$ and observe $t = \sqrt{\frac{2s}{g}} = \sqrt{\frac{2 \cdot 1.0\,\mathrm{m}}{9.8\,\mathrm{m/s^2}}} = 0.45\,\mathrm{s}$

The moving observer measures a time given by equation 47-1. For small v/c the best way to make this calculation is to use a binomial expansion of the radical (see the Introduction, Mathematical Background).

$$\Delta t = \frac{\Delta t_o}{\sqrt{1 - v^2/c^2}} = \Delta t_o\left(1 - v^2/c^2\right)^{-1/2} = \Delta t_o\left[1 + \frac{1}{2}\frac{v^2}{c^2} + \frac{3}{8}\frac{v^4}{c^4} + \ \dots\right] \qquad \textbf{(47-2)}$$

The first, and largest, relativistic correction term is $\frac{1}{2}\frac{v^2}{c^2} = \frac{1}{2}\left(\frac{25}{3.0 \times 10^8}\right)^2 = 3.5 \times 10^{-15}$.

This means that the correction is $\Delta t = \Delta t_o(1 + 3.5 \times 10^{-15})\,\mathrm{s}$. Time measurement this precise is beyond the capability of the cesium clocks. This is an unobservable effect.

47-2 For the situation of the previous problem take the moving observer from the train to a $2500\,\mathrm{m/s}$ jet plane and then to a rocket ship traveling at $0.90\mathrm{c}$. (It is common, and very convenient in relativistic calculations, to express velocity as a fraction or decimal times c.)

Solution: Again using the binomial expansion and looking at the first term

$$\frac{1}{2}\frac{v^2}{c^2} = \frac{1}{2}\left(\frac{2500}{3.0 \times 10^8}\right)^2 = 3.5 \times 10^{-11}$$

This small a time difference is observable but it is a difficult experiment.

For the rocket ship traveling at $0.90\mathrm{c}$, use equation 47-1 directly.

$$\Delta t = \frac{\Delta t_o}{\sqrt{1 - 0.90^2}} = \frac{\Delta t_o}{0.43} = \frac{0.45\,\mathrm{s}}{0.43} = 1.0\,\mathrm{s}$$

The observer in the rocket ship measures the time for the falling coin as $1.0\,\mathrm{s}$, over twice what the stationary observer sees. These two problems illustrate how relativistic effects are observable only as relative velocities approach c.

47-3 A "strange particle" is observed to move at $0.96c$ and have a lifetime of 2.0×10^{-8} s. What is the lifetime of the particle in its own reference frame?

Solution: In equation 47-1 Δt_o is the (proper) time an observer traveling with the "strange particle" would measure so find

$$\Delta t_o = \Delta t\sqrt{1 - v^2/c^2} = 2.0 \times 10^{-8}\,\text{s}\sqrt{1 - 0.96^2} = 0.56 \times 10^{-8}\,\text{s}$$

Length Contraction

The time dilation leads to a length contraction with velocity. Consider a vehicle moving at velocity v between two stars with one observer in the vehicle and another at a fixed (with respect to the stars) point. The external observer measures the (proper) distance between the stars as L_O, observes the velocity of the vehicle as v, and writes the time interval as $\Delta t = L_O/v$. The time interval Δt is not a proper time because measurement of Δt would require synchronized clocks at both stars. The observer in the vehicle sees the stars moving at v and measures a (proper) time interval Δt_O. This observer determines the distance between the stars as $L = v\Delta t_O$. Write these two equations as a fraction

$$\frac{L}{L_O} = \frac{v\,\Delta t_O}{v\,\Delta t} \quad \text{and with} \quad \Delta t = \frac{\Delta t_O}{\sqrt{1 - v^2/c^2}}$$

$$L = L_O\sqrt{1 - v^2/c^2} \qquad \textbf{(47-3)}$$

The length measured by a moving observer is contracted by a factor equal to $\sqrt{1 - v^2/c^2}$.

47-4 In a soaring space ship (a vehicle capable of both space flight and conventional aircraft flight) you return from a space journey to find a new landing runway. Passing over this runway at $0.92\,c$ you measure its length as $1960\,\text{m}$. What is its length at your landing speed of $200\,\text{m/s}$?

Solution: At $200\,\text{m/s}$ there is no observable relativistic effect. Use equation 47-3 to find the length of the runway at your landing speed.

$$L_O = \frac{L}{\sqrt{1 - v^2/c^2}} = \frac{1960\,\text{m}}{\sqrt{1 - 0.92^2}} = 5000\,\text{m}$$

47-5 How fast must a meter stick be traveling relative to your reference frame for you to observe a 2% contraction?

Solution: Use equation 47-3 with $L = 0.98 L_o$

$$0.98 L_o = L_o \sqrt{1 - v^2/c^2} \quad \text{or} \quad v = 0.20c$$

Relativistic Momentum

At "low velocities" a force applied to a particle produces an acceleration according to the familiar $F = d/dt(mv) = m(dv/dt)$. As the velocity of the particle approaches c, the momentum becomes velocity dependent with the force statement taking on the form

$$F = \frac{d}{dt} \frac{mv}{\sqrt{1 - v^2/c^2}}$$

The most convenient interpretation is to associate the $\sqrt{1 - v^2/c^2}$ with the mass and say that the effective mass of a moving particle is

$$\frac{m_o}{\sqrt{1 - v^2/c^2}} \qquad \textbf{(47-4)}$$

where m_o is the rest mass, measured with the mass not moving in the inertial frame where the measurements are made.

47-6 What is the effective mass of an electron moving at $0.80c$?

Solution:

$$m = \frac{m_o}{\sqrt{1 - v^2/c^2}} = \frac{9.1 \times 10^{-31}\,\text{kg}}{\sqrt{1 - 0.80^2}} = 15.2 \times 10^{-31}\,\text{kg}$$

47-7 Intergalactic space travelers need to know the relative velocities and masses of their space ships. Each ship, therefore, has a $1.0\,\text{m}$ long bar painted on the side of the ship alongside their rest mass. As you pass by a ship you measure this $1.0\,\text{m}$ bar as $0.93\,\text{m}$. What is your relative velocity? You also observe their rest mass printed as $365{,}000\,\text{kg}$. What is their mass relative to you? What does an observer in the other ship measure for your $1.0\,\text{m}$ bar?

Solution: The length you observe is $L = L_o\sqrt{1 - v^2/c^2}$ where L_o is the $1.0\,\mathrm{m}$, the length an observer at rest with respect to the vehicle would measure, and L is the length you measure so

$$0.93 = 1.0\sqrt{1 - v^2/c^2} \quad \text{or} \quad v = 0.37\,c$$

The relativistic mass you observe is

$$m = \frac{m_o}{\sqrt{1 - v^2/c^2}} = \frac{365{,}000\,\mathrm{kg}}{0.93} = 393{,}000\,\mathrm{kg}$$

Observers in the other space ship measure your bar as $0.93\,\mathrm{m}$ and relative speed as $0.37\,c$.

As an exercise find the effective mass of an electron at $0.999c$ and at $0.99999c$. As the velocity approaches c, the effective mass approaches infinity implying the necessity of an infinite force to reach c. This shows the theoretical impossibility of material objects traveling at c or beyond.

PHYSICAL CONSTANTS

Name	*Symbol*	*Value*
Acceleration due to gravity	g	$9.8\,\mathrm{m/s}$
Speed of light	c	$3.0 \times 10^{8}\,\mathrm{m/s}$
Electron-volt	eV	$1.6 \times 10^{-19}\,\mathrm{J}$
Electronic charge	e	$1.6 \times 10^{-19}\,\mathrm{C}$
Gravitational constant	G	$6.7 \times 10^{-11}\,\mathrm{N \cdot m^2/kg^2}$
Boltzmann's constant	k	$1.4 \times 10^{-23}\,\mathrm{J/K}$
Avogadro's number	N_A	6.0×10^{23} molecules/mole
Gas constant	R	$8.3\,\mathrm{J/mole \cdot K}$
Mass of electron	m_e	$9.1 \times 10^{-31}\,\mathrm{kg}$
Mass of neutron or proton	m_n or m_p	$1.7 \times 10^{-27}\,\mathrm{kg}$
Mechanical equivalent of heat		$4.2\,\mathrm{J/cal}$
Permittivity of free space	ε_o	$8.8 \times 10^{-12}\,\mathrm{C^2/N \cdot m^2}$ (F/m)
	$1/4\pi\varepsilon_o$	$9.0 \times 10^{9}\,\mathrm{N \cdot m^2/C^2}$
Permeability of free space	μ_o	$4\pi \times 10^{-7}\,\mathrm{Wb/A \cdot m}$ (H/m)
Speed of sound		$343\,\mathrm{m/s}$
Standard atmospheric pressure	1 atm	$1.0 \times 10^{5}\,\mathrm{Pa}$

CONVERSIONS

$1\,\mathrm{kg} = 1000\,\mathrm{g} = 0.068\,\mathrm{slug}$

$1\,\mathrm{m} = 100\,\mathrm{cm} = 3.3\,\mathrm{ft}$

$1\,\mathrm{nm} = 10^{-9}\,\mathrm{m} = 10\,\mathrm{\mathring{A}}$

$1\,\mathrm{m/s} = 3.3\,\mathrm{ft/s}$

$1\,\mathrm{rad} = 57.3^o$

$\pi\,\mathrm{rad} = 180^o$

$1\,\mathrm{N} = 10^5\,\mathrm{dyne} = 0.22\,\mathrm{lb}$

$1\,\mathrm{cal} = 4.2\,\mathrm{J}$

$1\,\mathrm{eV} = 1.6 \times 10^{-19}\,\mathrm{J}$

$1\,\mathrm{hp} = 750\,\mathrm{W} = 550\,\mathrm{ft \cdot lb/s}$

$1\,\mathrm{J} = 10^7\,\mathrm{erg} = 0.24\,\mathrm{cal} = 0.74\,\mathrm{ft \cdot lb}$

INDEX

A
Absolute pressure, 159
Acceleration, 30
 angular, 111, 120
 center of mass, 94, 99
 centripetal, 111
 due to gravity, 37, 139
 radial, 66, 121
 tangential, 116
Alternating current generator, 313
Ampere, 272
Ampere's law, 318
Ampere-Maxwell law, 370
Amplitude, 209
Angle:
 for polarization, 384
 of incidence, 380
 of reflection, 380
 of refraction, 380
Angular acceleration, 111, 120
Angular frequency, 111, 120
Angular momentum, 123, 125
Angular velocity, 111
Antenna, 372
Antinode, 216
Archimedes' Principle, 162
Atomic mass unit, 327
Average Value of a Function, 18
Avogadro's number, 118

B
B-field, 301
Ballistic pendulum, 104
Banking of curves, 69
Barometer, 161
Battery, 277
Bernoulli's equation, 166
Binomial expansion, 5
Biot-Savart law, 326
Boltzman constant, 187
Boyle's law, 181
Brewster's law, 385
Buoyant force, 162

C
Calorie, 175
Capacitance, 260
 dielectric, 268
Capacitive circuit, 357
Capacitive reactance, 357
Capacitive time constant, 296
Capacitor:
 charging, 294
 cylindrical, 262
 discharging, 298
 displacement current, 370
 energy storage, 267
 in parallel, 265
 in series, 263
 parallel plate, 260
 spherical, 263
Center of mass, 93, 100
Charge, 230
Charles' law, 181
Circuit:
 junction rule, 286
 loop rule, 284
 multi loop, 284
 RC, 294
 RL, 346
 RLC, 354, 356
Circular motion, 66
Collision, 102
 conservation of momentum, 107
 elastic, 106
 inelastic, 103
Concave mirror, 386
Conductivity:
 electrical, 271
 thermal, 178
Conical pendulum, 69
Conservation:
 angular momentum, 123, 125
 energy, 82
 linear momentum, 93
 in collisions, 97, 102
Constant volume process, 195
Constants, physical, 403
Convex mirror, 387
Coordinate systems, 6
 Cartesian, 7
 cylindrical, 8
 spherical, 7
Coulomb's law, 230
Critical angle, 382
Cross product, 25
Current:
 alternating, 352
 direct, 272
 displacement, 370
 induced, 336
Current density, 272
Cyclotron, 303

D
Damped oscillator:
 electrical, 350
 mechanical, 157
d'Arsonval meter, 316
Decibel, 225
Density, 159
Derivatives, 12, 30
Determinants, 3
Dielectrics, 268
Diffraction, 392
 double slit, 392
 Fraunhofer, 393
 grating, 394
 single slit, 393
Dipole, electric, 234
Dispersion, 383
Displacement current, 370
Doppler effect, 226
Dot product, 23
Drift speed, 274

E
Electric field, 234
Elastic collision, 102, 106
Electric current, 271
Electric dipole, 234
Electric field:
 energy density, 267
 induced, 339
 lines of force, 234
 charged disk, 239
 electromagnetic wave, 372
 line of charges, 237
 group of charges, 234
 sheet of charges, 238
 point charge, 234, 240
Electric flux, 243
Electric generator, 335
Electric potential, 250
 disk, 258
 group of charges, 253
Electromagnetic induction, 277
Electromagnetic oscillations, 372
Electromagnetic waves, 372
 energy density, 375
 polarization, 370
 Poynting vector, 377
 speed, 373
Electromotive force, 277
Electron, charge to mass ratio, 308
Electron-volt, 256

EMF, 277
Energy, 125, 151
  conservation, 125
Energy density, 351
  electric, 375
  magnetic, 316, 344
Entropy, 207
Equilibrium, 130
Equations, 1
Equation of state, 181
Escape speed, 140
Expansion, thermal, 174
Exponential, 12
  decay, 298
  growth, 294

F
Farad, 260
Faraday's law, 332
First law of thermodynamics, 161
Flow, fluid, 165
Fluid, pressure, 159
Flux:
  electric, 243
  magnetic, 332, 341
Focal length, 390
Force, 56
Forced oscillations:
  electric, 356
  mechanical, 157
Frequency:
  and period, 210
  angular, 112
Friction, 60, 84
Functions, 10
Fusion, heat of, 177

G
g, 37, 139
Gas constant, R, 181
Gas thermometer, 172
Gas, ideal, 171
Gauge pressure, 159
Gauss' law:
  electricity, 243
  magnetism, 301
Gaussian surface, 261
Generator, ac, 335
Grating, diffraction, 394
Gravitation, 83, 139

H
Hall effect, 309
Harmonic motion:
  damped, 157
  forced, 356
  simple, 145
Heat:
  capacity, 195
  conduction, 178
  pumps, 206
Heat of fusion, 177
Heat of vaporization, 177
Henry, 341
Hertz, 145
Hooke's law, 179

I
Iceberg, 165
Ideal gas, 181
Image, 386
Impedance, 362
Incidence, angle of, 380
Index of refraction, 380
Induced electric field, 339
Induced emf, 277
Inductance:
  mutual, 342
  self, 341
  solenoid, 343
Inductive circuit, 359
Inductive reactance, 350
Inelastic collision, 102, 103
Integral, 15
Intensity, 223, 376
Interference, 392
  double slit, 392
  grating, 394
  single slit, 393
Internal energy, 193
Inverse square law, 223

J
Joule's law, 275
Junction law, 286

K
Kelvin scale, 172
Kepler's laws, 142
Kinematics, 28
Kinetic energy, 82, 103
  definition, 82
  in oscillator, 151, 352
Kinetic theory, 181
Kirchhoff's laws:
  junction, 286
  loop, 284

L
LC circuit, 352
LC oscillator, 352
Length contraction, 399
Lens, thin, 386, 390
Lenz's law, 333
Light, 372, 380
  polarization, 384
  speed, 373
Linear motion, 28
Line integral, 318
Logarithms, 12
LR circuit, 346

M
Magnetic field:
  and circulating charges, 303
  and electromagnetic wave, 373
  energy density, 316, 344, 375
  flux, 332, 341
  of earth, 302
  of solenoid, 324
  of toroid, 325
Magnetic flux, 332, 341
Magnetic force:
  between wires, 312
  on charge, 302, 308
Magnetic moment, 314
Magnification, 388
Manometer, 161
Mass spectrometer, 306
Maxwell's equations, 368
Mean free path, 271
Mean free time, 274
Mirror, 386, 387
Mole, 192
Moment of inertia, 156
Momentum, 93
  and Newton's laws, 93
  angular, 123
  linear, 93
  relativistic, 400
Motion in one dimension, 28
Motion in a plane, 47
Mutual inductance, 342

N
Newton's second law, 56
Node, 216
Normal force, 60

O
Ohm, 274
Ohm's law, 273
Optical fiber, 380
Optics:
  geometric, 386
  physical, 392
Organ pipes, 219

Oscillations, 352
  damped, 157, 354
  electromagnetic, 352
  forced, 361
  LC, 352
  mechanical, 145

P
Parabola, 11
Parallel connection:
  capacitors, 265
  resistors, 279
Parallel plate capacitor, 260
Partial derivative, 211, 223
Pascals, 159
Pendulum:
  ballistic, 104
  conical, 69
  simple, 155
Period and frequency, 210
Permeability, 318
Phase change, 177
Phase velocity, 211
Phasors, 356
Plane mirror, 386
Plane, motion in, 47
Polarization, 380
  by reflection, 384
Potential energy:
  and work, 82
  capacitors, 267
  simple harmonic oscillators, 151
Power:
  electric, 365
  electromagnetic wave, 375
Poynting vector, 377
Pressure, 159
  absolute, 159
  gauge, 159
Pressure amplitude, 223
Projectile motion, 47
Proper time, 396

Q
Quadratic formula, 2

R
R (gas constant), 181
Radian, 9
Ray tracing, 387
RC circuit, 294
RLC circuit, 356
Reactance:
  capacitance, 357
  inductive, 359
Real image, 387
Reference frames, 396
Reflection, 380
  law of, 380
  total internal, 382
Refraction, 380
Relativity, 396
Resistance, 273
  equivalent, 280
  parallel, 279
  series, 279
Resistive circuit, 357
Resistivity, 273
  temperature dependence, 276
Right-hand rule, 25
RL circuit, 346
RLC circuit, 354, 361
rms (root-mean-square), 366
Rotational inertia, 123
Rotational motion, 110

S
Satellites, 141
Scalar product, 24
Second law of thermodynamics, 202
Series connection:
  capacitors, 263
  resistors, 279
Simple harmonic motion, 145
Simultaneity, 396
Single slit diffraction, 393
Snell's law, 381
Solenoid, 324
Sound:
  level in dB, 225
  speed, 227
Space station, 114
Special relativity, 396
Specific gravity, 159
Specific heat, 176
Spring constant, 79
Standing waves, 216
Stored energy:
  capacitor, 267
  inductor, 351
Superposition principle, 231
Synchrotron, 340

T
Temperature, 171
Temperature scales:
  Celsius, 172
  Kelvin, 172
Tesla, 302
Thermal conductivity, 179
Thermal expansion, 174
Thermometer, 171
Thin lens formulas, 390
Time constant:
  capacitive, 296
  inductive, 347
Time dilation, 396
Toroid, 325
Torque, 130
  angular acceleration, 110
  angular momentum, 122
  on electric dipole, 241
Total internal reflection, 380
Transformation, heat of, 177
Transformers, 366
Trigonometric functions, 8
Trigonometric identities, 10

U
Uniform circular motion, 110
Unit vector, 21, 35

V
Vaporization, heat of, 177
Vectors, 19
Venturi tube, 167
Velocity, 30
Virtual image, 387
Volt, 250
Volume flow rate, 168

W
Wave equation, 212
Waves, electromagnetic, 372
Weber, 332
Weight, 71
Wire, magnetic force on, 320
Work, 76
  as integral, 79
  for variable force, 81
Work-energy, 76

X

Y
Young's experiment, 392

Z

ABOUT THE AUTHORS

Dr. Robert Oman received the B.S. degree from Northeastern University and the Sc.M. and Ph.D. degrees from Brown University all in physics. He has taught mathematics and physics at several colleges and universities including University of Minnesota, North Shore Community College, Northeastern, and University of South Florida. He has also done research for Litton Industries, United Technologies, and NASA, where he developed the theoretical model for the first pressure gauge sent to the moon. He is author of numerous technical articles, books, and how-to-study books, tapes, and videos.

Dr. Daniel Oman received the B.S. degree in physics from Eckerd College. He received his M.S. degree in physics and his Ph.D. in electrical engineering both from the University of South Florida. He has done research on CO_2 lasers and solar cells, and he has authored several technical articles. Daniel is currently working on the manufacturing of microelectronics at Lucent Technologies.

(X)

p. 56 $\| -\vec{a} \| = 4/7^2 \leftarrow 580^2$

p. 58 (sec)^2 or s^2